Popular Development

Rethinking the Theory and Practice of Development

For my family, Holly, Jay, Isabella, Ted, Florence, and Mary Ellen

Popular Development

Rethinking the Theory and Practice of Development

John Brohman

First published 1996

Blackwell Publishers Ltd
108 Cowley Road
Oxford OX4 1JF

Blackwell Publishers Inc.
238 Main Street
Cambridge, Massachusetts 02142
USA

HC

59.7

.B6887

1996

British Library Cataloguing in Publication Data
A CIP catalogue record for this book is available from the British Library.

ISBN 0 155786 3156 (hardcover)
ISBN 0 155786 3164 (paperback)

Library of Congress Cataloging-in-Publication Data
Library of Congress data has been applied for.

Printed in Great Britain by T. J. Press Ltd, Padstow, Cornwall
This book is printed on acid-free paper

Contents

CONTENTS

Acknowledgments

Research for this book was in part made possible by grants from Simon Fraser University and the Social Sciences and Humanities Research Council of Canada. Much of the book was written at Simon Fraser University and was facilitated by the helpful staff of the library there. For their assistance in many different ways, I am grateful to: Robyn Adamache, Alison Atkinson-Hill, Bob Besterd, Alexa Cartwright, Paul DeGrace, Fiona Jeffries, Shamira Jetha, Heather Milne, Dick Peet, Kerry Preibisch, Natasha Silva, Jessie Smith, John Toye, John Vosbourgh, and Joan Waringer. I would also like to acknowledge my commissioning editor at Blackwell, John Davey, for his unfailing patience and many helpful suggestions, as well as Anthony Grahame, the copy editor. Finally, I want to especially thank my partner and compañera, Holly Keller-Brohman, not only for her editorial assistance but also for her constant encouragement in spite of many months of sacrifice while I was working long hours on the manuscript.

The author and publishers gratefully acknowledge the following for permission to reproduce copyright material: *CLAG Yearbook* (parts of chapter 3), *Economic Geography* (parts of chapter 2), *Review of Income and Wealth* (figure 3.1), The World Bank (figures 3.2 and 3.3), and *World Development* (table 3.6).

Abbreviations

Asian NICs	Asian newly-industrializing countries (Hong Kong, Singapore, South Korea, Taiwan)
CACM	Central American Common Market
CEPAL	Comisión Económica para América Latina, or Economic Commission for Latin America (United Nations)
chaebol	large, family-owned conglomerates (South Korea)
CIDA	Canadian International Development Agency
dirigiste	state-led or Keynesian development model
EEC	European Economic Commission
EOI	export-oriented industrialization
EOZs	export-oriented zones
FAO	Food and Agriculture Organization (United Nations)
GATT	General Agreement on Tariffs and Trade
GDP	gross domestic product
IDB	Inter-American Development Bank (branch of World Bank)
ILO	International Labor Organization
IMF	International Monetary Fund
IRD	integrated rural development
IRDP	integrated regional development planning
ISI	import-substitution industrialization
latifundio	large rural estate (Latin America)
minifundio	small farm (Latin America)
MITI	Ministry of International Trade and Industry (Japan)
NAFTA	North American Free Trade Agreement
NGOs	nongovernmental organizations
NICs	newly industrializing countries
NIDL	New International Division of Labor
NIEO	New International Economic Order

North   First World, developed countries, advanced industrialized
                  countries, capitalist core
NTEs   non-traditional exports
OAU   Organization of African Unity
OECD   Organization for Economic Cooperation and Development
SAPs   structural adjustment programs
South   Third World, developing countries, underdeveloped coun-
                  tries, capitalist periphery
TNCs   transnational corporations
UNCHS   United Nations Center for Human Settlements
UNCTAD   United Nations Conference on Trade and Development
UNCTC   United Nations Conference on Transnational Corporations
UNDP   United Nations Development Program
UNEP   United Nations Environment Program
UNICEF   United Nations Children's Emergency Fund
UNIDO   United Nations International Development Organization
UNRISD   United Nations Research Institute for Social Development
USAID   United States Agency for International Development
WCED   World Commission on Environment and
                  Development (United Nations)
WHO   World Health Organization (United Nations)
WID   'Women in Development' approach

Introduction

In recent years, development studies have entered a period of crisis. The mainstream frameworks dominant in postwar development studies seem unable to meet the South's most compelling challenges. Yet, newer alternative frameworks appear equally impotent or remain relatively undeveloped and unexplored. While a crisis of uncertainty and inertia grips the field of development studies, Third World countries are desperately seeking solutions to their mounting development problems.

This book addresses the need for a critical evaluation of development strategies at both the theoretical and practical level. It reviews mainstream development models to determine what, if anything, each has to offer. Alternative development frameworks are also examined to establish whether they might present viable options and, if so, under what circumstances. Innovative ideas from various development strategies, which might be reformulated and combined within a new hybrid development approach, are given particular attention. The book proceeds from an initial critical deconstruction toward the eventual positive reconstruction of a new development strategy, called 'popular development,' a strategy designed to meet the diverse needs and interests of the popular majority in Third World countries.

In addition to providing a critical evaluation of mainstream and alternative development frameworks, this book integrates a sharp theoretical analysis with empirical research of practical development experiences occurring in all of the South's major regions. Coverage focuses on the theory and practice of development in the capitalist periphery and omits consideration of socialist Third World countries (e.g., China, Cuba), which might form the subject of another quite lengthy volume. Although radical development theories (e.g., dependency and world systems theory, the modes of production approach) are not covered per se, many elements from these frameworks are incorporated into the analysis.

Part One of the book covers mainstream theories and practices of development. Chapter 1 begins this analysis with an examination of the postwar evolution of mainstream development studies, including the now dominant framework of neoliberalism. Many central elements of the neoliberal approach were ostensibly developed in reaction to widely acknowledged shortcomings of other mainstream development paradigms and can only be fully understood within this context. However, a detailed examination of mainstream development frameworks reveals not only many areas of contrast, but also many areas of similarity, that neoliberals have overlooked in their rush to portray the framework as a fresh approach to Third World development. Many of the most important issues and debates in contemporary development studies revolve around these similarities and contrasts. A review of the major postwar development paradigms is therefore fundamental to understanding the direction of current development thinking. The review begins with an examination of the modern origins of mainstream development theory immediately after the Second World War, proceeds to an analysis of the development of Keynesian frameworks such as growth theory and the modernization approach, and culminates in a discussion of the more recent rise of neoliberalism.

Chapter 2 examines the transformation of these abstract theories into the practical strategies that have dominated postwar development in the South. It first looks at the agroexport or primary export model, which was the prevailing strategy in most smaller, rural-based countries, especially during the early postwar period. Next, analysis shifts to the Keynesian strategy of import-substitution industrialization, which has been the centerpiece of postwar development in most larger, urban-oriented countries. It then covers the recent rise of neoliberal strategies of export-led growth. Finally, there is a detailed study of non-traditional exports, a new outward-oriented growth sector which a growing number of analysts believe offers good development prospects for many countries. Although the analysis finds contradictions and shortcomings in all these development strategies, it also uncovers some positive aspects which should not be overlooked in the formulation of new development approaches.

[Additional material that appeared here in the original manuscript was deleted due to considerations of length. However, this material is available in three journal articles. The first, Brohman (1996), complements the section on nontraditional exports in chapter 1. It addresses issues and debates over international tourism, another outward-oriented growth sector. Emphasis is placed on linking tourism-led development to the long-term interests of the popular majority in local communities instead of the short-term goals of an elite transnational minority.

The second and third articles examine common theoretical problems of mainstream development frameworks, ranging from the older modernization approach to neoliberalism. Brohman (1995a) explores tendencies toward universalism, Eurocentrism, and ideological bias in development studies. It points out that these have resulted in many unconventional or theoretically challenging ideas being ignored, particularly those based in forms of indigenous, popular knowledge from the South itself. Brohman (1995b) addresses the problem of economism in development studies, which limits the conceptual space afforded other key areas of development, including sociocultural and political relations, the intersubjective realm of meanings and values in development, and environmental relations and issues of sustainability.]

Chapter 3 examines the postwar development experience of the Asian NICs (Newly Industrializing Countries), which have been the focus of much recent attention in the development literature. The realities of this development experience are compared with neoliberal explanations. Two of the most pressing questions in contemporary development studies are addressed: does the neoliberal explanation of development in the Asian NICs stand up to serious scrutiny? Is it transferable as a new model of development for other Third World countries? These questions are answered by examining the following aspects of NIC development: the role of the state in development, the compatibility of inward- and outward-oriented elements of development, the influence of internal historical and sociocultural conditions on development, and the impact of external geographical and historical factors on development.

Chapter 4 looks at the practical application of neoliberal development strategies in the rest of the South. The rush to embrace the neoliberal approach in the South, particularly through increased reliance on structural adjustment programs (SAPs), presents an opportunity to examine the actual performance of this development approach. Although it is recognized that many factors beyond the immediate control of Third World states have profoundly influenced the outcomes of SAPs, the effect of variations in state policies is emphasized. The importance of ideological considerations in framing neoliberal policies, especially within the IMF and World Bank, is also stressed. Many of the shortcomings of outward-oriented policies and other liberalization measures are revealed, and alternative development strategies are suggested.

Chapter 5 continues analysis of the neoliberal development experience in the South. Stress is placed on the social and political nature of many neoliberal contradictions uncovered in the previous chapter. Attention is given to variations in both state–society relations and the internal composition of Third World states. Such variations may strongly affect the political feasibility of neoliberal measures – with important implications for state legitimacy and social stability. These factors have had a profound impact on the social and environmental sustainability of SAPs. An alternative approach would emphasize sociopolitical factors and the need to construct a broadly based social consensus behind any process of structural change.

Part Two of the book examines alternative development theories and practices. It begins in chapter 6 with an analysis of the major alternative frameworks, especially those focused on broadening equality and meeting human needs. Most of these development alternatives represent rather isolated and small-scale efforts, which have yet to make a substantial impact on the overall development trajectory of Third World countries. Nevertheless, they have introduced many important ideas, concepts, and methods for development, which have been largely overlooked in the mainstream literature. Analysis of the alternative frameworks provides valuable new insights into development, although contradictions and anomalies also remain which must be resolved for these approaches to become more globally important.

Chapter 7 furthers analysis of alternative development strategies by examining derivative frameworks of regional and spatial planning. These alternative planning approaches stress the contribution of a well-balanced, efficient, and locally suitable spatial organization to more equitable growth and democratic participation. Decentralization is often seen as critical to promoting more appropriate development and assisting in local resource mobilization. Debates over decentralization and participation have shed light on the interwoven nature of social and spatial structures in development processes. In addition, the complex role that place and locality play in creating specific development contexts has been revealed. However, like alternative strategies in general, these planning approaches must overcome a series of theoretical and practical shortcomings to play a larger role in development.

Chapter 8 focuses on one of the most important new emphases of alternative development: popular participation and empowerment. Progress in this area is essential for any development strategy to meet the South's most pressing development problems. Nevertheless, it has been given inadequate attention in both the theory and practice of development during most of the postwar period. Recently, however, popular participation and empowerment have received more attention, assisted by many new ideas and methods from a range of social sciences and humanities. At the same time, though, ventures into unfamiliar conceptual and methodological territory have created new problems which must be overcome for progress to continue.

Chapter 9 explores a related area of development that has also received much belated attention: women and gender relations. Until quite recently, the role of women in development was all but invisible within various theories and strategies. A variety of new approaches, however, has made considerable progress in addressing this problem. Such efforts have been greatly assisted by the emergence of women's organizations and movements in many countries. Although much work remains to be done, recent theoretical and practical advances have been impressive in areas related to the role of women and gender relations within processes of both production and social reproduction.

Chapter 10 analyzes a third recent emphasis of alternative development: environmental relations and sustainable development. Worsening environmental problems have prompted a rethinking of the development agenda, placing more stress on issues of sustainability alongside more traditional concerns for growth and equity. As a result, a variety of notions of sustainable development have been introduced which have quickly made an impact on development debates. Moreover, as with women's issues, the recent rise of various environmental movements has greatly influenced alternative development theories and practices. Although such work has barely begun in most countries and contains many contradictions and shortcomings, it also promises to have a profound influence on future directions of development.

Finally, chapter 11 discusses the principal elements of a new approach to Third World development, which is called 'popular development' here. Third World societies by their very nature are complex and multifaceted and so is the concept of popular development. This chapter summarizes the diverse components of popular development and links them together to provide a clear overall picture of the approach. Whereas popular development addresses many of the central issues and questions that have frequently occupied postwar development studies, it also employs new concepts and methods designed to overcome the common shortcomings of our development frameworks. In this way, popular development seeks to advance the rich alternative development tradition that has evolved in recent years. As the term implies, popular development focuses on a central concern of all the alternative approaches – creating development which will prove more appropriate to the needs and interests of the popular majority in Third World countries.

Popular development rejects formalistic models and preconceived theories in favor of a more flexible approach which stresses the contextuality of development, which involves a complex and ever-changing interplay of objective conditions and subjective concerns. One of the central themes of this book is that postwar strategies have contorted the meaning of development to serve the narrow interests of an elite minority rather than the broader and more diverse concerns of the popular majority. Indeed, many recent trends in development seem, if anything, to be deepening this problem. In both the theory and practice of development, the time for top-down models administered by an elite minority needs to end. Positive change in the lives of most Third World people awaits their empowerment, which can only take place through prolonged political struggle of the popular sectors in cooperation with like-minded social forces. Recently there has been progress in this area, mainly through the actions of various popular movements, but in most Third World countries these struggles have barely begun.

Part One

Mainstream Theories and Practices

1

The Postwar Tradition in Theory

The profound economic crisis that has recently shaken much of the South has contributed to a dramatic change in mainstream theories and strategies of development. Over the past two decades, neoliberalism has moved from the fringes to the center of development studies, thereby displacing Keynesianism as the dominant development orthodoxy. Many of the central elements of neoliberalism were ostensibly developed in reaction to some widely acknowledged shortcomings of other mainstream development approaches and are only fully comprehended within this context. However, an examination of mainstream development paradigms reveals many areas not only of contrast, but also of similarity, that have been overlooked by neoliberals in their rush to portray their framework as a fresh approach to Third World development. Many of the most important issues and debates of contemporary development studies revolve around these areas of similarity and contrast among the different development approaches. A review of the major postwar development frameworks is therefore fundamental to understanding the direction of current maintream development thinking. The review begins with an examination of the modern origins of mainstream development theory following the Second World War, proceeds to an analysis of Keynesian frameworks such as growth theory and the modernization approach, and culminates in a discussion of the more recent rise of neoliberalism.

The Early Postwar Era of Mainstream Development Theory

The field of development studies emerged shortly after the Second World War and rapidly became focused on a series of macroeconomic problems, particularly those concerning global inequalities between rich and poor countries. While economists played a leading role in the elaboration of postwar development theory, the field has also been notably interdisciplinary with contributions from a variety of other fields, including sociology, anthropology, political science, social psychology, and geography. In fact, the recent trend in development studies is decidedly away from dependence on the conceptual apparatus of a single discipline and toward development of hybrid theories that incorporate elements from a number of related fields.

In contrast to recent mainstream development literature, a sense of enthusiasm and optimism imbued the early postwar era of development studies (Chakravarty 1991; Hirschman 1982; Sutton 1989). The technical and conceptual advances associated with the rise of positivist science in fields such as economics would, it was hoped, provide the tools that state planners needed to bring about a rational process of rapid development: this type of intervention might accelerate economic growth and development and hence the closing of the gap between rich and poor countries.

Links with US Geostrategic and Economic Interests

Development studies became the field by which the South's problems could be rationally assessed and technically managed with the aid of Western guidance. Its conceptual apparatus and analytical techniques quickly became more than intellectual tools to understand development; within the context of the Cold War, they rapidly also became powerful instruments of international relations. The idea that Third World development was a process and goal for which the First World, especially the US, should provide assistance grew out of the auspicious experience of the Marshall Plan in Europe, as well as Cold War fears and ambitions (Black 1991: 48).[1] Europe's speedy recovery under the Marshall Plan lent credence to the notion that planned intervention, if rationally conceived and scientifically managed, could stimulate rapid development. However, not any kind of development would do; the essence of the 1947 Truman Doctrine was that any political or economic change not sanctioned by the US would be arrested.

The mainstream frameworks of growth and modernization theory both arose in the context of Cold War competition between the superpowers for

---

[1]   In the early postwar era, US official development assistance (ODA) dwarfed that of all other states as well as international and nongovernmental entities. As recently as 1982, the US continued to be the largest single contributor, accounting for 22 percent of ODA – almost double the portions of either Japan or Saudi Arabia, the two second largest contributors. However, in 1989 Japan overtook the US as the world's largest donor of nonmilitary aid (Black 1991: 63–5). Much US aid has been closely tied to American geostrategic interests. From the late 1950s to the mid-1970s, the majority of US aid was directed to Asia as a corollary of the wars in Korea and Vietnam. Since the late 1970s, most aid has been directed to the Middle East, about half of it to Israel and Egypt. By the late 1980s, more than 60 percent of the total US foreign assistance budget of about $15 billion was allocated to military and security (ESF) categories (ibid.: 63).

influence in the South: the US and other OECD countries offered capitalist growth and modernization to counter the Soviet Union's proposal for socialist development.[2] Mainstream development theory may be regarded, at least in terms of its political rationale, as the 'ideological child of [Communist] containment' (Preston 1986: xv). As Slater (1993: 419) explains, the extension of mainstream development frameworks into the Third World 'reflected a will to geopolitical power. It provided a discursive legitimation for a whole series of practical interventions and penetrations that sought to subordinate and assimilate the Third World Other.' Indeed, some of the later disillusionment with development studies arose from the realization that much of mainstream theory had served as an instrument for American foreign policy and that this had never been dealt with satisfactorily (T. Smith 1985: 543).[3]

Growth Theory and Keynesian Developmentalism

Growth theory as a school within development studies was popularized from the late 1940s to mid-1950s, after which it was gradually absorbed into the broader framework of modernization theory. Growth theory was based on the transfer to the Third World of a series of Keynesian models for analyzing economic growth that had been developed in the US and Europe, including the Harrod (1948)–Domar (1957) model on savings and investment, Rostow's (1956) stages-of-growth theory, Rodan-Rodenstein's (1943) 'big push,' Scitovsky's (1954) externalities models, and Nurkse's (1953) work on industrialization. Although it later developed a broader conceptual base when it was extended into the modernization framework, growth theory was largely contained within the disciplinary boundaries of economics. The idea of development was conceived rather narrowly as economic growth; social and cultural factors received attention only in their role of facilitating (or hindering) appropriate societal changes that would accompany growth.

Because its intellectual roots were so extensively based in Keynesian economics, growth theory was sometimes termed International Keynesianism.

---

2 Oman and Wignaraja (1991: 3) note that concern by the US and other OECD countries that conditions of socioeconomic underdevelopment left poor countries 'ripe' for Communist takeover was reflected in a number of foreign-policy initiatives during the early postwar period, including 'Point IV' of the Truman administration's 1949 foreign-policy program, the United Kingdom's 1950 'Colombo Plan,' and the Kennedy Administration's 'Alliance for Progress' program in the early 1960s.

3 Although some authors such as W. W. Rostow and Samuel Huntington never felt that this was a problem and openly linked their work with the expansion of US interests into the Third World.

The Keynesian program of growth took a strongly interventionist stance toward Third World development, which stressed comprehensive development planning by reformist states in cooperation with foreign donors. In terms of ideological affinities, growth theory was closer to US New Deal or European social democratic programs than to the laissez-faire tradition associated with neoclassical economics.

The Economic Bases of Growth Theory

Growth theory envisions development as a process of capital formation which, in turn, is largely determined by levels of savings and investment. Domestic savings ought to be directed toward productive investment, especially in high-growth sectors such as manufacturing industries. In instances where market imperfections prevent this process from reaching a successful conclusion, intervention may be required from the state and/or external sources. Growth is regarded as a unilinear process which will endure once momentum is gained; no conceptual space is afforded for possibilities of subsequent decline or underdevelopment.

As income levels increase with development, the marginal propensity to save also rises, freeing capital for further investment and giving growth a more self-sustaining character. Growth becomes essentially market-driven, but during the critical initial stages of development, breaking free from the inertia of long-term stagnation might require extensive state intervention. An often quoted statement by Lewis (1950: 36) captures the image of the difficult initial stage around which development efforts should be concentrated, and after which self-sustaining growth will take place:

> Once the snowball starts to move downhill, it will move of its own momentum, and will get bigger and bigger as it goes along . . . You have, as it were, to begin by rolling your snowball up the mountain. Once you get it there, the rest is easy, but you cannot get it there without first making an initial effort.

The rise of growth theory marked a rift in mainstream development studies between a new, more Keynesian interventionist school and orthodox neoclassical theory, which continued to stress the importance of a 'pure' market and export-led growth based on principles of comparative advantage. Growth theorists turned steadily toward Keynesianism, doubting the ability of neoclassical theory to translate its microeconomic base of individualized, short-run decision-making into a dynamic macroeconomic theory for long-term development. Three principal areas of criticism of neoclassical economics arose from this era (see Myint 1987: 110–12):

1   Neoclassical theory is static and is focused on allocations of given resources,

while development problems are dynamic and should concentrate on increasing investable resources through stimulating savings and investment. Whereas a 'big push' is needed to initiate the development process, neoclassical theory offers only marginal adjustments and piecemeal improvements to market forces.

2   Neoclassical models neglect structural rigidities common to developing countries that prevent markets from responding to price changes in the 'normal' theorized manner. Such rigidities might range from the sociological (e.g., 'irrational' peasant responses toward economic incentives) to the more technical (e.g., limits to labor- or capital-intensiveness under differing historical conditions).

3   The neoclassical emphasis on development based in comparative advantage and free trade is inappropriate to the late industrializers of the South. Direct static losses from state intervention to support industrialization may be more than offset by dynamic gains (e.g., external economies associated with technological change and improved skills, linkages with other economic sectors, long-term benefits from promising 'infant' industries). Unlike the nineteenth century, export-led growth is no longer a viable option; countries must turn to alternative 'engines' of growth such as import-substitution industrialization (ISI) or balanced internally-oriented development between agriculture and manufacturing.

The Image of Growth as a Sequence of Stages

Although growth theory rejected some aspects of neoclassical economics as inappropriate for Third World development, its models and theoretical framework nevertheless remained firmly rooted in Western economic history. This meant that growth theory was structured by a Eurocentric vision of development based in a Keynesian interpretation of the unique, albeit historically important, experience of core industrial capitalism. The 'new nations' of the postcolonial world were required to follow the Western model in order to develop; this represented a 'modernization imperative' (Nayar 1972) based on a stylized version of Western economic history whereby countries pass through a sequence of stages on a unilinear path toward higher Western-style development.

Perhaps the most popular of these stage models (at least among laymen, if not necessarily among development specialists) was Rostow's (1956) 'stages of growth' model. Rostow's model was committed both to Westernization and to industrialization. As was generally characteristic of growth theory, it focused on the problem of raising the savings and investment rates in order to direct investable surplus toward manufacturing industries. Rostow theorized that countries pass through a sequence of development stages leading from a traditional society to modern capitalism based on mass consumption. However, the most critical period within this development transition was termed the 'take-off' stage to self-sustaining growth, in which investment capital must be mobilized by a 'big push.' For this critical juncture, marginal adjustments in the mold of neoclassical economics would not be sufficient to prompt the structural transformation or 'big push' upon which subsequent development depended. If circumstances prevented domestic capital from fulfilling this historic task, the state should intervene and, if necessary, ought to be supported by direct foreign investments and/or official development assistance.

Although some research continued to be devoted to analyzing development stages and related subjects,[4] by the late 1950s growth theory had fallen out of fashion with most development theorists. Looking back, Streeten (1972) contends that the rather rapid demise of growth theory was inevitable in that it attempted to address complex problems of Third World socioeconomic and political change with Keynesian concepts that had been developed primarily for core industrial economies.

Similarly, Preston (1986) concludes that growth theory's 'basic metaphor' of a unilinear path of development was false and consequently produced an inappropriate set of reworked concepts and analytical tools. Problems of Third World development could not be equated with those in First World societies and were, therefore, not susceptible to simple Keynesian remedies.

However, despite these basic contradictions, growth theory did prepare much of the conceptual groundwork for subsequent mainstream development research, especially that associated with the modernization approach. In fact, many of growth theory's central ideas (e.g., development as economic growth based on industrialization, the critical role of savings and investment, the need for state intervention in development planning, sequential development based on the historical experience of the West) have continued to survive in various forms long after growth theory[5] per

---

4   For example, studies by Lewis (1954) and Kuznets (1955) into changes in income distribution, as economies pass through different development stages, stimulated much related theoretical and empirical research. Results generally suggested, following the history of the West, that wealth and income are less evenly distributed during early stages of growth, but that once an economy matures, distribution will become more balanced. Other research translated stages theory from economics to related disciplines concerned with development, such as political science. A study by Organski (1965), for instance, extended stages of Western political development to Third World areas. As was characteristic of Rostow's framework, all societies were said to pass through a fixed sequence of stages, each of which would be dominated by similar issues and problems.

5   Although growth theory is usually associated with subsequent Keynesian-inspired development research such as modernization theory, some of its ideas (e.g., development as growth, the importance of rising rates of savings and investment, development based on the Western model) have also survived in slightly altered form within anti-Keynesian frameworks such as neoliberal development theory.

se fell out of fashion. It is mainly because of this lasting influence that it is difficult to dismiss growth theory out of hand.

Modernization Theory and Continuity in Mainstream Development

In many ways modernization theory represented a deepening and extension of the basic conceptual apparatus of growth theory. Economic factors continued to occupy a central place in modernization theory, and development remained a process by which societies passed through a sequence of stages derived essentially from the history of the West. However, modernization theory extended the analysis of development into a more interdisciplinary realm with the addition of theories of social and institutional change to complement those focused on economic transformation.

Basic Characteristics of Modernization Theory

The models used in the modernization approach became more complex and less narrowly economic. The incorporation of non-economic elements such as social practices, beliefs, values, and customs required an extension of conventional ideas concerning development beyond the economic dimension. Although modernization and development remained closely tied to economic performance, this was, in turn, dependent on social, cultural, and political factors. According to the theory, these non-economic factors should evolve in a manner consistent with the logic of capitalist economic growth. The question of which combinations of norms and institutions would facilitate, or impede, the modernization process became central.

Modern society was thought to be composed of typical economic patterns (e.g., capitalist work rhythms, mass consumerism, high savings and investment rates), typical social patterns (e.g., high literacy and urbanization rates), and typical psychological attributes (e.g., rationalism, achievement motivation). Modernization theory claimed a high correlation existed in Third World societies between the degree of modernization and the diffusion of Western-style cultural and attitudinal traits. Modernization thus involved pressing into service institutions in critical areas, such as education and the mass media, to support the diffusion process. The interlinking of changes in both economic and non-economic factors (e.g., in attitudes toward work, wealth, savings and risk taking) could then take place in a mutually reinforcing manner to support development.

Given the social context of Cold War competition in the South within which modernization theory unfolded, there was also a tendency toward optimism. Some rather pessimistic, somewhat uncomfortable elements of growth theory (e.g., initial problems overcoming economic stagnation in the Harrod–Domar model) were erased in favor of a more positive view, in which capitalist development was made freely available to all who would have it (Preston 1986: 89–90). Rather than being difficult to attain, growth and development were made easier both to achieve and to sustain. Clearly, in the context of aid-donor competition and superpower rivalries, this benign, inclusive image of development offered a much more attractive package to sell.

It has been noted that modernization theory offered no monolithic structure; differences in stress, specificity, and interest produced considerable variations within the overall approach (Hulme and Turner 1990: 39). The theory's diverse intellectual origins, resulting from its multidisciplinary roots, accentuated such differences. Although some of the dominant themes of modernization arose from classical sociology, these were also subjected to different interpretations, which combined to 'form the mish-mash of ideas that came to be known as modernization theory' (Harrison 1988: 1).[6] It is therefore only possible to identify some common, but not intrinsic, elements of modernization theory. It should be remembered that these were not equally shared by all who employed the general approach, even though modernization theory has often been critiqued as an undifferentiated whole.

The main features of modernizaton theory, at least in its classical form up to the mid-1960s, can be briefly summarized:

1   Modernization involves a mixture of development factors (e.g., technological change, capital accumulation, changing values and attitudes) that may be analyzed from a variety of disciplinary perspectives. However, most accounts gave highest priority to inducing social changes (e.g., in values, norms, beliefs, customs) that would prompt related change in other spheres of development. The theorization of such social change came principally via the conceptual apparatus of classical sociology.

2   Societies and their component parts (e.g., values, institutions, social groups, regions) can be divided into modern and traditional spheres. These two spheres are antithetical and, for the most part, exist separately. Although 'dual societies' might coexist in the short term, the modernization process implies the dissolution of the traditional sphere.

3   The direction of change that the modernization process induces is broadly similar for all Third World societies and closely resembles the history of the industrial capitalist world. Although rates of change

6   Among the sociological frameworks that helped to shape the modernization perspective were evolutionism, diffusionism, structural functionalism, systems theory, and interactionism (ibid.: 1).

may vary and temporary social 'dislocations' may occur among some groups, in the end modernization is inevitable and assumed to be beneficial for all. Modernization via capitalist industrialization is not a unique experience of the West but offers a general blueprint for development. Modernization is synonymous with Westernization.

4 The internal diffusion of key factors of development (e.g., changes in values and attitudes, technological innovations, investment capital) originally carried forth modernization within the West. However, for contemporary Third World societies, diffusion of these factors normally originates from the outside. In many instances, this should speed the modernization process, allowing the gap between rich and poor societies to narrow quickly.

5 The speed of diffusion, and of the overall modernization process in general, are critically dependent on the modernizing elite of each society. These 'change agents' act as innovators and diffusers. Therefore, development policies should target them, especially in the initial stages of modernization, to facilitate rapid structural transformation.

6 Although diffusion originates from the outside, modernization essentially depends on factors internal to each society. Especially important for the overall process is the removal of various social and cultural 'barriers' to modernization, many of which are linked to the continuing existence of a traditional sector. Deficiencies resulting from backward internal structures (rather than external factors, such as dependency or geopolitics) are the fundamental causes of underdevelopment. If structural change can be induced, then growth and modernization will follow quickly.

The Roots of Modernization Theory in Durkheim and Weber

Much of the social theory underlying the modernization approach, particularly that associated with the traditional/modern dichotomy, flows from the ideas of the nineteenth-century classical sociologists Max Weber and Emile Durkheim. Durkheim's (1984) *The Division of Labour in Society* differentiated between two basic types of society, each founded on distinct forms of social cohesion. Traditional societies were held together by 'mechanical solidarity' based on group similarity, with their members adhering to a rigid pattern of traditional norms and values. By contrast, modern societies developed more specialized functions and institutions, thereby facilitating the rise of 'organic solidarity' and prompting increasing social differentiation. Weber, like Durkheim, sought to explain the key factors, particularly those connected with industrialization, that were responsible for making 'modern' Western societies distinct from others. In *The Protestant Ethic and the Spirit of Capitalism* (1958), he concentrated on the emergence of a Protestant 'work ethic' and frugality based not in economics, but in religious strictures against waste and extravagance. In addition, he stressed the appearance of rationalization, a cultural process that he believed was peculiar to Western society.

It has been pointed out that attempts by modernization theorists to combine the original ideas of Durkheim and Weber into a single grand theory of development do an injustice to both, expecially to the work of Weber (Webster 1990: 48–9). Because each author conceived of the contrast between modern and traditional societies in quite different terms, attempts to simplify and combine their work have been conceptually problematic. Moreover, this simplification process also removed many of the difficulties that both authors believed would accompany modernization processes. While modernization theory exuded a sense of optimism and rejoiced at the prospects of Western-style modernity, neither Durkheim nor Weber regarded modernization so benignly (Hulme and Turner 1990: 37). Durkheim stressed the anomie and dislocation that often accompanies the destruction of traditional societies by modernization, while Weber took a very pessimistic stance toward the concentration of power that he believed was inevitably linked with modern, large-scale development.

The Creation of Dualistic Development Models

Possibly because they were caught up in the wave of postwar optimism over the prospects of rationally managed development, or perhaps because they felt the pressures of formulating a rapid Western response to the threat of Soviet expansionism, modernization theorists simplified and reconstructed the work of these classical sociologists to give theoretical support for their models of Western-style development. The traditional (*gemeinschaft*) versus modern (*gesellschaft*) dichotomy was fitted into the mainstream sociological framework of structural functionalism to produce a series of dualistic development models which extoled the virtues of Western-led modernization and were amenable to new positivist methodological techniques and statistical procedures. These models dominated the development literature over much of the late 1950s and 1960s. Among the best known examples of this type of modeling are the works of Lewis (1954), Ranis and Fei (1961), and Jorgenson (1961).

The transition model of W. Arthur Lewis illustrates the basic characteristics of the dualistic development model. Following its classical sociological predecessors, this model divides Third World economies into traditional and modern sectors. The traditional sector is composed of peasant agriculture and other occupations employing labor-intensive production methods and primitive technologies. The decision-making process in this sector is irrational in the economic sense that it is not profit maximizing. Because of the presence of 'disguised' unemployment and low productivity in the traditional sector, it can also act as a labor reserve for the modern sector without decreasing output. This labor transfer will take place if the modern sector offers wage rates above the (institutionally determined) wage in the traditional sector.

By contrast, the modern sector is composed of capital-intensive and technologically-advanced industry and agriculture. Firms in this sector operate rationally and are assumed to maximize profits by hiring labor up to the point that their marginal product equals the wage rate. As the economy grows these firms reinvest their profits, expand output, and draw labor from the traditional sector. Wages will not rise in the modern sector until the labor reserve of the traditional sector has been exhausted (i.e., so long as there exists labor with zero marginal productivity in the traditional sector). This 'unlimited supply of labor' for the modern sector provides a powerful economic stimulus, especially for the early stages of industrialization, which allows profits, savings, and investable surplus to rise quickly.

From this perspective, the process of modernization and development in Third World societies essentially depends on the displacement of the traditional (largely peasant) sector by a modern (mostly industrial) sector. The traditional sector is non-dynamic; it does not represent a source of development, but is merely a reservoir from which the modern sector can extract labor and other resources. Basically, the modern sector acts as a 'pole of development' from which various sociocultural and economic elements of modernization radiate outward.

The Focus on Values and Pattern Variables

In this formulation, the principal barriers to development are found in the traditional sector, but should gradually disappear as this sector is displaced by the forces of modernization. Particularly for sociologists associated with modernization theory, changing values were critical for the transformation from traditional to modern society. This focus on changing values was initiated by Parsons' (1951) formulation[7] of distinct sets of 'pattern variables' linked with tradition and modernity; it subsequently formed an important

[7]   A large number of sociologists and others employed variations of Parsons' pattern variables to conceptualize the role that changing values play in the modernization process. For example, Hoselitz (1960) maintained that the transition from a traditional to a modern society essentially depends on changing pattern variables. Lipset (1967) claimed that traditional Latin American societies preserve values that are antithetical to accumulation and growth. Lerner (1958) stressed that development in the Middle East was linked with the spread of modernity and modern values. McClelland (1961) connected development with the rise of entrepreneurship based on changing values associated with 'achievement motivation.'

theoretical buttress for much of the sociocultural work that took place in modernization theory.

In formulating his conception of pattern variables, Talcott Parsons fitted the dualistic notion of modernization developed by Weber and his successor Ferdinand Tönnies[8] into a structural-functionalist approach to social change. Parsons' pattern variables were used to distinguish the essential forms of social interaction and organization in traditional and modern societies. It was assumed that social values and related norms and institutions play a critical role in determining the potential for development of various societies. Development involved much more than simply initiating economic changes; new values, norms, institutions, and organizations had to be introduced to transform the old social order. Elements of traditional societies that constitute obstacles to modernization had to be replaced. Modernization was viewed as a broad process that required changes in social values toward the modern 'ideal' of individualistic capitalism. The 'collective' values and all-encompassing interpersonal relationships that characterized traditional societies had to give way to the 'individual' values associated with the more complex division of labor and increased socioeconomic mobility of advanced capitalism. Resultant changes in individual thought processes and behavior would lead to broader systemic transformations in institutional orientation which would, in turn, promote more rapid change at the individual level.

Modernization studies based on Parsons' dualistic framework paid particular attention to three sets of variables that were thought to differentiate traditional from modern values. The first was the determination of status by achievemental versus ascriptive criteria. In modern societies, positions were thought to be achieved because of qualifications and experience – in contrast to traditional societies in which familial ties, ethnicity, or other unachieved or ascribed attributes dominate. The second concerned the governance of patterns of interaction by universalism versus particularism. Modern universalistic organizations were thought to subject their members to similar rules and regulations, thereby contributing to systemic efficiency and legitimacy, whereas the particularism of traditional organizations discriminated in favor of certain individuals and groups over others. The third focused on the specificity-diffuseness dichotomy concerning the extent to which roles are either narrowly or broadly defined. It was thought that the expectations and obligations associated with a modern system of role relationships have greater specificity – in contrast to the more diffuse system that characterizes a traditional order.

[8]   Like Weber, Tönnies (1855–1936) developed a dualistic conception of modernization that separated traditional from modern societies. The former were said to be bound together by traditional community ties or *gemeinschaft*, while the latter tended to evolve cohesion from more functional associations or *gesellschaft*.

A large number of modernization studies applied derivatives of Parsons' pattern variables to outline a path of Western-style development by which Third World societies could overcome problems of stagnation and under-development rooted in the forces of tradition. One of the best known was a study by Lipset (1967) that linked the preservation of feudal values in Latin America, derived from fifteenth- and sixteenth-century Iberian mores and institutions, with the failure to engender the entrepreneurial spirit necessary for capitalist growth. Lipset maintained that particularistic and ascriptive patterns of behavior characterized traditional Latin American societies; as a consequence, diffuseness, elitism, and weak achievement motivation prevented Latin Americans from adopting more capitalistic traits such as pragmatism, materialism, and a stronger work ethic. Because of this sociocultural burden, the development of modern, rational, capitalist enterprises, which might propel growth through organizational efficiency and calculated risk taking, had failed to take place in Latin America. Therefore, the path to Latin American development lay in the adoption of more 'modern' structural patterns (i.e., an achievement orientation, universalism, and functional specificity) to facilitate capitalist growth.

The Evolutionary Nature of Modernization Theory

The modernization approach envisioned development as a process of rapid induced changes that cumulatively would result in linear progress toward an end point closely resembling the contemporary advanced capitalist world. Much of the theory became either implicitly or explicitly evolutionary. Although not all evolutionists among development theorists adopted the modernization approach, 'there does appear to be some kind of "fit" between structural functionalist and evolutionist perspectives' (Harrison 1988: 59). The common indices used to measure progress along the road to development were the factors thought to be responsible for growth among 'modern' Western societies (e.g., the growth of formal rationality and complex bureaucratic organizations, increased social differentiation and industrial specialization).

Development involved passing through a set sequence of stages based on Western history. As in its dualistic conception of social change, modern-ization theory's evolutionism can be traced to nineteenth-century philoso-phers and social theorists, particularly Comte and Spencer (Klarén 1986: 9).[9] Comte theorized social evolution as a series of stages of human development beginning with a traditional (theological and military) soci-

---

[9]   Auguste Comte (1798–1857) was a French philosopher who is regarded by many as the founder of the positivist philosophy of science. Herbert Spencer (1820–1903) used many of the ideas of Charles Darwin to develop theories of social evolution that became influential in the latter part of the nineteenth and the early twentieth centuries.

ety and culminating in a modern (scientific and industrial) society. Likewise, Spencer maintained that a gradual evolution of human society had taken place through a series of increasingly complex stages. These ideas on social evolution were combined with the dualistic traditional–modern conception to produce a theory of social change in which development was conceived as sequential 'stages of growth' leading toward industrial capitalism.

The Political Implications of Modernization

Most modernization theorists regarded the middle and upper classes as crucial to this process. Progressive members of these classes would play the role of a 'modernizing elite' composed of 'change agents' who would be the bearers of modern (particularly entrepreneurial) values. The middle classes were expected to expand their population share as modernization proceeded, providing a stabilizing force for the overall process. In addition, an expanding middle class should facilitate upward mobility for the lower classes, establishing conditions not only for industrialization based on mass consumption, but also for liberal democratization.

In most modern societies these classes are overwhelmingly urban and industrially oriented; they are generally also well educated and believe strongly in universal public education. The spread of a westernized educational system was considered especially important to solidify the growing influence of this modernizing elite (Johnson 1958, Lipset 1967). Generally, the lower classes were virtually excluded from consideration, or were regarded negatively as bearers of (backward) traditions rather than modernity. The stress placed on developing individual rationality and entrepreneurship, as well as the concepts of innovation and diffusion by which the modernization process would spread, were all aimed at inducing a top-down, center-outward process of capitalist development via a modernizing elite.

In its original version modernization theory took the process of political change that would accompany Western-style development more or less for granted. Economic growth based on mass-consumption industrialization would generate more egalitarian socioeconomic structures upon which liberal democratic institutions could be erected. However, as modernization theory began to evolve in the 1960s, it soon became apparent that the process of political change accompanying modernization in many Third World countries was far from smooth. For some theorists (e.g., Black 1966; Huntington 1968; Smelser 1963), modernization began to be seen as an uneven process that could prove to be quite disruptive and might sometimes serve as a source of sociopolitical conflict. Neither traditional nor transplanted Western political institutions were proving sufficiently strong or flexible to withstand the new sociopolitical tensions unleashed by modernization.

Given the rise of revolutionary pressures in many countries, political theorists such as Huntington adopted the position that maintenance of stability rather than promotion of democratic institutions should be the primary goal of political modernization. This new emphasis on order and stability tended to favor strong (military) regimes, instead of experiments in democratization, as best able to provide the necessary underpinnings for sustained capitalist growth (Klarén 1986: 12). It also provided theoretical justification for US intervention in support of authoritarian regimes that could provide the 'strong government' necessary to sustain the modernization process (Hulme and Turner 1990: 42).

While some political theorists chose to support authoritarianism as a necessary antidote to the social and political unrest that often seemed to accompany modernization, others (e.g., Eisenstadt 1970) regarded such instability more pessimistically, as symptomatic of the fundamental failure of modernization. Rather than offering upward social mobility and increasing equality, Eisenstadt noted that modernization had widened socioeconomic cleavages and had failed to provide any functional basis for the harmonious reintegration of newly differentiated social groups. His analysis was particularly damning of the modernizing elites who, rather than performing their prescribed roles as progressive leaders and bearers of modernity, had more often deepened social divisions by vascillating over crucial issues or using their influence to manipulate policies in their own interests. Hulme and Turner (1990: 42) regard Eisenstadt's pessimistic assessment of the prospects for Western-style development in the Third World as 'an appropriate obituary for the modernization paradigm.' By the turn of the 1970s, classical modernization theory was under increasing intellectual attack. It was apparent that the traditional–modern transformation had failed to take place in the prescribed manner and, instead, had manifested itself in stagnant economies, widening inequalities, and political repression over much of the South.

The Intellectual Legacy of Modernization

For much of the 1970s and 1980s, the modernization paradigm, if not totally moribund, was relegated to the margins of development theory under a sustained barrage of criticism from both the Right (i.e., neoliberals) and Left (i.e., dependency and neo-Marxist theorists). In recent years, however, there has been a move to revive modernization theory, albeit in a substantially renovated form that takes note of the many criticisms to which it has been subjected. This 'rethinking' of modernization generally acknowledges the devastating theoretical critiques levelled in the 1970s (e.g., by Smith 1973; Roxborough 1979) against its pro-Western ideological bias and its adoption of a deterministic and formalistic evolutionism based on Western history.[10]

Thus, Nash (1984) attempts to delink modernization from westernization, thereby recasting the modernization process in a more universal manner. Similarly, Roxborough (1988) maintains that modernization theory does not require a unilinear approach to development but may be used to analyze a variety of societal structures and development paths. According to this perspective, the extrinsic features of modernization theory can be stripped away, leaving development theorists with important conceptual tools for analyzing historical change, seen in terms of the increasing rationalization of human action (in the Weberian sense of formal rationality). This could result in a reformulation of modernization theory that might incorporate concepts from different bodies of development thought. Some analysts have recently adopted the position that considerable potential exists for the creation of a hybrid theory that might blend elements of the modernization approach with those of other development frameworks, such as dependency or world systems theory. Apter (1987: 29), for example, claims that both the modernization and dependency frameworks suffer from 'crucial absences, omissions, and blindnesses' that could be overcome through a carefully constructed process of theoretical convergence. The broad focus of modernization theory might lend itself to analyses of sociocultural factors of development that have often been overlooked by more economically oriented theories, while dependency concepts might correct some of the ethnocentric and bourgeois liberal biases of the modernization perspective.

Similarly, Harrison (1988) argues for the 'commensurability of perspectives' between the modernization school's emphasis on internal development factors and the dependency–world systems schools' stress on external forces. Development theorists are urged to pay more attention to cultural and political structures and to incorporate both objective and subjective elements into their analyses. Neither modernization nor dependency theory has been able to account for recent changes in Third World development (e.g., the rapid rise of the East Asian NICs) notwithstanding that modernization theory might contribute concepts needed to incorporate cultural traditions and processes of social change in a new hybrid framework. Bossert (1987) sees a 'convergence of thought' taking place in Latin American development theory, prompted by rising dissatisfaction with the traditional approaches of the Right and Left. Higgot (1983: 103) urges dependency theorists and other members of the 'radical' camp to reconsider the core propositions

[10]   A lengthy theoretical critique of the modernization approach, neoliberalism, and other frameworks within the mainstream development tradition appears in Brohman (1995a and 1995b).

of modernization theory, while resisting 'the temptation to throw out the modernization baby with the bath water.'

The movement toward convergence between a renovated modernization theory and other development perspectives has gained momentum in recent years. The realization among many theorists and practitioners of development that neither mainstream nor radical frameworks offer viable solutions to critical Third World problems has prompted this movement. Tired old conceptual dichotomies (e.g., modernization–dependency, internal–external) seem no longer to afford needed explanatory power. On a more practical level, none of the traditional approaches seems capable of offering a coherent alternative strategy or viable way out of the present crisis. While development theoreticians bemoan the apparent stagnation of their work, the practitioners of development become increasingly demoralized, as the old ideas and methods seem unable to cope with changing conditions, new contradictions, and unfamiliar pressures.

However, while some movement has recently been discernible toward a convergence between modernization theory and other development approaches, sizable obstacles remain. Over and above substantial ideological differences, many modernization theorists and proponents of more radical approaches continue to insist that their respective theoretical frameworks are mutually exclusive. Accordingly, many of these analysts believe that a hybrid approach can evolve only at the cost of theoretical coherence, producing a hodgepodge of incompatible concepts and ideas. Others are prepared to accept a certain lack of coherence, at least in the short term, as an acceptable trade-off for advancing more meaningful research into pressing development problems. However, any movement toward a hybrid alternative involving modernization theory will have to carefully unpack a considerable amount of conceptual baggage associated with that framework.

Ian Roxborough (1988: 755–6) offers a good example of what this process might entail by exploring alternative conceptualizations of the traditional–modern dichotomy. If these terms are taken, as in much of classical modernization theory, to refer to concrete types of society, they will obscure a wide range of variations in social structures that remain central to any serious analysis of historical change. However, replacing the term 'traditional' with 'precapitalist' or 'premodern' societies might alleviate some conceptual problems. The notion of a transition from precapitalist (or premodern) societies to modern, capitalist societies may dispense with at least some of the objections that have been levelled at the traditional–modern dichotomy. The processes of modernization and development become synonymous, and the notion that 'traditional' or 'premodern' has any meaningful content is abandoned. The concept of modernization, then, is devoid of much of its formalistic and ideological content; it becomes simply the capacity for social transformation. This transformative capacity is, in turn, linked with social differentiation and an increase in the formal rationality of social action. The transformative potential of societies increases as they become more complex. Following Parsons and others, this represents a core concept of modernization theory for explaining historical change.

However, one should note that the usefulness of the theory likely to result from such a process is closely related to the meaning and use of key concepts and terms. This, in turn, depends on whether the core theoretical propositions of modernization, many of which can be traced to the social theory of Weber and other classical thinkers, can be liberated from the ideological baggage that they acquired during the postwar rise of mainstream development theory. In the end, this baggage might only represent some extrinsic features of modernization theory that can be stripped away without affecting its core propositions, which view history in terms of increasing rationalization, in the Weberian sense. It is difficult for any theory of historical change to dismiss this focus completely. However, its inclusion in any hybrid development approach will necessitate discarding classical modernization theory's distorted image of Third World progress following the footsteps of the West. This might allow a respecification of the boundaries and the elaboration of mutually compatible concepts between hitherto antithetical approaches (e.g., modernization and dependency theories) to the study of development.

The Rise of Neoliberal Development Theory

While some analysts have recently called for the development of a hybrid approach incorporating the modernization and dependency–world systems perspectives, many other theorists have turned away from the mainstream 'developmentalist' paradigm in favor of neoliberalism. The neoliberal 'counterrevolution' is dedicated to counteracting the impact that Keynesianism has had on mainstream development theory since the rise of growth theory and the modernization approach in the early postwar period. As we saw in the previous two sections, most mainstream theorists, particularly in the 1950s and 1960s, rejected the neoclassical emphasis on market forces and outward-oriented growth as inadequate for the special Third World needs of rapid development. Key problems of Third World development were thought to be essentially dynamic and, therefore, not amenable to solutions based on 'static' neoclassical models of market-directed resource allocation. Especially during the critical 'take-off' stage in the transition to sustained growth and modernization, it was believed that the state should supply a 'big push' through massive investment programs and comprehensive development planning, rather than simply relying on market forces to eventually overcome the inertia of tradition.

However, by the end of the 1960s there were clear signs that this developmentalist paradigm had gone into decline and was rapidly being supplanted by a more orthodox neoclassical approach in major development sub-fields such as international trade, agricultural economics, and development planning. By the late 1970s, a major change was underway within mainstream theory as a whole, prompted both by the rise of anti-Keynesian conservatism in Europe and the US and by the seeming inability of the developmentalist approach to offer viable solutions for mounting Third World problems. The election in 1979–80 of conservative political parties espousing stridently anti-Keynesian monetarist policies in the US, the UK, and West Germany gave confidence to neoclassical theorists that they had finally returned to the centerstage after a prolonged period at the margins of development theory. Moreover, it afforded them access to important domestic and international platforms, as well as prospects for greater recognition and financial reward. The rapid decline of much of the South into economic crisis by the late 1970s provided additional impetus for the neoliberal counterrevolution. As many countries began to suffer the disastrous effects of economic stagnation, declining per capita incomes, rising indebtedness, and fiscal insolvency, a reappraisal of mainstream development theory took place that called into question the effectiveness of traditional policies. This resulted in a profound change in the accepted economic, financial, and sociopolitical orthodoxy, which placed a new emphasis on supply-side factors, private initiative, market-led growth, and outward-oriented development, while turning away from older developmentalist policies based in demand stimulation, import-substitution, state intervention, and centralized development planning.

Monetarism versus the Political Economy of Keynes

Perhaps the central figure in the early development of the anti-Keynesian framework in development studies was Harry Johnson, a technically proficient economist and specialist in monetary theory who held chairs at both the University of Chicago and the London School of Economics (Toye 1987: 23–4). Monetarism has subsequently played a dominant role in shaping neoliberal theory and the structural adjustment programs of the IMF and World Bank. Monetarist economics can be traced back to the equilibrium theorists of the late nineteenth and early twentieth centuries, particularly Irving Fischer (Preston 1986: 112–13). For Fischer, the mechanism of interest-rate adjustments basically sustained economic equilibrium: high rates decreased growth and low rates increased it. Moreover, the quantity of money in an economy, which is directly derived from government policies, absolutely determined the prices of goods. This 'Quantity Theory of Money' has formed a key element in the anti-Keynesian approach of Johnson and other neoliberals. The growing macroeconomic problems (e.g., inflationary pressures, indebtedness) of many countries are viewed basically as monetary phenomena resulting from excessive government spending and other demand stimulation that has driven up the quantity of money in their economies to unsustainable levels.

Rather than spurring growth and development, Keynesian state intervention into Third World economies is held responsible for a series of interrelated problems (e.g., stagnation of production, rent-seeking and inefficiencies in the state sector, inflationary pressures, indebtedness, fiscal insolvency). The present crisis in the South is not attributable to external factors, such as neocolonialism or global structural inequalities, but fundamentally is the result of misguided internal policies based in a deluded belief that Keynesianism would foster development. According to this perspective, the South cannot resume growth and development until it rejects Keynesian state interventionism in favor of more rational policies based in orthodox neoclassical theory. Such policies would basically reduce the development role of the state to providing the framework within which a laissez-faire market could operate efficiently.

Monoeconomics versus Duoeconomics

In addition to its anti-Keynesianism and adherence to monetarist principles, the neoliberal counterrevolution also proposes a replacement of the so-called 'duoeconomics' of the developmentalist paradigm by 'monoeconomics' based in neoclassical theory. According to Hirschman (1982), development economists may be divided into two basic types. On the one hand, there are those who take a monoeconomic perspective, whereby orthodox neoclassical theory is equally applicable in both the North and the South. Although the South may have peculiar characteristics, such as higher levels of uncertainty, its economic agents make decisions and its markets function essentially according to the same logic as in the North. Most importantly, individuals or firms respond similarly to structures of incentives, whether they are (properly) established by the economic market or (improperly) set by the political market.

On the other hand, there are the followers of a duoeconomic or mainstream Keynesian approach who believe that standard market economics is of little relevance to the special problems of development in the South. Without substantial theoretical modification, they see the neoclassical paradigm as incapable of addressing problems associated with underdevelopment that arise from a series of particular Third World structural conditions. These include: massive unemployment and underemployment, despite instances of high growth and industrialization; frequent market failures based on poorly developed circulation systems, financial networks, and other economic structures; the continuing influence of culture and tradition on common forms of behavior that detract from utility maximization; the persistence of extreme societal polarization and inequalities, which are aggravated by poorly articulated social, sectoral, and regional structures; and a pronounced vulnerability to externally generated crises due to high levels of foreign dependency and economic concentration within the external sector.

Neoliberals contend that these problems do not require a different approach to development specific to the South, but are amenable to general solutions based in standard economic principles. Moreover, neoliberals charge that incoherent development strategies, resulting from the mainstream developmentalist paradigm's perversion of neoclassical economics, have aggravated many of these problems. According to Haberler (1987), development economists initially acquired a series of anti-market, *dirigiste* biases by way of basic misconceptions within Keynesian theory concerning the roots of the Great Depression in the excesses of laissez-faire capitalism. Haberler and other neoliberals maintain that these misconceptions caused development economists to adopt strategies that stressed comprehensive planning and other forms of state intervention, which have prevented markets from functioning properly. The economic performance of most Third World countries has subsequently been poor, especially when compared with those few countries that followed policies based on neoclassical principles of market-directed growth, minimal state intervention, and an open trade regime (e.g., the East and Southeast Asian NICs).[11] As Theodor Schultz (1980: 639) noted in his Nobel lecture: 'A major mistake of much new development economics has been the presumption that standard economic theory is inadequate for analyzing the economic behavior of low-income countries.' Accordingly, future growth and development in the South await the replacement of policies based in the Keynesian duoeconomic approach by neoliberal policies based in the monoeconomics of neoclassical theory.

Classical Liberalism and the Marginalist Revolution

Following the early research of Harry Johnson, neoliberals have given monetary theory a central place in their paradigm. However, the broader roots of their monoeconomic approach to development can also be traced back to eighteenth- and nineteenth-century classical liberalism, particularly the work of Smith and Ricardo, and to the marginalist revolution in the late

11 A detailed analysis of this neoliberal depiction of development in the Asian NICs appears in chapter 3.

nineteenth century, which established much of the conceptual basis for the subsequent rise of neoclassical theory.

The neoliberal school generally stems from Adam Smith and his laissez-faire principle that was outlined in his most influential work, *The Wealth of Nations* (1880). Developed in reaction to eighteenth-century mercantilism, which held that the exercise of strict state controls over key economic activities such as international trade and investments maximizes a country's wealth, the laissez-faire principle instead calls for a minimization of state intervention in economic transactions. Liberal arguments against state intervention were later reinforced by David Ricardo's theory of comparative advantage, which maintains that countries have a common interest in the free flow of goods, services, and capital across borders, and calls for the elimination of tariffs, quotas, and other forms of state interference in trade. The law of comparative advantage states that a country may raise its consumption level above that which would be possible in a state of autarchy by specializing in the production and export of commodities for which it has the lowest production costs, derived from relative advantages in key factors of production (e.g., raw materials, labor costs, technological expertise). Each country (and the world as a whole) will obtain more goods, at a constant level of factor input, by specializing in and exporting those commodities that it can produce most efficiently within the international division of labor. Moreover, free trade will eventually lead to factor price equalization among countries so that, for example, wage differences between the North and South will be gradually reduced, thereby promoting more equal global income distribution. Thus, the liberal legacy maintains that the road to greater development for both rich and poor countries lies in specialization, free trade, and global integration rather than in autarchy, self-sufficiency, and state intervention.

While classical liberalism supplied the intellectual roots and ideological agenda for the neoliberal program of market-led, outward-oriented development, the major theoretical and analytical components of neoliberalism can be traced back to the origins of neoclassical economics in the marginalist revolution. Marginalist or neoclassical economics, as it became known, essentially views human behavior as the outcome of rational choice. Economic theory is focused on relationships between a basic psychological end, utility, and a set of scarce resources. Individuals and firms must make rational choices in a world of unlimited desires, but scarce means. The role of the 'rational actor' or *homo economicus* (economic man) becomes one of making the 'best' choices, that is, those that maximize one's ends given the limited means available. Barnes (1988: 477) notes that *homo economicus* plays a dual role in neoclassical economics, supplying it with both a unifying theoretical concept and a methodological agenda:

The first [role] is theoretical. By defining the best choices, economic rationality allows the neoclassical vision of the economic problem, couched in terms of means and ends, to be realized. The second role is that of providing neoclassicism with a methodological agenda. It is an agenda based upon reducing the complexity of economic events at any time or place to the universal trait of rational choice making; a trait that, because of its determinist nature, is easily represented in a formal model.

The neoliberal development model is based on a neoclassical reading of the economic history of the industrialized capitalist world. It thus stresses elements of development such as market-led growth; increased savings and private investment based on high profits and, at least initially, low wages; gradual industrialization, beginning with light industries; innovation diffusion and technological advancement through increasing global economic integration; and the progressive 'trickle down' of the benefits of growth to all social classes, economic sectors, and geographic regions.

The neoliberal model provides a powerful means of simplifying complex social processes of development so that an agenda for policymaking can be established based in the microeconomic theory of neoclassical economics. In the early twentieth century, the marginalist revolution established microeconomics as the new paradigm for economics as a whole, thereby giving the discipline a much more unified core model and more parsimonious approach to development studies than other fields offered. Moreover, neoclassical economics exemplified the rise of positivism across the social sciences with its employment of the 'scientific method' of hypothesis testing, its ontological focus on empirical facts and events, its derivation of law-like generalizations, and its promise of predictable results based in the replicability of its models.

Neoclassical (and, by extension, neoliberal) models focus on the sphere of market exchange and, via aggregation, derive society-wide conclusions from an individual level of analysis by employing sophisticated statistical techniques. Market mechanisms are seen to promote general welfare most efficiently by maximizing utility to the individual. An assumed set of fixed consumer preferences reduces human behavior to 'a remarkably parsimonious postulate: that of the self-interested isolated individual who chooses freely and rationally between alternative courses of action after computing their prospective costs and benefits' (Hirschman 1985: 7).

This 'rational-actor' approach, coupled with an emphasis on gradual, marginal, and equilibrating change, narrows development economics to a field in which calculus and other quantitative procedures can be employed to yield more scientifically certain results. However, this process also leaves aside many non-economic elements of development. In addition, it restricts economic analysis to (directly observable and measurable) factors that can most comfortably fit its new modeling techniques, thereby excluding from study much of the broader traditional subject matter of classical political economy.

Public Choice Theory and the Minimalist State

In fact, a new type of neoliberal political economy has recently evolved in the form of public choice theory,[12] which focuses on the allocation of public resources in the political market and emphasizes redistribution to powerful interest groups (e.g., Bates 1987; Bauer 1972, 1984; Becker 1983; Niskanen 1971; Olsen 1965, 1982). Using the assumptions and methods of neo-classical economics, public choice theorists construct simple models that yield testable hypotheses which can be used to predict, and supposedly explain, political behavior in the real world (Dearlove 1987: 7). Within development studies, cross-disciplinary influences emanate from (neoclassical) economics rather than from the mainstream tradition in political science or political sociology. Gone are traditional concepts of political modern-ization and development with their characteristic terminology of systems, functions, structures, and cultures. Instead, the focus has turned to raising the scientific level, analytical rigor, and precision of political studies to match the supposedly more sophisticated economic analyses of development offered by neoclassical theory (Hettne 1990: 67). In effect, public choice theory challenges neoclassical economists to remake political economy. Eco-nomic science, it is argued, is relevant to all major aspects of development, including those related to politics, because it provides a generalized theory of rational behavior (i.e., choice and decision-making) that is supported by a rigorous method of scientific enquiry and explanation.

While they call for more scientific and rigorous political analysis, many public choice theorists are also keenly aware of the normative and ideo-logical implications of their work. What they are attempting is a paradigm shift in development studies which will lead away from the old Keynesian emphasis on market failure and the benevolent state toward a new neoliberal focus on political failure and the predatory state. Within the mainstream Keynesian tradition, the state operates in the common interest of society as a whole. The state is called upon to correct frequent market failures and attend to societal objectives (e.g., income redistribution, cultural and environmental protection) that are believed to lie beyond the scope of the market. By contrast, the public choice framework is characterized by the neoclassical belief that real-world markets may sometimes be imperfect,

---

[12]   Sometimes also variously termed rational choice, neoclassical political economy, new institutionalist economics, new political economy, or the political economy of rent-seeking (see Sridharan 1993).

but real-world governments are even more imperfect – imperfect markets are, therefore, generally regarded as preferable to imperfect governments.

Public choice theorists contend that mainstream development theory has been characterized by a Keynesian bias under which the state can do no wrong. However, some analysts contend that, in reaction to this perceived bias, public choice theory itself has swung to the opposite extreme, in which the state can do no right (e.g., Killick 1989; Streeten 1989). The state is regarded simply as the executor of discriminatory policies in favor of self-interested pressure groups upon whose support it depends. Planners, bureaucrats, politicians, and other members of the state use their privileged positions within the policymaking process to extract 'rents' in return for influencing policy outcomes. This 'rent-seeking' behavior by the state distorts economic activities away from efficient and equitable resource allocations.[13] Rent-seeking is turned into an essentially political phenomenon,[14] and the best way to reduce rent-seeking is to place strict limits on state activities. State intervention is not the solution, as in Keynesian theory, but the problem. The state does not optimize the welfare of society as a whole, but only that of special-interest groups. State intervention causes, rather than cures, market imperfections. Moreover, most Third World governments do not have the autonomy to pursue 'correct' policies (i.e., those supporting market-directed growth) because of interference from powerful pressure groups and their supporters within the state itself.

Public choice theory seeks to explain how politically rational states can follow irrational development strategies (Lal 1984). It explains policies that have inhibited economic growth in many countries (e.g., maintenance of

13 Rent-seeking can take a variety of forms. Sometimes it consists of illegal activities such as bribery, payoffs, or smuggling. It may also consist of surpluses generated through the creation of monopolies resulting from discriminatory government licencing arrangements, import–export regulations, and so on. Another common example is the restriction of jobs within the civil service to particular groups by various means, such as requiring excessive educational qualifications or selective advertising. In many countries, rent-seeking is also associated with selective distribution of commodities that are in short supply. Whatever form it takes, such rent-seeking is associated with considerable economic inefficiencies and waste of scarce resources, particularly in the South. Economic costs attributable to rent-seeking within the public sector in some Third World countries are estimated to be 10 percent or more of total GNP (Grais et al. 1986; Krueger 1990).

14 Ironically, within the classical political economy of Adam Smith, David Ricardo, and James Mill, a major source of inefficiencies and wasteful expenditures was not so much rent-seeking within the political arena, but the extraction of rents in the private sector resulting from monopoly controls by landlords, moneylenders, commercial intermediaries, and others. It has not escaped notice that public choice theorists, despite their intellectual heritage in nineteenth-century liberalism, have virtually ignored the common phenomenon of private rent-seeking in Third World countries in favor of an exclusive, and highly ideological, focus on the state (e.g., Bagchi 1993; Streeten 1993).

overvalued exchange rates, large inefficient bureaucracies) without resorting to assumptions of ignorance or willful misbehavior by politicians and bureaucrats. Instead, the state is seen as a rational actor which seeks to maximize its economic and political utility by extending its influence within powerful groups in society, albeit at the cost of slower long-term growth and development (Rausser and Thomas 1990: 372–3).

However, the question remains whether this neoliberal political economy model is any more relevant to problems of Third World political development than the Keynesian pluralist model of the state that it seeks to replace. It has yet to offer any sophisticated theory of the Third World state, whether based in an examination of the internal structures and mechanisms of the state itself or in an analysis of the social composition of the state within broader societal structures.[15] While public choice theory offers an analytical framework for understanding the seemingly irrational development strategies that have been followed by many Third World states, it provides no logical apparatus for political reforms that may produce more effective policies. In the end, this paradigm is limited by its narrow and cynical view of the political process. While this provides much of the ideological affinity between public choice theory and neoliberalism in general, it offers little analysis of real-world political complexities in which states may engage in a range of both productive and predatory behavior according to a variety of internal and external influences.

[15] A more detailed critique of mainstream and neoclassical political theory appears in Brohman (1995a and 1995b).

2

Strategies of Growth and Industrialization

Chapter 1 described the evolution of development theories. Chapter 2 now considers how these theories were transformed into the practical strategies that dominated postwar development in the South. It first looks at the agroexport or primary export model, the prevailing strategy in most smaller, rural countries. The analysis then moves to the Keynesian strategy of import-substitution industrialization, which, for most of the postwar period, has been the dominant development approach in most larger, urban countries. The recent rise of neoliberal strategies of export-led growth is then covered. Finally, there is a detailed study of nontraditional exports, a new outward-oriented growth sector which many analysts believe offers good development prospects. Although contradictions and shortcomings can be uncovered in all these strategies, they contain positive aspects that ought not to be overlooked in the formulation of new development approaches.

The Agroexport Model

Throughout the postwar era, nearly all mainstream development strategies (including, most recently, neoliberalism) have called on the majority of Third World countries to exploit their 'comparative advantages' in cheap land and labor by expanding exports of agricultural goods and other primary commodities. This advice was particularly aimed at poorer, smaller countries in regions such as Central America and sub-Saharan Africa, which had little or no history of export-oriented industrialization. Indeed, agroexport production did expand rapidly during the early postwar period in regions like Central America, becoming a 'motor' for outward-oriented development. But many analysts now contend that a number of inherent contradictions have gradually played themselves out during recent years to render the agroexport model, at least in its classical form, dysfunctional to future Third World development. At the same time, however, many countries possess few realistic alternatives to agroexport production in the short-to-medium term, especially to earn foreign exchange necessary for macroeconomic stability. Thus, a debate is now being waged over whether to retain agroexport production as a major axis of development and, if so, in what form. This section addresses these issues, first by examining the traditional agroexport model, and secondly by exploring possibilities for the establishment of alternative strategies which retain a significant agroexport component.

Comparative Advantage and Outward-Oriented Growth

As we saw in chapter 1, the principle of comparative advantage has been a key element of neoclassical development. This principle assures Third World countries that the road to higher growth and development lies in specialization and exchange – even with the advanced industrial countries of the First World. The classical nineteenth-century Ricardian conception of comparative advantage has since been modified and incorporated into neoclassical theory, especially through the work of Heckscher, Ohlin, and later Samuelson (1948). The resultant Heckscher–Ohlin–Samuelson (HOS) model of international trade assumes equal access to production technologies throughout the world, so that comparative advantages arise only from differential factor endowments (e.g., land, labor, capital). It follows that land- or labor-abundant Third World countries should specialize in and export land- or labor-intensive goods, while leaving the production of capital-intensive goods to those (First World) countries with greater endowments of capital.

The HOS model thus maintains that specialization and trade will increase levels of production and consumption in both developed and developing countries. The model also suggests that there is a tendency toward equalization of factor prices, including wage rates, following the development of trade. This would allow wage rates to rise gradually in many Third World countries, as specialization in labor-abundant goods eventually reduces their relative abundance of labor and drives up wages. Conversely, continuing capital-intensive production will steadily reduce relative labor scarcity in the First World, gradually decreasing its marginal productivities and wage rates. The neoclassical HOS model, therefore, holds out prospects for growth and development in the South on two fronts. During their initial stages of development, Third World countries can enjoy higher levels of growth and gradual wage increases through the production of primary commodities for the world market. During subsequent stages of development, many Third World countries will begin to gain comparative advantages in semi-manufactured goods as gradual changes in their factor endowments allow for specialization in more sophisticated products.

In addition to stressing the advantages of specialization and trade, neo-classical theory also suggests that Third World countries can move to higher stages of development more quickly by maintaining an outward economic orientation. Due to the supposed lack of entrepreneurship and technological skills in most traditional societies, private foreign investment is seen as the best means of providing the capital and expertise needed to employ more sophisticated production techniques. It is implied that developing countries can only harm themselves by placing artificial limits on the free flow of trade and investment, since benefits from both the comparative advantage principle and the theory of mobility of capital transfers depend on the adoption of policies promoting outward-oriented economic growth.

Basic Elements of the Classical Agroexport Model

Neoclassical development strategies have thus focused on both stimulating growth according to comparative advantages and attracting foreign capital through infrastructure projects and other programs designed to provide a profitable and stable environment for investment. In many Third World countries, this has led to increased exports of primary commodities, particularly from the agricultural sector, to exploit comparative advantages based on relatively cheap labor and land. Especially for many of the poorer and smaller Third World economies, growth in the agroexport sector was thought to be critical for attracting (foreign) investment capital, creating a positive trade balance, and expanding job creation through the operation of 'multiplier' and 'spread' effects.

Since the Second World War, export agriculture has indeed attracted substantial foreign capital and has been the principal source of growth for many of the smaller Third World economies. 'Traditional' export crops (e.g., coffee, tea, bananas, sugar, cotton), as well as newer agroexports (e.g., animal feeds, beef, fruit and vegetables), have claimed a growing proportion of cultivable land in the South, including large tracts of tropical rainforest and grassland savanna. In many areas of Latin America, Africa, and, to a lesser extent, Asia, the deterioration of fragile ecosystems has been linked to agroexport expansion and the accompanying displacement of peasants into environmentally sensitive areas (e.g., López 1992; Amin 1993). At the same time, the production of staple foods for the domestic market has stagnated or declined in most agroexport-dominated economies; indeed, in many of the South's main agricultural zones, it has been supplanted entirely. Moreover, society has become polarized through the exigencies of agroexport production, in which land concentration and absolute labor exploitation essentially generate comparative advantages. This has blocked possibilities for creating more broadly based development models stressing economic diversification for both internal and external markets.

The reorientation of agricultural production toward exports has also produced a profound structural transformation in many rural areas. The agroexport model, at least in its classical form, implies the destruction of small/medium peasant forms of production and enterprise, and of the rural village communities upon which they are based. Characteristically, the lateral expansion of large-scale agribusiness has squeezed peasants off their land in traditional food-producing areas. These former peasants have commonly been converted into a landless or near-landless floating reserve of labor which is often seasonally employed for peak periods of labor demand, such as to harvest crops for export. Peasant displacements have also contributed to growing rural–urban migration in many countries, thereby swelling the urban labor reserve, which exerts downward pressure on wages and undermines non-wage relations of production in urban areas. An enlargement and reconstruction of the overall surplus population takes place as members of rural families (notably youths, the elderly, and women) are pushed into the latent labor reserve, with many forced to eke out a bare existence in the nebulous informal sector.

For many Third World countries, particularly in Latin America, the agroexport model was associated with high rates of economic growth from the end of the Second World War until the mid-1970s. From 1950 to 1977, Central American agroexports increased by twelve times. During much of this period, some of Latin America's poorest and smallest countries (e.g., El Salvador, Guatemala, Nicaragua) sustained among the highest overall growth rates in the region, based on the dynamism of their agroexport sectors (Brohman 1989). By the end of the 1960s, however, many development theorists began to recognize that high rates of economic growth were not necessarily correlated with other basic objectives of Third World development. Deteriorating trends in employment, income distribution, and levels of poverty often accompanied impressive growth rates.

The Agroexport Model and Postwar Central America

Although much of the South has had some experience with the agroexport model, it is perhaps in Central America where the model has been most forcefully applied for the longest period of time. The Central American development performance therefore represents a good case study for assessing the model's long-term impact. Because economies that are concentrated on the export of primary commodities are extremely vulnerable to fluctuations in world market conditions, they characteristically follow a 'boom and bust' pattern. The Central American economies have traditionally been concentrated on a few agroexport sectors and, during the postwar era, their overall economic performance has closely followed this classic pattern of boom and bust (FitzGerald 1991).

Although there were some variations in the performance of individual economies, Central America generally experienced a boom period of relatively high growth from the early 1950s to the mid-1970s. This growth was based on a very limited number of agroexports: the rise of cotton in the 1950s, beef in the 1960s, and sugar in the early 1970s complemented more traditional crops of coffee and bananas. However, this boom period has given way to a pronounced recession that began in the late 1970s and has continued in most of these countries until the 1990s. Of the Central American economies, only Costa Rica has resumed a high rate of economic growth in recent years. However, its growth has essentially been based on the rise of non-traditional exports, tourism, and other forms of economic diversification, rather than continuing to depend on a reduced number of traditional agroexports.[1]

The agroexport model in Central America functions according to 'a logic of the minority' (Collins 1985: 108). The expansion of the model to serve the interests of a narrow economic and political elite has created the very conditions that are responsible for the marginalization and impoverishment of the majority. Investments and growth are concentrated among a few agricultural sectors controlled by foreign capital and allied fractions of the domestic bourgeoisie. Meanwhile, the remainder of the economy molders in neglect, unable to meet even the basic needs of the majority of the population (e.g., Whiteford and Ferguson 1991).[2] Typically, the benefits of economic growth are concentrated among a few large-scale

[1] A more detailed examination of new development strategies based on economic diversification into sectors such as tourism and non-traditional exports is carried out in the final section of this chapter.

[2] Within Central America, Costa Rica has represented an exception to this pattern of extreme polarization between social classes and economic sectors. Although agroexport production has traditionally formed an important part of the Costa Rican economy, the major agroexport sectors (at least until recently) have not been marked by excessive levels of concentration as in the other Central American economies. Historically, rural development in Costa Rica has included many medium-sized farms, in contrast to the more polarized latifundio–minifundio structure in much of the rest of Latin America. Many analysts contend that this has allowed a more diversified and internally articulated pattern of economic growth to evolve in Costa Rica which has, in turn, facilitated the rise of a social democratic political system. All of these features differentiate Costa Rican development from the classical agroexport model. However, as we shall see in the last section of this chapter, the recent rise of agroexport production in Costa Rica has been marked by many of the same characteristics (e.g., land concentrations, peasant displacements and growing rural–urban migration, widening rural inequalities) that have traditionally defined the agroexport model in the rest of Central America.

export producers, primarily in the dominant agricultural sector and related agro-industries. The economic structure closely conforms to the classical *Cepalista* model of Latin American peripheral capitalism (see Prebisch 1950),[3] with the dominant economic sectors oriented toward the reproduction and extension of a dependent capitalist mode of production. CEPAL (1983) characterized Central American export-led growth as 'superimposed development,' with a relatively modern agroexport sector superimposed on and independent of the remainder of the underdeveloped, internally oriented economy. Despite relatively high growth rates during the boom period of agroexport expansion, few internal 'multipliers' have been created that might generate jobs and income for the majority of the population.

Within this pattern of development, the evolution of the external sector 'determines the global behavior of the economy'; restrictions found in that sector 'mark the limit on the rate of domestic economic activity' (ibid.: 5–6). A direct relationship is established between the performance of the export sector, on the one hand, and overall rates of economic expansion, investment and capital accumulation, levels of employment, external balance of payments and import capacity, and the principal sources of government revenue, on the other. The export sector not only determines economic patterns, but also conditions the evolution of the social structure and configurations of political power. Torres-Rivas is among the many analysts who have stressed the fundamentally flawed nature of the agroexport model in Central America:

> The export-oriented economy notably retarded national and social integration and contributed to the extreme rigidity of political and social relations . . . development has been determined by an externally-oriented dynamic whose essential nature has remained unchanged in spite of efforts (after World War II and especially after 1955) . . . to implant a new productive base dependent on the growth of an internal market. (Torres-Rivas 1980: 25)

State Intervention to Support Agroexport Capitalism

In contrast to the rather haphazard involvement of the state within the Central American economies before the Second World War, state intervention in support of the accumulation requirements of agroexport capitalism has become much more pronounced in the postwar era. In addition to strengthening repressive labor policies and various other methods of coercion

---

3   During the 1950s and 1960s, the United Nations CEPAL or *Comisión Económica para América Latina* (Economic Commission for Latin America) based in Santiago, Chile, was a center of criticism of the agroexport model and other forms of dependent development in Latin America. A more detailed analysis of CEPAL's alternative strategy follows in the next section, concerning import-substitution industrialization.

designed to expand the rural labor reserve, the state has employed a variety of mechanisms that have decisively influenced the overall pace and direction of the accumulation process in the dominant agroexport economy. Prominent among these state policies and programs have been: discriminatory credit, tariff, pricing, and exchange-rate policies; selective construction of roads and other infrastructure; and the provision of publicly subsidized irrigation, research and extension, storage, and processing/marketing facilities to favored producers (Brohman 1989).[4] State economic intervention has accelerated tendencies toward concentration and centralization of capital in key agroexport sectors and related processing, commercialization, and import/export activities. Patterns of regionally uneven development have also been accentuated as state resources are directed toward major concentrations of large-scale agroexport production, while areas dominated by peasant producers of basic grains and other foodstuffs are left to stagnate in abject poverty and isolation.

This development strategy has been described as a 'repressive agroexport model' (Barraclough 1982: 15). If state economic measures prove insufficient to meet the accumulation requirements of the agroexport bourgeoisie, armed force is brutally applied to bring recalcitrant social sectors into line. Peasants, rural workers, and the urban poor have no real role in the system other than providing a steady source of cheap labor. Because the markets for Central America's principal (agroexport) production sectors are located overseas, methods of absolute exploitation can be used to hold down labor costs without adversely affecting demand for the goods produced. Rather than promoting the rise of a relatively 'progressive' modernizing bourgeoisie whose profitability might be based on technological advance, state policies have strengthened patterns of absolute exploitation and ownership concentration that have traditionally supported agroexport production by a reactionary landholding oligarchy.

Land Concentration, Semiproletarians, and Absolute Exploitation

Agroexport profitability stems not from increases in relative surplus linked with rising productivity, but is based on the extraction of absolute surplus

---

4   Moreover, many discriminatory state programs that have favored large-scale agroexport production over small–medium food production in Central America have also been supported by international aid and lending organizations (e.g., US Agency for International Development, Inter-American Development Bank, World Bank). Particularly noteworthy in Central America (as well as in other areas such as the Brazilian Amazon, the Dominican Republic, and Paraguay) have been greatly increased amounts of investment capital supplied by international donors and lending institutions for expansion of the export beef industry since the 1960s.

derived from maintaining very low labor costs.[5] Land concentrations by the agroexport bourgeoisie have reinforced traditional precapitalist mechanisms of peasant exploitation and have provided impetus for the rise of newer, more capitalistic exploitative forms as land-poor peasants have been forced to seek seasonal wage labor in the agroexport sectors. The creation of a massive reserve of seasonal rural laborers has become a condition for meeting the demands of cheap labor upon which agroexport profitability and international competitivenesss are largely based. Low wages and poor working conditions associated with forms of absolute exploitation are related both to the seasonality of agroexport labor requirements and to the lack of alternative sources of steady income for masses of rural semiproletarians and itinerant proletarians (de Janvry 1981).[6] At the same time, the ability of these groups to meet their families' needs from sources other than temporary wage labor in the agroexport sector (e.g., other types of seasonal wage labor, activities in the informal sector, partial subsistence production) allows agroexport producers to keep labor costs at levels beneath those which would be required to maintain a permanent, fully proletarianized labor force.

Accordingly, Central American rural development has created masses of semiproletarians and itinerant proletarians, which may be regarded as 'peculiar forms of the proletarianization process of the capitalist agroexport model' (Nuñez 1980: 39). The expropriation and displacement of much of the rural population created a mobile labor force which had few alternatives but to respond to the requirments of cheap, seasonal labor within the

[5]   Profitability in most productive sectors of the advanced capitalist world during the postwar period has been based on increasing relative surplus value. Increases in relative surplus are linked with rising productivity, which also allows wages and labor costs to increase. In the agroexport economies of the South, however, profitability in the dominant agricultural sectors has traditionally been based on extraction of absolute surplus through maintenance of low labor and land costs rather than increasing productivity. For a more detailed discussion of forms of absolute surplus extraction linked with an analysis of the Central American agroexport model see Torres-Rivas (1981).

[6]   In his analysis of alternative roads of capitalist rural development in Latin America, de Janvry coined the term semiproletarians for land-poor peasants who are forced to seek seasonal wage labor (normally in agroexport harvests) to meet the social reproduction needs of their families. They are part peasant and part wage laborer, thus becoming rural semiproletarians. Itinerant proletarians are usually completely landless and have been forced into a cyclical migratory pattern of seasonal wage jobs. They are therefore fully proletarianized in that they normally work only as wage laborers; however, they are also itinerant in contrast to more sedentary permanent proletarians. These semiproletarians and itinerant proletarians, many of whom are former peasants displaced by postwar land concentrations, form the bulk of the (seasonal) labor requirements of the principal agroexport sectors in Central America and many other rural Third World economies.

dominant agroexport sectors around which capitalist growth revolved. Barry (1987: xiv) reports that, by the 1980s, rural landlessness had tripled since the 1960s and that about 80 percent of farmers possessed insufficient land to feed their families; at the same time, 85 percent of the best land was used for agroexports and 45 percent of total arable land was devoted to cattle grazing.

By the end of the postwar agroexport boom, all of the Central American countries (with the exception of Honduras)[7] had developed immensely expanded rural labor reserves, composed primarily of part-time peasants and migrant workers who supplied the bulk of the labor power used by their agroexport sectors (Brohman 1989). Because other sources of income have been largely blocked as a result of agroexport expansion, the peasantry throughout Central America has remained dependent on seasonal wage labor during agroexport harvests as its principal source of family income. In Guatemala, for example, Burback and Flynn (1980) found that temporary wage labor in the agroexport sector accounted for almost three-quarters of total peasant family income. In Nicaragua, three export crops (coffee, cotton, and sugarcane), with particularly heavy labor requirements during the harvest season, controlled almost 54 percent of the total rural workforce in the 1970s (Baumeister 1984).

Concentrated Land Tenure and Income Inequalities

Within Third World economies dominated by agroexport production, there is a strong correlation between concentrations of land tenure and unequal income distributions (e.g., Barraclough 1982; Enge and Martinez-Enge 1991). Because land is generally the principal means of production in such economies, extreme concentrations of land ownership commonly lead to equally skewed distributions of income. In Central America the acceleration of land concentrations in the postwar era greatly widened income inequalities (Vilas 1984: 74). Of all the countries in the region, by the 1970s, only in Guatemala did the poorest 50 percent of the population receive a larger share of national income than the wealthiest 5 percent, and even there the margin was quite slight (23.5 percent versus 21.8 percent). In both Nicaragua and Honduras, the wealthiest 5 percent of the population received roughly double the percentage of national income that the poorest 50 percent earned. In all five countries, the share of national income garnered by the wealthiest 20 percent exceeded that of the poorest 50 percent by a wide margin: 60.0 percent to 15.0 percent in Nicaragua, 50.7 percent to 20.8 percent in Costa Rica, 64.8 percent to 12.4% in El Salvador, 51.3 percent to 23.5 percent in Guatemala, and 58.9 percent to

7   In Honduras, the economic dominance of the 'banana enclave' along the Caribbean coast led to the creation of a more sedentary rural proletariat.

17.3 percent in Honduras. The index of income polarization (measuring the inequality of average income between the wealthiest 5 percent and the poorest 50 percent of the population) was most extreme in Honduras (20.0) and Nicaragua (18.6), but was also substantial in the other countries of the region: 11.2 in Costa Rica, 15.0 in El Salvador, and 9.2 in Guatemala.[8]

Postwar inequalities became particularly pronounced among Central American countries, owing to the absolute domination of their economies by agroexport capital. Meanwhile, a marked tendency toward more regressive income distribution also marked rural development in other areas of Latin America that were subjected to a rapid, if somewhat less concentrated, form of agroexport expansion. A number of studies carried out near the end of the agroexport 'boom' in the 1970s reveal a staggering rate of rural poverty for Latin America as a whole: a 1975 World Bank study found that 42 percent of the region's rural population had per capita incomes of less that $75 per annum, while a 1978 CEPAL study reported that 62 percent of the region's rural households could not satisfy their basic needs (de Janvry 1981: 85). As a direct consequence of growing inequities and impoverishment, malnutrition and associated problems also increased, especially among the most vulnerable sections of the population such as poor children. By the 1980s, it was estimated that about three out of four children in Central America were malnourished (Barry 1987: xiv), while in Mexico some 90 percent of the rural population suffered from a severe deficiency of calories and protein (Esteva 1983: 13). Nor is this pattern unique to Latin America – links between agroexport growth and problems of malnutrition and food scarcity have been uncovered in many other Third World areas, including Africa (Bryant 1988), the Pacific Basin (Schuh and McCoy 1986), and Asia (Chisholm and Tyers 1982).

Increasing land concentrations, new and more onerous forms of rural exploitation, and widening rural inequalities have accompanied postwar agroexport booms in a succession of Third World countries. As Grindle (1986: 7, 112) notes, increasing rural inequalities in postwar Latin America are caused not by the isolation or backwardness of the peasantry, as the dualist thesis of the modernization approach contends, but by the ways in which peasants have been inserted into the expanding capitalist economy in the countryside:

> The rural poor are not isolated or backward and have not simply been left behind by the modern sector, but the growth in their unemployment and underdevelopment, landlessness, wage dependence and migration is a direct result of developments in the modern capitalist sector and of state policies . . . At the same time that policies for agricultural modernization

8   For more Central American data on land tenure and income distribution, see Brohman (1989: 518–19).

increasingly dominated markets and profits, peasants were driven into greater debt, squeezed from their land, forced into wage labor, and pushed to migrate in increasing numbers.

Systemic Limitations and Contradictions of the Agroexport Model

The contradictory nature of the agroexport model in Central America has not only blocked possibilities of development for much of the peasantry and domestic agriculture, it has also limited growth within the industrial sector to a narrow branch of activities tied through forward/backward linkages to agroexport production. The limited industrialization that took place under the Central American Common Market (CACM) in the 1960s and 1970s actually strengthened the hold of the dominant agrarian-based export structure over the remainder of the economy (Torres-Rivas 1980: 28–30). The polarization that characterized agricultural development was replicated and extended into the industrial sector with the establishment of a reduced group of large capital-intensive enterprises linked to agroexport production alongside a large number of small, technologically backward operations aimed at domestic consumption. This type of truncated industrialization di 1 little to alter the international position of Central America's economy as an exporter of largely unrefined agricultural products and an importer of manufactured intermediate and final consumption goods. The concentration of industrial production within sectors supplying goods to the advanced capitalist world did not permit the broadening of patterns of development to other areas of the economy. Very partial internal processing of most agroexports meant that most of the value-added and employment associated with turning primary agricultural products into final consumption goods was exported to the developed world.

With the possible exception of Costa Rica, the basic characterisitics of the postwar Central American economies are defined by: the dominance of overall production by the agricultural sector; a chronic crisis within agricultural production for the internal market; the weakness and narrowness of an industrial structure based on agro-industrial processing; the bloated nature of the informal sector and unproductive commercial activities; and severe polarization of growth in sectoral, regional, and class terms. All of these characteristics can be seen as direct consequences of the particular logic of accumulation upon which the agroexport model is based. The manner in which agroexport production takes place, supported by diverse forms of absolute exploitation and the concentration of land and other means of production, conditions the entire internal socioeconomic structure and the nature of political power. The linking of economic growth to a subordinate position within the international division of labor, based on the comparative advantage offered to agroexport sectors by maintaining low labor and land costs, retards national socioeconomic and spatial integration and inevitably leads to an extremely rigid polarization of social and political relations.

The growth and pattern of overall development in an agroexport economy depends on an externally oriented dynamic in which the demand for the goods of its principal production sectors comes not from domestic consumption but from overseas markets. Because their sources of demand are external, the key agroexport sectors operate according to an independent logic of accumulation which has little correspondence to the necessities of broader development for other economic sectors. The lack of domestic demand for consumption goods blocks the spread of internally oriented growth. It is also directly related to the accumulation logic of agroexport production, rooted in property concentrations and deepening forms of absolute exploitation. Mechanisms which might stimulate internal demand (e.g., agrarian reforms, rising wage levels and income redistribution, improvements in the social wage) do not have a functional relationship to the dynamic of the accumulation process of the dominant agroexport sectors. Indeed, they would impede agroexport production, serving to undermine the bases of its comparative advantage (i.e., cheap labor and land) in the international commodity markets.

Growing systemic tension within the agroexport model has recently been noted in most Central American countries (e.g., Bulmer-Thomas 1988; LaFeber 1983; Pelupessy 1991b; Torres-Rivas 1981; Williams 1986).[9] Although the way in which systemic contradictions eventually manifest themselves is also dependent on indeterminate sociopolitical factors which may be particular to each social formation, there is widespread agreement among many analysts that the agroexport model, at least in its classical form, has recently become exhausted and offers no real future for development. As intractable societal problems have become more acute in Central America, the model is said to have entered the stage of its 'final crisis,' which increasingly calls into question the ability of the model to overcome its central contradictions in the absence of structural change. The focus of many recent accounts of Central American development has been on the economic and political difficulties involved in breaking with the agroexport model against the interests of the region's powerful landholding oligarchy and its domestic and foreign allies. Possibilities for broadening development based on industrialization and other forms of economic diversification have been blocked because the class fractions that dominate the major productive sectors and political arenas have opposed economic changes that might reduce their profits and power (FitzGerald 1991; Torres-Rivas 1981). At the same time, the logic by which the agroexport model operates has

9   Similar tensions have also been noted in other areas of the South in which postwar development has been dominated by agroexports (e.g., Barraclough 1982; Collins 1985; de Janvry 1981; Weisskoff 1992).

systematically impoverished large segments of the population and greatly accelerated environmental destruction in many rural areas (Whiteford and Ferguson 1991; Williams 1986).

From this perspective, the roots of the present crisis in Central America can be traced to the maintenance of conditions that permit an exploitative and exclusionary agroexport model to endure against the interests of the majority. Within agrarian-based societies such as those of Central America, strong links exist between patterns of land tenure, societal polarization, and the arbitrary, and often ruthless, exercise of political power. Many analysts portray the recent rise of political instability and military conflict in Central America as symptomatic of the exhaustion of the agroexport model. Williams (1986: 191), for example, argues that 'even if the wars in Central America ended today . . . a [continuation of the same] development program would produce the conditions for a resurgence of the conflict within ten or fifteen years.' Accordingly, if stability and development are to return to Central America, new development strategies must be found that operate according to a different logic than the outmoded agroexport model. On the one hand, the model relies on a world commodities market that has substantially changed in recent years. Stagnant global demand has combined with increased supply, resulting from the adoption of similar export-led strategies by all of the South's major agricultural producers, to send prices of many traditional agroexports into a protracted decline. On the other hand, the degree of political repression, economic exploitation, and environmental destruction required to maintain the 'comparative advantages' of the major agroexport sectors is no longer feasible.

Possibilities for an Alternative Agroexport-led Development Model

Most analysts agree that future development for agroexport economies is dependent on structural change which responds to the needs and interests of the popular majority rather than of a narrow elite. Nevertheless, in traditional agroexport areas such as Central America in which neither regional autarchy nor widespread economic diversification are realistic possibilities for the immediate future, agroexport production must be maintained at least in the medium term. Agroexports in small, narrow dependent economies represent the equivalent of a 'capital goods sector' (FitzGerald 1985), in that agroexports have the unique ability to generate foreign exchange, which in turn determines the availability of the producer goods needed to generate overall economic growth. The deterioration of key agroexport sectors would cause unacceptably high social and economic costs in terms not only of foreign-exchange earnings, but also of internal savings and investment capital, productive employment, income generation, and sources of government revenues. Given the existing structures of production and social-class formations in the region, there are no economic sectors which could readily replace agroexports. It follows that any program of economic revitalization must include a strategy to recover the agroexport dynamic.

However, while selected agroexports should continue to play a role in the creation of any viable alternative development strategy, the central mechanisms by which the old model operated need to be replaced to allow for more broadly based and sustainable development. A dynamic agroexport sector need not necessarily be based on an exclusionary and exploitative *latifundio–minifundio* model. Instead, it might be based on alternative forms of rural organization such as medium-sized farms and cooperatives that, if given proper state support, could combine economic viability with social equity. Small/medium farmers and cooperatives might complement their traditional focus on domestic food production with a carefully managed entry into selected agroexport markets. Indeed, many small/medium farmers' associations in Central America have indicated that their members would welcome the opportunity to diversify into export sectors if provided with proper conditions (Rosene 1990). Development policies to promote exports ought to be designed to meet the specific needs of small/medium producers (through risk minimization, export diversification alongside food production, use of labor-intensive production techniques). In cooperation with national farmers' organizations, mechanisms such as credit, service, and marketing cooperatives and other forms of producers' associations ought to be encouraged to enhance farmers' technical and marketing expertise and help to promote market diversification, especially into overseas areas.

An alternative, more broadly based development model for areas such as Central America will require a more flexible and diversified export platform in which comparative advantage is derived from an increasingly skilled workforce, an expanded domestic wage-goods sector, and other factors beyond those rooted in land concentration and absolute labor exploitation that propelled the old agroexport model. Mechanisms such as distributive agrarian reform and fiscal modernization need to be implemented to facilitate the transfer of profits to sectors and social classes beyond the agroexport elite. In addition, public investment priorities should shift toward providing a basic social and economic infrastructure (e.g., rural health care, education, technical assistance programs, roads, irrigation) that would permit wider economic participation and allow real living standards to rise via productivity gains without sacrificing international competitiveness. Unit costs of labor can decline for agroexports and other globally competitive economic sectors (allowing wages to rise alongside profitability) without reliance on capital-intensive technologies, if labor intensity improves (with income incentives) and the labor force becomes increasingly skilled (as a result of better social infrastructure). An additional advantage is that the labor force could also become increasingly flexible in its ability to adapt to new production sectors and techniques. This means that an expanded wage-goods sector (to provide income incentives) and social services sector (to heighten labor skills) should be seen as essential components rather than alternatives for a renovated export-led development model in areas such as Central America.

However, a new development model for Central America should promote enlarged and more equitable primary agroexport structures as well as opportunities for diversification into both domestic wage-goods and export sectors based in new areas such as non-traditional agriculture, manufacturing, and services. In fact, Costa Rica's relatively successful growth performance since the early 1980s (based largely on non-traditional exports) helps to illustrate the potential for increased economic diversification in the region. While there is nothing inherently wrong with increasing production from traditional agroexport sectors under the conditions outlined above, this does not mean that promising opportunities for diversification into non-traditional export sectors ought to be pushed aside.

In regions such as Central America, in which productive structures and class formations have historically been dominated by the agricultural sector, forms of economic diversification based on forward/backward linkages with primary agroexport production probably need to play a leading role, at least in the initial stages, in any viable development strategy. Moreover, in addition to increasing industrialization linked to agroexports, the development of internally oriented agro-industries should also be encouraged to supply a broader range of wage-goods and other basic needs for domestic markets. Research in other areas of the South has demonstrated that, among industrial sectors, agriculturally based manufacturing often has strong links with the local economy. For example, in an inter-industry study of different strategies of export promotion for India, Dholakia et al. (1992: M155) conclude:

> if our objective is to generate high income effects without sacrificing the linkage effects on the rest of the economy so as to achieve diversified high growth in the system, the agri-based manufacturing sectors are obvious candidates for intensive export promotion measures.

The Need for Regional Cooperation and State Assistance

It is widely accepted that economic diversification can help to reduce the instability in export earnings that has plagued many Third World economies dominated by primary commodity production. However, new Third World exporters attempting to gain a foothold in global markets often face substantial entry barriers and other marketing constraints within overseas trade. This is especially true for many trans-oceanic marketing channels in which scale of production is important and where transnational capitals and their Third World affiliates have historically dominated trade (van der Laan 1993). In addition to such constraints, monopolistic and oligopolistic firms based in the North control most international marketing chains, especially for agroexports and other primary commodities, and appropriate much of the surplus produced by exporters from the South. In the case of the coffee trade between Central America and West Germany, for example, Meister (1991) finds that Central American producers receive only one-quarter of the economic surplus that they generate and that their production costs represent only one-sixth of the coffee's consumer price. These figures would seem to indicate that Central American producers might better improve their position within global commodity markets via collective negotiation along international marketing chains (perhaps with the support of producer associations, the state, and/or regional trading bodies) rather than by taking measures to reduce production costs, such as by lowering wages. Moreover, as Meister (1991) suggests, much room exists for Central American countries to work together, in cooperation with producer associations, to increase their share of selected global markets. With state assistance, action could be taken in the fields of advertising, improving export services, inspecting and certifying the quality of exports, and so on. Actions on these fronts might improve the image of Central American products in selected markets and encourage Central American exporters to pay more attention to the international reputation of their products.

Problems such as excessive surplus extraction and steep entry barriers within global marketing channels underscore the need for more basic research into the functioning of international markets, as a critical initial step for any export-led development strategy. Such research might uncover methods for Third World countries to exploit new export opportunities, as well as exposing constraints and limitations that need to be overcome in sectors that may at first seem promising. In the case of Central America, for example, there is a critical need to examine the nature of the marketing channels for traditional and non-traditional agroexports and related agro-industrial products. Particular attention should be paid to the distribution of surplus between producers and merchants, on the one hand, and to opportunities for access to new markets, on the other.

In addition to exploring possibilities for expanding and transforming trade with the North, export diversification should also be encouraged by seeking new ways to stimulate South–South trading links. In many cases, creating means to strengthen regional trading blocs and common markets (e.g., the CACM in Central America) might further this goal. As is the case in Central America, such trading blocs have often existed for prolonged periods 'on paper' only, or in quite limited form. Finding methods to facilitate free trade within the regions of the South opens up possibilities for rapid market expansion for firms previously restricted to relatively small domestic markets. Moreover, enhanced regional economic cooperation that conforms to the 'logic of the majority' should allow for the mutually beneficial exchange of products according to various countries' factor proportions and areas of technological expertise. Trade ties with local NICs may prove especially important for opening up new opportunities for growth in many rural countries. For example, regional economic cooperation could facilitate the exchange of Central American agrarian-based products for capital equipment or new technologies from more industrialized regional economies such as Mexico and Venezuela. Many development analysts and organizations (notably UNIDO and UNCTAD of the United Nations) believe that enhanced cooperation among developing countries has an enormous and, as yet, largely unexplored potential for overcoming problems of scale in areas such as production, and research and development. Bagchi (1990: 412), for instance, cites the case of growing cooperation between Cuba and Mexico in selected fields of biotechnology as offering an appropriate example of mutually beneficial exchange between Third World countries with different factor proportions and areas of expertise.

In regions such as Central America where the market 'logic' of the old agroexport model has produced severe societal polarization, it should also be apparent that any revitalized export-led development strategy will require considerable state intervention, at least in the initial stage. The market will not guarantee the maintenance, let alone the increase, of export production. Neither will it permit, in its presently polarized form, increasing economic participation by disadvantaged classes and social groups. Without policies to assist, for example, disadvantaged small/medium rural producers, any new cycle of export-led growth in Central America will inexorably reproduce the exclusionary character of the old agroexport model. As in the past, export-led growth will conform to the narrow interests of an elite minority rather than to the broader needs of the popular majority.

However, the inclusion within a renovated export-led development strategy of measures designed to raise general levels of rural productivity and economic participation (e.g., improved primary and secondary education, technical assistance programs, health care), as well as more tailored programs to meet the special needs of small/medium producers (e.g., risk minimization, export diversification while maintaining food production, adoption of labor- rather than capital-intensive techniques) might encourage broader diversification into new, potentially profitable export sectors that until now have been the exclusive domain of large-scale capital. For most of Central America, the correct package of selective policies must include redistributive agrarian reforms that can be compatible with an overall strategy to stimulate agroexport production. Various forms of support will need to be extended to small/medium producers arranged in many diverse (individual and cooperative) forms of production and exchange. The effects of such support, in terms of productive employment and income generation, as well as access to land and other major means of production, are important considerations for this strategy. Finally, attention should be focused on creating new conditions for increased social harmony and political stability, without which any future development strategy, whether or not it contains a significant agroexport component, cannot be sustained.

Import-Substitution Industrialization

For much of the postwar period, import-substitution industrialization (ISI) occupied a prominent place within development theories, in addition to playing an important role in the practical development experiences of many Third World countries. However, in recent years ISI has been subjected to a withering attack from critics on various sides of development studies, particularly from many neoliberal strategists. They argue that the contradictions and shortcomings of previous ISI strategies have made it as a fundamentally flawed model for future Third World development. This critique, though, has not gone unanswered: a group of development theorists, especially from the Latin American neostructuralist camp, have begun to question the neoliberal depiction of ISI and whether this forecloses all prospects for the development of new ISI approaches. Such debate has once again opened up possibilities for the inclusion of a renovated ISI approach within the development strategy of countries that seek to balance export-led growth with a complementary inward orientation. Any renewed approach to ISI, however, must take into consideration the successes and failures of past ISI strategies, which will be examined in this section.

The Rise of ISI in Postwar Development Strategies

In the 1950s many mainstream development theorists, especially those who had adopted the new Keynesian growth models, began to turn away from primary exports and to see industrialization as the key vehicle to propel rapid development for the South, particularly in some of its larger, more developed economies. Industrialization was regarded as especially important for alleviating capital constraints, which were commonly viewed as the key roadblock to growth and development in most countries. Moreover, there was widespread pessimism over the ability of exports to generate earnings fast enough to keep pace with the rapidly increasing import requirements of a modernizing economy. Possibilities for achieving more rapid and self-sustaining growth became linked with the rise of import-substitution industrialization, which would promote needed economic diversification while attracting direct foreign investment and concessional capital (i.e., aid) to spur growth.

The ISI strategy called for increasing production of manufactured goods for domestic consumption to nurture national markets. This would decrease external dependency and heighten self-sufficiency; absorb surplus labor, especially from the traditional agricultural sector; reduce balance-of-payments problems; foster more advanced stages of industrialization; and establish linkages with related sectors to encourage economic diversification. A mix of policies including fiscal incentives (e.g., low-interest loans, tax concessions) and protection from foreign competition (e.g., tariffs, quotas, licencing, exchange controls) was frequently put in place to promote ISI. In many countries, state-owned enterprises were created, either to carry out import-substitution directly or to support private capitals involved in this enterprise. ISI was also often combined with efforts by regional development authorities to integrate lagging regions into national economies via industrialization and the creation of backward linkages with primary sectors supplying industrial inputs.

In a few larger countries, ISI was focused on the creation of heavy industries, but more often it was chiefly directed at establishing light industries to supply intermediate and final consumption goods. In addition to typically high levels of state involvement, ISI also commonly attracted investment by foreign capitals designed to improve their access to local markets and circumvent mounting trade barriers. Nevertheless, foreign capitals often proved to be highly resistant to state efforts to regulate and shape the direction of ISI (e.g., through policies intended to promote internal forward/backward linkages, local reinvestment of profits, shared ownership and control by nationals, higher levels of local employment). In some cases, disputes over such issues led to foreign capitals being threatened with nationalization of their local assets; in rare instances, such threats were actually carried out.

By the 1950s, ISI had become the dominant development strategy in much of the Third World, especially for the larger economies of Latin America and South Asia (e.g., Argentina, Brazil, India, Mexico, Pakistan). The social origins of ISI can be traced back to the dramatic downturn in international commodity markets caused by the Great Depression (and extended by the Second World War), which undercut previous development strategies based on export-led growth. In addition, ISI became associated in many countries with the increasing economic strength and ideological assertiveness of a new, modernizing, urban-based bourgeoisie which was seeking to wrest economic and political power from more reactionary elements of the traditional rural oligarchy that had been linked to the old agroexport model.

In some Latin American countries (e.g., Brazil under the Vargas and Kubitschek administrations, Argentina under Perón), ISI played a key role in facilitating the state-led transformation of society in the interests of a new power bloc composed of an urban-based class alliance (led by the industrial bourgeoisie accompanied by labor, technocrats, the military, and other urban middle-class elements) (Cardoso and Faletto 1969). ISI often provided a valuable tool for the centralization and consolidation of power by this new industrially oriented hegemonic bloc. In addition, the state frequently viewed ISI as strategically important for increasing national self-reliance, particularly in many of the larger, more assertive Third World countries. ISI also had strong symbolic value for many countries in that it represented a further step along the widely acknowledged path to modernization.

By the end of the Second World War, ISI had provided the major Latin American countries with most of their basic consumption goods (e.g., processed goods, textiles, footwear, pharmaceuticals) and construction materials (e.g., cement, lumber, paints). During the 1950s and 1960s, many of the larger countries in the region also began import substitution in heavy industrial sectors (e.g., steel, basic chemicals), while smaller countries initiated ISI in the consumer goods sectors. During these initial stages of ISI, growth rates were typically high. Investment capital flowed into a series of technologically simple and relatively cheap industries that had ready (and protected) markets for an array of products. However, once these relatively easy gains had been acheived, ISI strategies began to run into severe difficulties. Since the early 1970s, ISI has come under increasing attack from both Left and Right, especially from neoliberals who associate it with excessive state intervention and the undermining of market-led, outward-oriented growth (see Harris 1986).

Support by the US and Transnational Capital for ISI

Most analysts seek to explain the rise of ISI by examining domestic political factors in Third World countries and/or the effects of global economic changes on those countries. For example, neoliberals commonly portray ISI as a key component of protectionist development policies promoted by populist and nationalist coalitions in the South. However, this interpretation neglects the fact that in many countries the most dynamic (consumer durables, intermediate and capital goods) ISI sectors were dominated by foreign-owned transnational corporations (TNCs), often with the support of considerable state subsidization.[10] It also neglects the important role that US and other transnational capitals played in sponsoring ISI, particularly in the postwar period. Using case studies of the Philippines, Turkey, and Argentina to support their analysis, Maxfield and Nolt (1990: 78) find that 'US internationalist businessmen and government officials worked with local proponents of industrialization to shape the formulation and implementation of ISI policies.' Ironically, ISI was promoted by those sectors of US society commonly associated with advocacy of liberal trade policies: the executive branch of the state and transnational corporations. Moreover, these authors contend that ISI was an important US initiative and not merely a concession to Third World nationalists in the context of the Cold War. The US promoted ISI even in countries such as the Philippines where domestic support for the strategy was weak. In other countries, such as Turkey and Argentina, where some version of ISI would have been implemented in any case, the US encouraged the adoption of a limited version of ISI that would secure favorable conditions for US direct foreign investment. This ensured that any protectionism associated with ISI would not entail the loss of foreign markets for large US corporations, but would actually facilitate their globalization by providing preferential access to protected markets while excluding trade by other foreign competitors.

The Neglect of Agriculture

In addition to being promoted by the US government and transnational capital, ISI also received considerable theoretical support from First World academics. The principal focus of mainstream development models in the 1950s and early 1960s was almost invariably on industrialization as an essential aspect of long-run development.

Industrial expansion would protect developing economies from worsening terms of trade for primary products; it would also supply a more secure basis for steady growth based on economies of specialization and scale, technological transformation, and associated learning and demonstration effects. Third World development was basically seen as a transformative process from a traditional, agricultural, and rural economy toward a modern, industrial, and urban one. Industrialization was correlated with various benefits of development such as high employment and per capita income, while underdevelopment was seen as the legacy of insufficiently

10   Evans and Gereffi (1982: 138) report that in Latin America, for example, TNCs (especially from the US) rapidly consolidated their positions within the most dynamic ISI sectors of the largest countries (e.g., in Brazil in the sectors of automobiles, pharmaceuticals, rubber, nonferrous metals, electrical machinery and goods; in Mexico in the sectors of chemicals, rubber, nonelectrical and electrical machinery, and transportation equipment).

developed industrial sectors. Given the prevailing conditions in most developing countries, the transition to industrialization was normally thought to require an inflow of capital and technology from abroad, leading to increased global economic integration and higher levels of dependency by peripheral countries on the capitalist core. The principal source of growth in a succession of models was increasing capital stock based on industrialization, with a bias toward allocations to capital-goods production rather than the consumer-goods sector, including agricultural production. Since industrialization strategies were basically inward-looking, they spawned a whole generation of closed-economy growth models that demonstrated optimal capital deployment among economic sectors. The push was always on industry and away from agriculture.

If agriculture was given a role within mainstream development strategies, it was usually focused on providing support for the transition to industrialization. In addition to representing an almost limitless labor reserve upon which industries could draw, the peasantry was expected to contribute to national development through taxation and the provision of cheap foodstuffs for the urban population. Meanwhile, traditional agroexport sectors were heavily taxed in many countries by a series of state-initiated 'distortions' of both internal and external price structures (e.g., tariffs and exchange-rate policies, indirect taxes, quantitative and/or price controls, discriminatory interest rates) to prop up the ISI sector (e.g., Oliveira 1986; Thorp 1992).

The industrial bias of most mainstream strategies resulted in widespread neglect of agriculture except as a steady source of resources to be exploited for the industrial modernization process. In a study of postwar African development, Stewart (1991: 416) notes that 'the single most important policy mistake . . . was the neglect of agriculture, which received inadequate investment, research and development, infrastructure, and prices in most countries.' In the Middle East, Lawless (1988: 19) states that 'the emphasis placed on urban-industrial development as the path to modernization has been at the expense of agriculture, from which resources have been effectively transferred.' For Latin America, Kuczynski (1988) finds that the development bias toward industry and neglect of agriculture aggravated inequalities, contributing to an unusually skewed distribution of income, even by Third World standards. In India, the industrial bias of a succession of postwar development plans is seen by Singh (1987: 231) as 'a telling example of an attempt to develop on modern Western lines without considering the socio-cultural relevance of the model to Indian conditions.' Today, many analysts contend that the neglect of agriculture in favor of rapid industrialization in much of the South has contributed to a number of interconnected problems that many countries are experiencing, including growing inequalities in socioeconomic and regional terms, the stagnation of domestic food production, high levels of malnutrition and associated health problems, balance-of-payments shortfalls, and foreign indebtedness.

Ironically, while the agroexport sector was frequently subjected to discriminatory state intervention via exchange and tariff policies, it nevertheless remained an important source of capital for industrial investments within many ISI strategies. The dynamism of ISI in much of Latin America, for example, came to depend in large part on revenues generated by the major agroexport sectors. This intensified the vulnerability of many Latin American economies (e.g., in areas such as public-sector financing, industrial investment and employment) to external trade shocks resulting from fluctuations in the international commodity markets. Contrary to the intent of ISI to reduce dependency, this type of industrialization commonly functioned as a 'multiplier' of external shocks, thereby transmitting fluctuations in global market conditions to industries, the public sector, and other closely related activities such as construction (Rosales 1988: 32).

Principal Economic Problems of Traditional ISI Strategies

While structural transformation associated with industrialization may be an important goal of development strategies in many countries, this does not mean that any type of industrialization is appropriate under all circumstances. A review of postwar ISI strategies in the South clearly reveals that in many cases there were inherent problems in the very nature of the industrialization process. First, rather than diminishing dependence on imports, ISI often proved to be highly import intensive. New demands were created, especially among the middle and upper classes, for an array of industrial products with very high import coefficients. Many of these products were produced by transnationals and replaced goods manufactured by smaller domestic capitals that had been strongly linked with local rather than foreign suppliers. Moreover, overvalued domestic currencies, as well as discriminatory tariff and exchange-rate policies, tended to favor imported inputs for ISI industries while discouraging a broad range of exports. In many cases, this meant that ISI strategies actually consumed more foreign exchange than they generated, aggravating balance-of-payments and fiscal problems.

Secondly, ISI strategies often tended to concentrate on highly capital-intensive industries (e.g., automobiles, household appliances, petrochemicals). The technologies employed were, for the most part, borrowed directly from the North and were therefore also capital-intensive. Moreover, state policies linked with ISI commonly had the effect of making capital artificially cheap and thus further promoted the use of capital-intensive technologies. In the face of chronic shortages of hard-currency reserves in many countries, demands for major capital outlays to support industrialization led to heavy foreign borrowing, compounding problems of growing inflationary pressures and indebtedness. The labor-saving bias of capital-intensive production techniques in many ISI sectors also meant that employment creation failed to keep pace with rising output. Problems of broadening inequalities and increasing levels of unemployment and underemployment appeared in many countries, as the urban-industrial sector provided relatively few jobs for rural–urban migrants who were streaming into the larger cities. Furthermore, the concentration on capital-intensive technologies was inappropriate to the factor endowments of most Third World economies. Scitovsky (1984: 953) states that 'import-substituting industrialization seems to have meant concentrating on the activities in which the LDCs [Less Developed Countries] had a comparative disadvantage – as if the doctrine of comparative advantage had been stood on its head.' This contributed to a situation in which the bulk of the ISI sector in most countries had only minimal backward linkages with the remainder of the domestic economy, while, at the same time its industries were hopelessly uncompetitive in world markets and maintained persistently high levels of import dependence.

Thirdly, ISI was inherently limited in many countries by demand restrictions resulting from extreme income inequalities. Rural demand for industrial products was limited by the destabilization of small/medium farmers and rising exploitation associated with agroexport production. Urban purchasing power was held down by the lack of permanent industrial jobs and by the maintenance of low wages, both as a condition for attracting foreign investment and as a reflection of growing labor reserves. The lack of an indigenous capital-goods industry in most countries, coupled with the vertically-integrated and capital-intensive nature of many ISI sectors (especially those dominated by transnationals), further restricted employment provision, particularly of higher paying, more skilled jobs that could have fostered an industrial middle class to spur domestic demand. This meant that ISI production of consumption goods in many countries was either geared to supplying luxury items (e.g., automoblies, stereo sets, TVs, washing machines) for the fortunate few or found its internal markets for an array of more popular goods (e.g., shoes, clothing, bicycles, building materials) severely limited. Such demand restrictions were further compounded by the failure of various integration schemes (Andean Pact, Caribbean Common Market, Central American Common Market, Latin American Free Trade Area) to develop expanded regional and subregional markets for local products. As a result, ISI markets (especially in the smaller Latin American countries) remained much more socially and geographically circumscribed than was originally envisioned by Raúl Prebisch and other theorists at CEPAL and related development institutes.

Political Crisis and the Demise of ISI

In the end, the demise of the ISI approach in areas such as Latin America was probably based at least as much in mounting political contradictions as in economic incongruities. O'Donnell (1975) argues that the exhaustion of opportunities for ISI growth in the 1970s was closely tied to increasingly authoritarian rule and political repression by many regimes in the region. Black (1991: 82–3) adds:

> The real crisis of the ISI strategy was felt when it became clear that continual growth would be dependent upon expansion of the domestic market, and that such expansion implied a far-reaching redistribution of wealth and power. This recognition served to unite and mobilize elites – both domestic and foreign – who saw their interests threatened. The upshot was the suppression – in many countries through armed force – of effective demand and the adoption of a new strategy.

As ISI gained strength in Latin America, it became apparent that a new industrial fraction of capital would vie for state power in many countries with the traditional oligarchy based in agroexports and financial capital. Most development theorists, including those associated with CEPAL, believed that new urban-based populist coalitions led by the industrial bourgeoisie would gradually supplant the traditional elites. However, research into populist coalitions in, for example, the Southern Cone suggests that they were effective only when strategies were developed that accommodated or coincided with the interests of the old oligarchy (Cypher 1990: 49–50). The rise of ISI came during unusual historical circumstances that allowed for almost costless change. Rapid postwar economic growth supported structural change toward ISI without excessively affecting the profitability of the dominant (agroexport) sectors, and ISI provided benefits for important capitalist, military, and technocratic groups in many countries (Kaufman 1990: 129). In the 1960s and 1970s, however, when the postwar economic boom began to slow and ISI policies started to seriously impinge on the accumulation needs of important elements of the traditional oligarchy and foreign capital, populist coalitions supporting ISI were faced with disintegrating pressures that they could not withstand. Perhaps because of the optimism generated by the earlier period of relatively costless social transition, analysts at CEPAL and related institutes failed to realize the precariousness of the pro-ISI populist coalitions, due to changing conditions following the mid-1960s.

Critique and Defense of ISI in Development Studies

By the early 1970s, a growing number of theorists from both Right and

Left began to label ISI as an inherently flawed development strategy. From the Right, neoliberals linked ISI with excessive state interference in market mechanisms, and particularly with protectionist policies that had prevented countries from exploiting opportunities for outward-oriented growth based on comparative advantages. It was claimed that ISI had promoted neither efficiency nor equity. State policies associated with ISI were held fundamentally responsible for the economic stagnation and other related macroeconomic problems that were worsening in much of the South. From the Left, dependency and world-systems theorists contended that the neglect of needed changes in patterns of consumption and ownership fundamentally constrained ISI strategies. In addition, ISI was criticized for deepening the foreign penetration of Third World economies, encouraging the use of inappropriate technologies, and accelerating the net outward flow of capital toward the capitalist core (see Harris 1986).

Since the 1970s disillusionment with ISI strategies has spread across a broad range of ideological and analytical perspectives within development studies. However, a group of (largely neostructuralist) theorists have also presented a strong case that the failures of ISI were essentially attributable not to the thrust of ISI strategy itself, but to the contortion of its policies by the peculiar politico-economic structures that prevailed in many Latin American countries (e.g., Dietz and James 1990; Gereffi and Wyman 1990). According to this line of thought, many of the excesses and misdirected policies of ISI were inconsistent with ISI theory as represented, for example, by the work of Prebisch and others at CEPAL (e.g., Kay 1993; Rosales 1988; Thorp 1992). There is considerable evidence that these early theorists, rather than arguing for an extreme type of closed, inward-looking economy, advocated a moderately protectionist model with a strong emphasis on increasing efficiency and technical progress in ISI sectors that would eventually lead to enhanced international competitiveness and external openness.[11] Although the internal market was important to the CEPAL model of ISI, the model was not exclusively inward-looking. It was hoped that temporary protection would stimulate new export possibilities and that endogenous sources of productivity growth would allow for rapid development that would be compatible with goals of both increasing autonomy and enhanced competitiveness in global markets according to

[11]   Thorp (1992: 189) notes that the phrase *'desarrollo desde adentro'* (development from within) occurs in Prebisch's early writing, in contrast to the later description of *'desarrollo hacia adentro'* (inward-oriented development) that is usually attributed to him and other CEPAL theorists in the development literature. Rather than simply advocating an inward-oriented closed economy, the phrase *desarrollo desde adentro* expresses the idea that the Latin American countries ought to develop in a way that reinforces their internal capacities, respects their autonomy, and builds up their comparative advantages as gradual integration into the global economy takes place on terms favorable to Latin America itself rather than to the TNCs from the North.

long-run comparative advantage.

For many neostructuralists, the contortion of the original CEPAL model by Latin American countries and their consequent failure to sustain economic growth under ISI were based in the nature of political and economic power in the region. This was especially evident in the continuing dominance of ruling elites whose interests conflicted with an ISI-inspired social transition. Fajnzylber (1990: 335), for example, castigates the wasteful 'showcase modernity' of Latin America and compares it with the more progressive 'endogenous modernity' of the East Asian NICs. Greater societal polarization, less distributive equity, lower international competitiveness, and a short-term orientation toward consumption by urban-based elite groups characterize the former. By contrast, higher levels of societal integration and distributive equity, increasing international competitiveness, and a longer-term orientation toward savings and investment to meet strategic development goals characterize the latter. Calling into question the neoliberal assumption that 'excessive demand' by the subordinate classes in Latin America was chiefly responsible for the inflationary pressures and other macroeconomic problems that plagued the ISI model, Fajnzylber (1990: 325) contends that it was the ruling elite, not the popular sectors, that were mainly responsible for the wasteful consumption that distorted development in the region. For Fajnzylber and other neostructuralists, the eventual demise of ISI became a virtual certainty in the absence of a profound crisis in Latin America that would have forced change upon its reactionary elites and their overseas allies. The region's recent poor development performance 'was not caused by failing to "get policies or prices right," but rather by not "getting politics right"' (Dietz and James 1990: 203).

A Reconsideration of ISI

In retrospect, the branding of ISI as an inherently flawed strategy by many development theorists on both the Right and the Left has probably been an overreaction. Under ISI many countries achieved dramatic increases in industrial production, fostering the establishment of substantial industrial sectors to supply the local market. In many cases, ISI improved the economic prospects of an increasingly influential urban-based class alliance. This allowed state structures to be formed from a socially cohesive base that, for a time at least, generated rapid growth with relatively little class conflict. ISI was also chiefly responsible for the emergence of urban mass markets in some Third World countries, permitting more easily sustainable growth in the face of global economic downturns. In addition, substantial state outlays on social and economic infrastructure during the ISI period provided many countries with the foundations for future economic diversification and development.

For industrialization to proceed to more advanced stages, a base needs to be created through ISI (or some other equivalent program) of appropriate social, technological, and economic infrastructure. Modern industrial sectors are not established overnight, but are the end result of long periods of broad societal structural transformation. Considerable periods of learning and adaptation are normally required to make technology transfers and other aspects of industrialization appropriate to particular historical conditions prevailing in different countries. Import substitution may provide the means for countries to carry out this type of learning process. It may also offer domestic firms the opportunity to achieve scale economies upon which to eventually construct a viable export platform. Contrary to neoliberal opinion, much of the recent export-led development of, for example, the East Asian NICs such as South Korea and Taiwan grew from the industrial foundations originally established under ISI.[12] Thus, the way in which the inward-oriented ISI model is juxtaposed to the EOI (export-oriented industrialization) strategy in much of the development literature (e.g., see Chandra 1992) may be a false dichotomy that does little to further constructive debate. Rather than being an either–or proposition, the two approaches may in fact be complementary, as the development experience of the East Asian NICs demonstrates. Import substitution does not preclude rapid growth and the development of a strong exporting capability. Finding the appropriate mix of ISI and EOI to suit particular countries under changing global conditions might be a more appropriate focus for development.

Necessary Components of a Viable ISI Strategy

Hulme and Turner (1990: 105) claim that policy-makers currently face two principal alternatives concerning the future of ISI. On the one hand, they can follow the prescriptions of the neoliberals for relaxing trade restrictions and other forms of state intervention that have traditionally propped up the ISI sector, thereby forcing its industries to become more efficient. On the other hand, they can follow the advice of the dependency theorists who call for significant changes in economic structures to accompany the deepening of ISI (e.g., through agrarian reforms, measures to redistribute income). Although each country must find its own appropriate path to development based on its unique local conditions and changing global circumstances, development for much of the South depends on the adoption of a hybrid approach composed of elements of both these alternatives.

[12]   The transition of the East Asian NICs and some other countries from an initial phase of ISI to a later period of EOI (export-oriented industrialization) will be covered in more detail in the following section and especially in chapter 3, which focuses on the NIC experience.

In order to be sustainable, growth has to be based on the efficient use of capital and other scarce resources. Infant industries may require an initial period of protection but this should be gradually reduced within the framework of a strategic economic policy to ensure long-term efficiency and global competitiveness. A policy of rational and selective protection would allow for the takeoff of new ISI projects in sectors of strategic importance for structural economic change (e.g., capital-goods sectors), while eliminating problems of excessive, permanent, or indiscriminate protection. A process of selective ISI might also be used to create links within key productive chains that could promote intra-industrial and inter-sectoral articulation. In many countries an endogenous industrial sector could be strengthened and better articulated to the rest of the local economy, which could ensure production of basic goods and widely used inputs at the same time as it acts as a springboard for expanded exports. The build-up of a viable ISI sector should be recognized as particularly important for countries with relatively large internal markets (e.g., Brazil, India, Mexico). In these countries, development strategies oriented almost exclusively toward the external market (as in Hong Kong and Singapore) are inappropriate. Production for the domestic market may permit the reduction of unit costs through the achievement of scale economies, allowing a country to establish a solid foundation from which to enter export markets. It may also foster socioeconomic articulation, permitting more efficient use of small/medium enterprises, which is vital for job creation and income distribution.

In general, measures to facilitate the economic participation of different classes and social groups, economic sectors, and geographic regions must also accompany industrialization so that it conforms to majority interests rather than to those of a narrow elite, as has so often happened within ISI strategies. For the great majority of Third World countries, either individually or collectively within regional trading blocs, ISI ought to play an important role in promoting goals of both economic efficiency and distributive equity. The establishment of a viable ISI sector to meet majority interests requires accompanying programs that simultaneously increase internal demand and raise standards of living among the popular sectors. Production ought to be particularly encouraged in sectors (e.g., basic goods, housing) that have strong multiplier effects on other industrial and primary sectors. In addition, various programs need to be implemented to open opportunities for expanded participation of small/medium firms in the industrial sector. This might involve, for example, initiatives to expand subcontracting arrangements between such firms and larger enterprises, both public and private. Finally, however, it should be remembered that this type of broadly based ISI program oriented toward majority interests may present a fundamental challenge to the status quo in many highly polarized societies. Given the past record of ISI, one might expect under these circumstances to encounter strong opposition from the ruling elite, which will feel its interests threatened and will throw its support behind contrasting (at present neoliberal) development strategies.

The Neoliberal Stress on Export-led Growth

As was related at the beginning of the previous section, a strong current of 'export pessimism' pervaded mainstream development theory during the early postwar period. Many influential development theorists and policy-makers contended that global trade, especially for primary commodities, was too erratic to form the principal 'engine of growth' for Third World economies. Instead, it was believed that ISI would offer a more secure and orderly basis for the generation of sustained growth. However, since the late 1960s support for ISI among mainstream theorists has gradually given way to a renewed emphasis on export-oriented industrialization (EOI) and other forms of outward-oriented growth. This shift in development thinking has paralleled the resurgence of neoclassical economics as the centerpiece of the neoliberal counterrevolution in development studies.[13] It has also accompanied increasing interventionism by the IMF and World Bank into Third World policy-making via mechanisms such as structural adjustment lending. Generally, continued access to such lending, as well as to most other external sources of financing, has been made conditional on the adoption of policy reforms designed to reduce state economic intervention and generate market-oriented growth. In many countries such pressures have contributed to a decisive shift in development strategy away from export pessimism and ISI toward an optimism for the prospects of EOI and other forms of export-led growth.

The Neoclassical Theory of Export-led Growth

Rising support for EOI and export-led growth within mainstream develop-ment theory is based on seven interrelated arguments based in neoclassical theory. First, given low levels of domestic demand in many developing countries, growth in a range of (especially industrial) sectors is believed to be largely dependent on gaining access to global markets via export-oriented

[13]   Among the first to criticize ISI were a group of neoclassical theorists (e.g., Vinner, Haberler, Bauer) who argued that ISI interferes with the 'natural process' of devel-opment based on comparative advantage. Their view, which has remained popular among neoliberals, was that Third World countries, at least during their initial stages of development, should uniformly specialize in primary exports rather than attempt to develop more sophisticated industrial sectors through state intervention that would not conform to comparative advantages based on factor proportions.

trade strategies. Second, export-oriented policies are regarded as normally the least damaging in terms of microeconomic efficiency – in that they benefit total factor productivity more than any other popular policy option. Third, foreign trade multipliers associated with exports are thought to play an important part in facilitating long-term growth by expanding overall production and employment. Fourth, earnings from exports may foster macroeconomic stability by contributing to a more favorable balance of trade and external accounts, which is important for attaining better ratings in international financial markets (and thus easier access to foreign loans and investment capital). Fifth, export earnings may also provide foreign exchange for imported goods, particularly capital goods needed to increase the production potential of an economy. Sixth, rising export volume and competition within global markets are believed to create economic efficiencies associated with increasing scale economies and technological diffusion. Seventh, given these theoretical arguments, rapid economic growth among (especially East Asian) export-oriented NICs, as well as a series of country studies showing strong correlations between exports and economic performance, is interpreted as empirical evidence supporting the export-led growth hypothesis.

The Rise of Industrial Exports in Some Third World Countries

In recent years, export-led growth has been especially strong among Third World countries that have managed to erect industrial export platforms. Reversing earlier trends, both the output and export of manufactures have grown more rapidly since the 1960s in the developing countries than in the industrialized countries (see tables 2.1 and 2.2). Growth in manufacturing production among developing countries was 9.0 percent in 1965–73 and 6.0 percent in 1973–85, while among the industrial market economies it was 5.3 percent and 3.0 percent in the same two periods, respectively. This allowed the developing countries to increase their share of global manufacturing production from 14.5 percent in 1965 to 18.1 percent in 1985, while the share of the industrial market economies slipped from 85.4 percent to 81.6 percent over the same period. Similarly, growth of manufacturing exports among developing countries was 11.6 percent in 1965–73 and 12.3 percent in 1973–85, while among the industrial market economies it was 10.6 percent and 4.4 percent in the same two periods, respectively. As a result, the developing countries increased their share of global manufacturing exports from 7.3 percent in 1965 to 17.4 percent in 1985, while the share of the industrial market economies fell from 92.5 percent to 82.3 percent over this period. However, such growth has been quite unevenly distributed within the Third World. A relatively small number of 'middle-income' developing countries (especially the newly established NICs) have dominated growth in manufacturing exports, while

Table 2.1 Shares of production and exports of manufactures by country group, 1965, 1973, and 1985 (percent)

Country Group	Share in Production 1965	1973	1985	Share in Exports 1965	1973	1985
Industrial market economies	85.4	83.9	81.6	92.5	90.0	82.3
Developing countries	14.5	16.0	18.1	7.3	9.9	17.4
Low-income	7.5	7.0	6.9	2.3	1.8	2.1
Middle-income	7.0	9.0	11.2	5.0	8.1	15.1
High-income oil exporters	0.1	0.1	0.3	0.2	0.1	0.3

Source: World Bank (1987) table 3.1, p. 47

Table 2.2 Growth in production and exports of manufactures by country group, 1965–85 (percent)

Country Group	Growth in Production 1965–73	1973–85	1965–85	Growth in Exports 1965–73	1973–85	1965–85
Industrial market economies	5.3	3.0	3.8	10.6	4.4	6.8
Developing countries	9.0	6.0	7.2	11.6	12.3	12.2
Low-income	8.9	7.9	7.5	2.4	8.7	6.0
Middle-income	9.1	5.0	6.6	14.9	12.9	13.8
High-income oil exporters	10.6	7.5	8.4	16.2	11.5	16.0
Total	5.8	3.5	4.5	10.7	5.3	7.4

Source: World Bank (1987) table 3.2, p. 47

a much larger number of 'low-income' developing countries have lagged behind. Growth in manufacturing exports among middle-income countries was 14.9 percent in 1965–73 and 12.9 percent in 1973–85, while among low-income countries it was 2.4 percent and 8.7 percent in the same two periods, respectively. Consequently, the middle-income countries dramatically increased their share of global manufacturing exports from 5.0 percent in 1965 to 15.3 percent in 1985, while the share of the low-income countries fell slightly from 2.3 percent to 2.1 percent over this period.

Export-led growth has been fueled in an increasing number of Third World countries by the attraction of TNCs and other capitals to export-oriented zones (EOZs),[14] which permit goods to be shipped to receiving (usually First World) countries without being subjected to prevailing tariffs, duties, and other forms of trade interference. In 1970, there were about 20 EOZs in ten Third World countries; by the late 1980s, more than 260 EOZs had been created in over 50 countries (UNCTC 1988: 169–72). The growth of EOZs has been particularly dramatic among the NICs of East and Southeast Asia and Latin America. In Mexico, for example, employment by *maquiladoras* (TNC branch-plants located in EOZs) jumped from 123,000 in 1982 to 412,000 in 1989, representing one of the swiftest rates of growth of manufacturing employment in the world during the 1980s (Harris 1991: 121). By 1990, Mexico had approximately 1,500 *maquilas* producing annual export earnings of about $2 billion (Sklair 1990: 112). In the early 1980s, most of these *maquilas* were relatively small plants owned by smaller US firms. Production was concentrated in low-skill operations in labor-intensive sectors and employed a high proportion of unskilled (particularly female) workers. Recently, however, average plant size has grown with the increasing participation of larger TNCs (including those based not only in the US, but also in Europe, Japan, and other Asian countries). Production has also begun shifting to more capital-intensive sectors that require a more highly skilled workforce in order to enhance the quality of output and allow more sophisticated '0–error' delivery systems to be implemented (Harris 1991).

Industrial Phases and the Internationalization of Production

Although the neoliberal development literature commonly contrasts (successful) EOI strategies with the (failed) experience of ISI in the Third World, in most of the NICs ISI and EOI are better conceptualized as two phases within the same dynamic process of internationalization of

---

14   These zones are sometimes also called 'free-trade zones' (FTZs) because they allow exporters freer access to normally restricted overseas markets. Another term for these zones is 'export-processing zones' (EPZs), which emphasizes their common orientation toward activities of final assembly or processing of inputs produced elsewhere.

the circuit of productive capital (Bina and Yaghmaian 1988).[15] Existing evidence indicates that ISI, rather than promoting 'self reliance,' has further integrated developing countries into the network of global production. However, because ISI was restricted to separate local markets, it eventually proved to be a constraint to the unification of the world market and the ability of capital to accumulate on a global scale. EOI has been the response of international capital to the need to develop a more unified network of capitalist production and exchange. According to this theory, ISI represents a prelude to EOI and the attainment of a higher level of global integration. In a sense, it set the stage for EOI and the internationalization of production for the world market. Many of the most successful Asian and Latin American exporters of manufactures (e.g., Brazil, Mexico, South Korea, Taiwan) developed their export platforms on the foundation of industries established during the import-substitution period. The basic difference between ISI and EOI, then, lies in the location in which their circuits of money capital are completed. For ISI the circuit is completed in the local market; for EOI it is completed externally by the realization of the value of domestically produced manufactures on the world market.

The Focus on Primary Exports in Most Third World Countries

Some highly publicized export-led growth strategies have focused on industrial exports by a small group of mostly middle-income developing countries, particularly the NICs. However, for a broader group of Third World countries, export-led growth has been concentrated in a few primary products. In many countries, the export structure remains essentially the same as it was in the (neo)colonial era under the agroexport model. This is especially true for sub-Saharan Africa, where only a few countries (notably Botswana, Mali, Mauritius, and Zimbabwe) have managed to diversify exports away from primary products during the last two decades. In Africa's low-income countries, the proportion of primary products within exports actually rose from 92 percent to 94 percent between 1965 and 1987, while it declined only slightly from 95 percent to 90 percent for the region's middle-income countries (Stewart 1991: 425).

Neoliberal development strategies associated with SAPs have encouraged export specialization in primary products for many Third World countries by expanding incentives for traditional exports. However, continuing export specialization in primary products historically has severely restricted efforts

---

[15]   These authors theorize that the progressive internationalization of capital has characteristically followed three stages: an early stage of primitive accumulation, a more advanced stage of primitive accumulation and ISI, and a final stage of export-led industrialization and global integration. ISI and EOI thus represent the final two stages of this process.

to broaden development in social, sectoral, and regional terms. Trade strategies concentrated on primary products have also commonly suffered from low demand and supply elasticities and a long-term decline in terms of trade in global markets. In the early postwar period, Prebisch (1950) and others linked worsening terms of trade for Third World primary commodities to an asymmetry on the demand side: exports from the South have a lower income elasticity relative to those from the North. However, more recent research by Krugman (1987) has also uncovered asymmetries on the supply side. The industrial sector in the North has increasing returns to scale (based on superior capital endowments and lower average costs in producing industrial goods), while the primary sector in the South has only constant returns to scale. Free trade between the North and South, then, will not only produce erratic growth and worsening terms of trade for the South, but will also lead to its systematic deindustrialization, as it finds itself caught within a low-level equilibrium trap that prevents entry into more sophisticated economic sectors (see Eswaran and Kotwal 1993).

Countries that fail to diversify exports beyond unrefined primary products are denied access to expanded employment opportunities and value-added, as well as vital learning processes associated with the production of higher-level goods. Experience gained in the production of more sophisticated goods results in the accumulation of labor skills and human capital, as well as general improvements in production techniques and organizational methods. Moreover, the effects of this learning process tend not to be confined to the sectors in which they originate, but spread through spin-off effects to other sectors as well. In recent years, an international division of labor based on the relative sophistication of production sectors has grown not only between the North and South, but also among developing countries themselves. A few (middle-income) countries have been able to stimulate development by diversifying production, especially of their leading export sectors, into more sophisticated and profitable economic activities. The bulk of (poorer) countries, however, remain specialized in production and export of a restricted number of primary products that, for the most part, have performed poorly and offer few prospects for breaking out of an inferior position within an increasingly rigid international division of labor.

Exploitation, Polarization, and Repression Accompanying Export-led Growth

In many cases, increasing exploitation and repression have accompanied export-led development, particularly of primary commodities. In this way the new wave of export-led neoliberalism closely resembles more traditional outward-oriented development strategies such as the agroexport model: the comparative advantage of their major export sectors is commonly derived from similar processes of absolute exploitation and monopolization of land and other means of production. The social marginalization that normally accompanies such processes means that repression is commonly applied on a massive scale to enforce labor discipline and ensure overall systemic stability. Within this context, Hettne (1990: 31–2) reports that: 'The comparative advantage will be with the most aggressive and repressive nation states, at least until their own disintegration sets in.' Indeed, programs of liberalization and export-led growth in areas such as Latin America have characteristically attacked workers' rights and benefits accumulated over previous decades. Institutions and mechanisms that had been established to protect workers and other members of the popular sectors have been systematically undermined in many countries. Banuri (1991: 193) argues that these policies represent an effort to resolve the incompatibility between the logic of export-led development and a persistently high degree of worker mobilization in areas such as Latin America. Beginning in the early 1970s, attempts to spur export-led growth through liberalization measures have been marked by increasing societal polarization and confrontation in many countries (e.g., Argentina, Chile, Peru) leading to the frequent employment of brutal repression by the security apparatus.

If stess is not placed on the creation of local linkages to spread the benefits of growth in social, sectoral, and regional terms, neoliberal export-led strategies risk replicating the vicious cycles of polarization and repression so commonly associated with past export-oriented development models. Under the guise of the new export-led models of the neoliberals, countries risk creating new and more sophisticated forms of polarized and dependent development. What is missing from strategies that focus only on increasing exports is a concern for the broader development goals of raising living standards of the popular majority and promoting more balanced growth among different economic sectors and geographic regions. In the absence of well-developed linkages between the export sectors and the rest of the economy, a limited and polarized form of development takes place that cannot act as a stimulus for overall growth and development to serve majority interests.

Issues and Questions for Export-led Growth Strategies

A set of criteria may be developed to evaluate the effects of export-led growth on overall development. These might include: the extent of linkages to the domestic economy; the creation of employment and value-added; the effect on external accounts and balance of payments; the fostering of genuine and appropriate technology transfer rather than merely technology relocation; the generation of jobs for skilled labor as well as for local managers, technicians, and other highly trained personnel; the establishment of favorable wages and working conditions relative to those prevailing in the country; and the rise of a relatively equitable social, sectoral, and regional distribution of the costs and benefits of growth. This would mean that maldevelopment accompanying export-led growth might be associated with some combination of: the destruction of internal linkages in the domestic economy; the failure to create satisfactory levels of local employment or value-added; the worsening of balance-of-payments problems and foreign indebtedness; the transfer of inappropriate (often capital-intensive) technologies developed for First World rather than Third World factor intensities; the loss of local skills and the failure to create skilled jobs for the local population; the intensification of labor exploitation; and the inequitable distribution of the costs and benefits of growth.

Reports of many of these problems appear with disturbing frequency in the development literature devoted to analyzing export-led growth in developing countries. At a general level, Black (1991: 85), for example, notes 'the failure of the [export-led growth] strategy to promote balanced and equitable growth in most Third World countries.' Similarly, Fröbel, Heinrichs and Kreye (1980) argue that export-led growth, especially that associated with EOZs, has produced only a truncated, severely circumscribed type of development that has excluded the majority from participating in the benefits of growth. Following an analysis of EOI in Mexico and China, Sklair (1990: 124) concludes that 'open-door strategies seem to offer a way out of the awful dilemma between dependency without development and capitalist development without social justice but . . . there is little evidence to suggest that this is anything more than a false promise in the interests of transnational capital and its partners, capitalist or otherwise, in the Third World.'

In Africa, Saha (1991: 2760) finds that liberalization and structural adjustment measures designed to promote primary exports have deepened the underdevelopment of other economic sectors and, most troublingly, have hastened a destructive process of deindustrialization in many countries. For India, Krishnaswamy (1991: 2417) reports that recent liberalization policies threaten 'the very fabric of the Indian nation' through an excessive centralization of economic decision-making, the distortion of democratic institutions, and the neglect of the bulk of the economy that lies outside of a few modern industrial sectors. Indeed, the relatively successful experience of a few (especially East Asian) NICs with export-led development is the exception rather than the rule for the developing world. The reality for the rest of the Third World is much more problematic.

Development in general and export-led development in particular can only be understood within a specific historical context. Absent from neoliberal development studies, particularly those based on IMF/World Bank models of structural adjustment, liberalization, and export-led growth, are country-specific analyses of class and other social relations. This omission is a product of the limited scope of neoclassical theory, and of the methods of positivist science in general, which prevent a detailed consideration of questions of class, gender, ethnicity, and other social relations. This represents a serious deficiency in the neoliberal approach since various forms of development ultimately are not just about abstract policies and empirical variables but, more fundamentally, are about class and other social relations within particular historical circumstances. Not all types of export-led development (e.g., agroexports, EOI) entail the same consequences for different social sectors. Indeed, similar export products may be produced according to different social relations, some of which may foster balanced development in majority interests, while others may promote polarized development for an elite minority. The impacts of a particular development strategy may differ dramatically as change takes place within the social structure of individual countries and in the position of these countries within an evolving international division of labor. Policies which may have had a positive impact in a particular country at a certain time may produce quite different results in another country or in another era. Likewise, policies that may produce satisfactory results at a macroeconomic scale, may have adverse effects on various social sectors and non-economic aspects of development. In order to make such distinctions and understand the underlying social dynamics, we need to deepen our analysis well beyond the narrow confines of neoliberalism based in neoclassical theory and positive science.

Non-traditional Exports: A New Growth Sector

With the renewed emphasis on outward-oriented growth which has accompanied the rise of neoliberalism, increasing attention has been focused on non-traditional exports (NTEs) as an important potential growth sector for many countries.[16] Indeed, recent strategies to promote growth of NTEs have proved remarkably successful in a growing number of countries, prompting much imitative behavior in many others. However, there have also been some common problems linked to NTE promotion, which call into question its usefulness as a major component of development strategies. These include foreign domination and dependency, socioeconomic and spatial polarization, environmental destruction, cultural alienation, and

[16]   Another important new growth sector for many countries is tourism, which is analyzed in Brohman (1996).

low levels of popular participation. This section analyzes such problems and explores ways in which they may be overcome by introducing changes in development policies. The design of alternative policies that call for increased popular participation and more coordinated state involvement in various aspects of development planning is emphasized.

Basic Elements of a Non-traditional Export Strategy

The theoretical case for the promotion of NTEs is based on the contribution that they can make to the development of a large, diversified trade sector to propel outward-oriented growth (e.g., Derosa 1992; Lücke 1993). Small trade sectors limited to a narrow range of exports based on resource-intensive products have severely circumscribed development in most Third World countries. Many of these products (especially traditional agroexports) have suffered a long-term decline in international terms of trade and have experienced severe limitations with respect to both supply and demand, affording few possibilities for export expansion at the margin (Eswaran and Kotwal 1993). At the same time, export expansion is necessary in most countries to maintain macroeconomic balance and to offset the cost of necessary imports, particularly of capital goods needed to keep industries and other domestic sectors functioning. A large, diversified export sector is also critical for generating growth in developing economies with small or underdeveloped internal markets. In addition, it permits countries to adjust more easily to periodic trade shocks caused by fluctuations in global commodity markets. If a country can diversify into a broader range of exports so that the variability of earnings from one subset of exports is largely offset by that from another, then that country will tend to face less uncertainty in its ability to finance imports and other necessities for development.

In global terms the strategy focuses on enlarging and diversifying the export sector, especially by exploiting niches in international markets according to a country's particular comparative advantages based on its factor proportions. For a few NICs (especially in East and Southeast Asia) this has meant rapid industrialization via export substitution; growth has been based on a growing number of increasingly sophisticated manufactures. However, for the bulk of Third World countries this strategy has concentrated on complementing more traditional resource-based exports with non-traditional agricultural exports (e.g., off-season vegetables and tropical fruits, ornamental flowers, specialty nuts) and/or low-level manufactured goods assembled in 'final touch' industries. Common sources of comparative advantage contributing to low production costs and high profitability among these export sectors include: inexpensive and abundant land and natural resources; a relatively cheap, unorganized, and compliant labor-force; the lack of labor, environmental, and other state regulations concerning production; and relatively low rates of taxation.

Given these sources of comparative advantage, the strategy calls for specialization in low-wage, labor- and land-intensive export sectors. In order to exploit opportunities for growth based on the principle of comparative advantage, new neoliberal development strategies maintain that trade restrictions ought to give way to liberalization and that macroeconomic policy should provide incentives to move resources from non-tradable sectors to tradable (i.e., export) sectors. Essentially, growth is to be export led; production of manufactures, food, and other goods for the internal market should occur only when domestic producers can successfully compete against importers without subsidies, duties, or other forms of state protection.

Examples of NTE Growth

Among African countries, Zimbabwe has enjoyed especially rapid growth in agricultural NTEs during the 1980s and 1990s. The growth of Zimbabwe's NTEs has been led by flower exports to Europe, which are based on Zimbabwe's comparative advantage of a (southern hemisphere) growing season that coincides with the European winter when flower prices there are at their highest. By the end of the 1980s, horticulture production in Zimbabwe had risen from almost nothing at the start of the decade to about 9 percent of total agricultural output. The value of Zimbabwe's flower exports increased sixfold between 1985 and 1989, easily representing the most rapid rate of NTE expansion in the country (Smith 1990: 160–1). The Zimbabwean government has been quick to recognize the potential of this sector not only as a source of foreign exchange, but also as a creator of rural employment. It is estimated that horticultural crops create an average of 0.7 full-time and 2.0 seasonal jobs per hectare, one of the highest employment : land ratios in Zimbabwe's agricultural sector (ibid.:162).

In Latin America, Chile and Costa Rica have been among the most successful non-traditional agricultural exporters during the 1980s and 1990s. In contrast to the historical domination of their economies by traditional resource-based exports (copper in Chile, and bananas and coffee in Costa Rica), much of these countries' export growth since the early 1980s has been based on NTEs. Like Zimbabwe, both countries began exploiting comparative advantages derived from their southern locations to ship high-value tropical vegetables and fruits, nuts, and horticultural products to North America during its winter months. Other important exports, primarily also for the North American market, have been wine from Chile and pharmaceuticals and seafood from Costa Rica. From 1984 to 1989 NTE growth was 348 percent in Costa Rica and 222 percent in Chile, representing an annual rate of growth of 28 percent and 17 percent, respectively (Barham et al. 1992: 49). Rapid growth of NTEs was a key factor in propelling the economies of these countries to among the highest rates of growth in the region during this (recessionary) period: the annual rate of GDP growth for 1984–89 was 4.0 percent in Costa Rica and 6.4 percent in Chile.

Common Problems of NTE Strategies

Given the success that a growing number of countries have enjoyed with NTEs, many Third World governments have begun turning to non-traditional exports in order to restimulate the growth that both traditional exports and domestic market expansion have failed to generate in recent years. NTE expansion has been made a key element in the neoliberal growth strategies of many countries and, as such, has often been actively promoted (in terms of inspiration, resources, managerial direction, etc.) by bilateral (e.g., US Agency for International Development (USAID), Canadian International Development Agency) and multilateral (e.g., World Bank, Inter-American Development Bank) aid and lending agencies. Intellectual support for the inclusion of NTEs as a critical part of most development strategies has also come from the community of development scholars (e.g., Hiemenz 1989; Paus 1989; Pelupessy 1991b).

However, there is also growing evidence that a series of problems commonly associated with the more traditional agroexport model is being replicated by NTE expansion. These problems include: the progressive concentration of land and other major means of production among a narrow minority; rising inequalities and dislocations, especially in rural areas; the inability of small/medium farmers to participate in the programs without state support; and increasing dependence on First World markets that are subject to wide fluctuations.

Many of these problems seem to be appearing in the very places that are being portrayed as successful cases of NTE expansion in the neoliberal development literature. A good example is in Costa Rica, where NTE expansion has been propelled by a new program of rural development, called *Agricultura de Cambio* (Agriculture of Change), supported by the state and external organizations such as USAID and the World Bank. In announcing the program, former Costa Rican President Arias said: 'We are concerned about the situation of the small producers. We know they need more help, they are the base of our democracy' (Desanti 1988). Under the program, however, poverty has increased dramatically in Costa Rica's main agricultural zones. One report states that over 80 percent of the rural population now lives below the poverty line (Rosene 1990: 371). Land concentrations, many of which have accompanied property purchases by foreigners, have displaced increasing numbers of peasants from traditional production areas of basic grains and other foodstuffs for the domestic market. NTE growth has been concentrated among larger, more affluent farmers that enjoy considerable advantages in important areas such as access to capital and bank credit, technical assistance programs, and expertise in marketing and import/export activities. State provisions of rural credit and technical assistance have increasingly been shifted away from encouraging peasant diversification, toward the further expansion of large-scale NTE operations (Lowder 1990: 97–8).

Without sufficient access to credit, technical and marketing assistance, or other forms of state support, small/medium farmers who have tried the *Agricultura de Cambio* have had very high failure rates. In an article quite critical of NTE development and the *Agricultura de Cambio* program in Costa Rica, Rosene (1990: 373) states:

> The Agricultura de Cambio program seems to be working directly against the small producers with the apparent intention of forcing them off the land to become cheap laborers in agribusiness farms or in the assembling and service industries . . . According to the farmers, the program will end up creating a dangerously skewed land tenure system. It will destroy Costa Rica's ability to feed itself, thus creating more dependency. It will increase social tension, threatening the stability that Costa Rica has enjoyed for so many years . . .

Evidence points to a similar process of concentration and centralization of capital accompanying the rise of NTEs in other Third World countries. In Zimbabwe, for example, differential access to capital, technological expertise, and forward/backward linkages has accentuated the concentration of horticultural exports among large-scale producers. The dominant position of the agrarian bourgeoisie in NTE expansion was solidified in 1986 with the establishment of the Horticultural Promotion Council by the Commercial Farmers Union, which represents the interests of large-scale farmers. Among its various activities, the HPC officially represents horticulture at the government level, organizes technical and marketing assistance, coordinates requests for foreign exchange to purchase inputs, and negotiates shipping arrangements with air carriers (Smith 1990: 161–2). Without a similar association, small and medium producers face considerable technical, organizational, and financial barriers that prevent entry into expanding NTE markets.

In Chile, the benefits of rapid increases in NTEs such as wine, tropical fruits, and vegetables have similarly been monopolized by the agrarian bourgeoisie to the exclusion of small/medium producers. In a process reminiscent of the agroexport boom in Central America three decades earlier, land concentrations, peasant displacements, and an increasing seasonality to patterns of rural labor demand have accompanied the rapid growth of NTE production in Chile (Carter and Mesbah 1993). The inability of the peasantry and other small/medium producers to participate in the country's NTE boom has led a growing number of observers to describe recent Chilean agroexport growth as 'exclusionary' (e.g., Ortega 1988; Cox, Niño de Zepeda, and Rojas 1990; Gomez and Echenique 1988). At the same time, the strong bias against state intervention in Chile's neoliberal development program has prevented the adoption of policies that might target groups of small/medium producers for assistance in establishing a presence in NTE markets.

Elements of an Alternative, Broadly Based NTE Strategy

If policies are not implemented to assist disadvantaged producers, the new cycle of export diversification based on NTEs threatens to reinforce the exclusionary character of more traditional postwar development models focused on agroexport production. As has happened so often in the past, export-led growth will serve the narrow interests of an elite minority rather than the broader needs of the popular majority. Conversely, however, the inclusion within NTE strategies of policies designed to meet the specific needs of small/medium producers (e.g., risk minimization, diversification into exports while maintaining food production, adoption of labor- rather than capital-intensive techniques) might encourage them to move into potentially highly profitable NTE sectors. This might create new sources of employment and accumulation that could help to reverse tendencies toward social, sectoral, and spatial polarization that have marked previous export-oriented development strategies. It might also help to lay the social foundations for increased political stability, without which any future development strategy, whether or not it contains a significant NTE component, cannot be sustained.

While many NTE sectors have been highly profitable in recent years, they have also often been quite risky, especially for smaller producers during the initial stages of production. Some of this riskiness is due to the uncertainties of adapting new production methods and technologies to diverse and often harsh local conditions. However, other important elements of risk stem from the vagaries of producing 'luxury' goods for a severely restricted group of First World countries. In most Third World areas, NTE production is aimed at a quite limited market (e.g., at the US from Latin America and East Asia, at Europe from Africa and the Middle East). Like more traditional exports, NTEs often compete with similar products from many other developing countries. While these exports may dominate their production sectors within a particular Third World country, they normally represent only a small fraction of similar imports by a large economy such as the US. Costa Rican flowers, for example, represent only 1 percent of all US flower imports; at the same time, flower exports to the US comprise 91 percent of all Costa Rican flower production (Rosset 1989). Under these circumstances, NTEs from the South often have only minimal demand stability in the large markets of the North. Moreover, because many NTEs are discretionary or 'luxury' items subject to sudden cutbacks during economic downturns, demand instabilities for these products are further accentuated.

Faced with such fluctuations, a transnational corporation may readily switch production to other goods or shift exports to another country. However, these types of changes are often much more difficult for small/medium producers who normally lack the capital, technical expertise, marketing arrangements, and knowledge of global market conditions to make such shifts smoothly. For all of these reasons, it is critical that policies encouraging the formation of support organizations for small/medium producers be put in place to minimize the considerable risks inherent in NTE production and assist in penetrating overseas markets. These support organizations might take on different forms according to the particular historical conditions prevailing in individual countries. In some countries, they might take the form of credit, service, and marketing cooperatives operating under the auspices of a national peasant organization. In other countries, independent producer associations focused on particular NTE sectors might be created.

Nevertheless, whatever type of producer-support organizations might be created for each country, they will normally require considerable state support, at least in their initial stages. The mechanisms by which such state assistance might be made available may, once again, be variable enough to conform as much as possible to particular historical conditions. At the same time, however, there are a number of areas that, given the nature of NTE production in most countries, will almost certainly have to be addressed in any program to heighten the participation of small/medium producers and spread the benefits of growth in NTE sectors to the popular majority. These include: increasing access to land and other major means of production; extending short- and long-term credit and other forms of financial support; providing agricultural extension programs and other technical assistance targeted at specific NTE sectors; improving provisions of more general social and economic infrastructure (e.g., education, health care, transportation systems), particularly in outlying rural areas; facilitating backward linkages, both to increase articulation with other domestic economic sectors and to increase access to needed imports at fair world prices; promoting forward linkages (e.g., in processing, refining, packaging) to capture more value-added and increase local multipliers; and, finally, assisting in penetrating foreign markets.

It should be emphasized again that the mechanisms by which various forms of state assistance might be provided may differ among countries according to local conditions. For example, increased access to land has successfully been provided for many small/medium rural producers via land reforms in some countries (e.g., Nicaragua, South Korea, Taiwan). However, given the present political realities of many other countries, less politically contentious methods of land redistribution might have to be employed, at least in the short term, in order to gain necessary support from powerful economic and political interests. For example, following the analysis of Carter and Mesbah (1993: 1085) in Chile, such methods might focus on the extension of 'self-financing land market reform policies – which include progressive land taxation, and the creation of land banks or land financing institutions such as mortgage banks – [that] will achieve the traditional land reform goal of linking agrarian growth with poverty reduction.'

Likewise, the state may choose different means to offer assistance in areas such as foreign market penetration or export financing. In a recent study of export promotion among East and Southeast Asian countries, the World Bank (1993b: 143–5) found that various states successfully employed different methods in both of these areas in order to meet goals for export diversification and expansion. Virtually every country in the study had some programs to ensure access to credit, often at subsidized prices. But there was also an impressive degree of variety within successful programs of export financing, including in the types of credit (long- versus short-term), the degree of subsidization (guaranteed access versus subsidized rates), the selectivity (all exports versus targeted export activities), and the means of delivery (specialized state-controlled financial institutions versus market subsidies). Similarly, nearly all of the states in the study recognized the difficulty that new exporters often face in penetrating foreign markets. But, once again, various means were chosen to encourage these exporters to overcome such problems. Some states directly subsidized export activity (direct income tax incentives), some subsidized market penetration (through exporter associations), some subsidized small/medium exporters to offset their particular difficulties in market penetration, and some promoted the creation of international trading companies.

Various means may be chosen, but active state involvement is needed in most Third World countries if NTE growth is to avoid reinforcing tendencies toward societal polarization that have accompanied previous export-led development strategies. If managed efficiently (and there are a growing number of examples of successful state-directed export promotion, particularly among the Asian NICs), the dividends that this would pay in terms of stimulating more broadly based and socially sustainable economic growth would far outweigh any short-term costs involved in state intervention. There is nothing inherently wrong with development strategies that seek to increase production of NTEs, or of agroexports in general, if they are consistent with larger societal goals (e.g., promoting growth with equity, maintaining access to affordable food and other basic needs).

Indeed, many peasants and other small/medium producers are not opposed to diversifying into NTE sectors. However, they wish to do so on terms by which they can effectively compete with larger producers and which allow them to minimize risks to acceptable levels. This normally entails putting mechanisms in place (e.g., technical assistance, credit programs, forward/backward linkages) to allow for true productive diversification – involving the continuing production of basic grains and other internal consumption goods alongside NTEs, rather than a complete switchover to the more risky export sectors. It also entails the creation of institutions (e.g., different forms of cooperatives, producer associations) to enhance producers' technical and marketing expertise and to assist, when necessary, in promoting market diversification. As with traditional agroexports, import substitution, or any other recent development approach, NTE strategies must include such considerations if they are to move beyond the immediate objectives of an elite minority toward the longer-term interests of the popular majority.

3

The Asian Newly Industrializing Countries

How has a group of small, resource-poor Asian countries sustained rapid development over the last three decades, while most other Third World countries have slipped into stagnation and crisis? Neoliberal development theorists claim they have an explanation, and argue that it should form the centerpiece of a new development model for the rest of the South. However, does the neoliberal explanation of development in the Asian newly industrializing countries (NICs) stand up to serious scrutiny?[1] Is it transferable as a new model of development that other Third World countries should emulate? These are some of the most pressing questions in development studies today. They are addressed in this chapter through analysis of the following aspects of NIC development: the role of the state in development, the compatibility of inward- and outward-oriented elements of development, the influence of internal historical and sociocultural conditions on development, and the impact of external geographical and historical factors on development.

Comparing the Asian NICs and Latin America

In recent years, proponents of neoliberal development strategies have often buttressed their arguments by pointing to the performance of the Asian NICs as an empirical illustration of the superiority of their outward-oriented, market-led development model. The development experience of the Asian

[1] The original Asian NICs are the 'Four Tigers' of Hong Kong, Singapore, South Korea, and Taiwan. The recent development performance of these countries is also sometimes compared with that of Japan in the early postwar period. In addition, a number of other Asian countries (e.g., Indonesia, Malaysia, Sri Lanka, Thailand, Turkey) have recently been given the status of NICs in much of the development literature.

NICs over the past three decades is seen to be the result of an evolutionary process of industrially induced modernization and socioeconomic structural transformation which the remainder of the South could replicate by adopting similar policies. Successful emulation of the NIC experience is thought especially to depend on locating an appropriate development niche within the global capitalist economy, which may be exploited by implementing sound development policies based on conventional neoclassical economic principles. Growth and development in the NICs are viewed as natural, inherent properties of their open capitalist economies, in which market forces have been allowed to operate freely with little state interference. According to Riedel (1988: 1), the NIC experience confirms a basic insight into development made by the classical liberal economist Adam Smith some two hundred years ago:

> . . . little else is requisite to carry a state to the highest degree of opulence from the lowest barbarism, but peace, easy taxes, and tolerable administration of justice; all the rest being brought about by the natural course of things. (Smith 1880)

Neoliberal Policy Lessons Derived from the Asian NICs

Accordingly, neoliberals stress general policy lessons that can be derived from the supposedly laissez-faire elements of NIC policies. Other Third World countries are called upon to drop their obsolete 'dirigiste' or state-centered development strategies in favor of a new neoliberal development program based on policies that supposedly reflect the successful market-led development experience of the Asian NICs. These policies include: the virtual elimination of restrictions on international trade, removal of controls on exchange rates, overall deregulation and internationalization of the financial sector, privatization of state enterprises, de-unionization and the creation of an unregulated labor market, specialization according to 'comparative advantage,' market-driven resource allocations by 'getting the prices right,' elimination of various regulatory mechanisms, and defining a generally 'minimalist' role for the state in development (e.g., Balassa 1981, 1991; Bhagwati 1986; Krueger 1986; Lal 1983).

Policies derived from the common NIC experience are given further coherence by their common theoretical focus on neoliberal and, by extension, neoclassical economic principles:

> . . . neoclassical economic principles are alive and well, and working particularly effectively in the East Asian countries. Once public goods are provided for and the most obvious distortions corrected, markets seem to do the job of allocating resources reasonably well, and certainly better than centralized

decision-making. That is evident in East Asia, and in most other parts of the developing and industrial world, and is after all the main tenet of neoclassical economics. (Riedel 1988: 38)

According to the neoliberal literature, adherence to basic neoclassical economic principles by the Asian NICs has especially been responsible for accelerating development resulting from increased economic integration into global capitalist markets. This was supposedly accomplished by policy changes that hastened movement toward the adoption of an open, market-led economic regime of export-oriented industrialization (EOI) based on trade liberalization, direct foreign investment, and the export of goods for the world market. Banuri (1991: 7–8) has examined in some detail the neoliberal version of market-led, outward-oriented growth in the NICs. In the early years of the neoliberal counterrevolution, the title 'export promotion' was commonly affixed to NIC development policies. Neoliberal prescriptions for import liberalization and currency devaluation within structural adjustment packages routinely contrasted the faster growth promised by new policies of export promotion (based on NIC performance) with the slower growth resulting from strategies of import substitution (characteristic of Latin American and many other Third World economies). Later, neoliberals invented the new title of 'outward-oriented policies' to describe a broader range of measures that, in addition to trade liberalization, also included financial liberalization and the removal of capital controls. From this 'outward orientation' it was only a short step to the more recent and still broader neoliberal notion of 'economic liberalization,' which calls for additional laissez-faire policies promoting privatization, deregulation, and de-unionization. Neoliberals currently use economic liberalization to mean the removal of controls in all markets including markets for foreign exchange (both current and capital account transactions), financial markets, labor markets, and markets for agricultural goods and other commodities.

Juxtaposition of the Asian NICs with Latin America by Neoliberals

Economic liberalization has frequently been linked in the neoliberal literature to the (successful) export-oriented growth strategies of the Asian NICs. By contrast, excessive state intervention is said to have characterized the (failed) import-substitution strategies followed by many Latin American countries. On the one hand, the export-led growth strategies of the NICs have supposedly been facilitated by realistic laissez-faire policies (e.g., on wage, exchange, and interest rates) and a reduced role for the state, which have allowed the NICs to 'get the prices right' and let their markets work. On the other hand, the import-substitution strategies of the Latin

American countries have supposedly depended on a larger role for the state and greater market intervention, resulting in distorted prices and severe macroeconomic imbalances. Neoliberals contend that while the Asian NICs were creating the conditions for sustained export-led growth based on stable prices and enhanced international competitiveness, Latin American countries were attempting to sustain flagging domestic growth through expanded international borrowing and increased state intervention to prop up an obsolete inwardly-oriented development model.

This divergence in development strategies is essentially believed to explain the contrast between the high growth rates and rising per capita real incomes enjoyed by the Asian NICs over the last two decades and the vicious circle of indebtedness, inflationary pressures, stagnant economic growth, and declining standards of living that Latin America has suffered during the same time period (e.g., Balassa 1991; C. Lin 1988, 1989; World Bank 1983, 1985, 1987). Therefore, the neoliberals contend that, if the Latin America countries (as well as much of the rest of the Third World) are to overcome their current economic malaise, they should drop their outmoded state-centered, inward-oriented development strategies in favor of a market-led, outward-oriented model that reflects the successful experience of the Asian NICs.

The Macroeconomic Development Record of the Asian NICs

The development performance of the Asian NICs has been spectacular.

Table 3.1   GDP growth rates in the Asian NICs and major Third World regions, 1960–1990 (% per year)

	1960–70	1970–80	1980–90
Asian NICs			
Hong Kong	10.0	9.2	7.1
Singapore	8.8	8.3	6.4
South Korea	8.6	9.6	9.7
Taiwan			
Third World Regions			
East Asia & Pacific	5.9	6.7	7.6
Latin America & Caribbean	5.3	5.4	1.7
Middle East & North Africa	n.a.	4.6	0.2
South Asia	3.9	3.5	5.6
Sub-Saharan Africa	4.2	3.6	1.7

Sources: World Bank (1982, 1992 and 1993a); Personal correspondence with World Bank

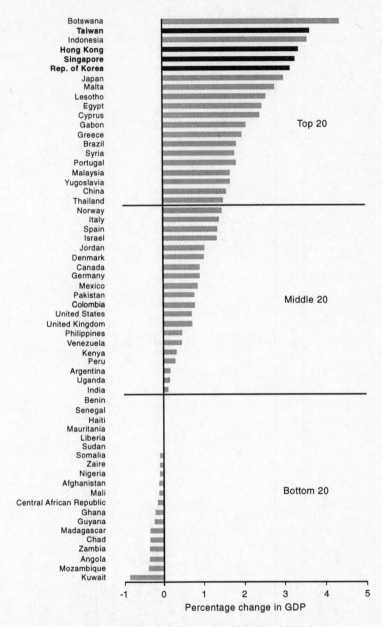

Source: Summers and Heston (1988), in World Bank (1993b) figure 1, p. 3

Figure 3.1   Change in GDP per capita for the Asian NICs and other selected countries, 1960–1985

Table 3.2 Share of exports in GDP (%)

	1960	1970	1980	1990
Asian NICs				
Hong Kong	70.9	92.2	88.0	133.9
Singapore	163.1	102.1	207.2	189.0
South Korea	3.4	14.1	34.0	31.0
Taiwan				
Third World Regions				
East Asia & Pacific	6.4	6.1	19.1	25.1
Latin America & Caribbean	14.8	12.6	16.0	16.8
Middle East & North Africa	n.a.	n.a.	42.2	31.5
South Asia	6.8	5.4	7.7	9.3
Sub-Saharan Africa	23.6	20.6	30.4	28.3

Sources: World Bank (1982 and 1992); Personal correspondence with World Bank

Table 3.3 Growth rate of exports in the Asian NICs and major Third World regions, 1965–1990

	Avg. Annual Growth Rate of Exports (%)	
	1965–80	1980–90
Asian NICs		
Hong Kong	9.1	6.2
Singapore	4.7	8.6
South Korea	27.2	12.8
Taiwan	18.9	12.1
Third World Regions		
East Asia & Pacific	8.5	9.8
Latin America & Caribbean	−1.0	3.0
Middle East & North Africa	5.7	−1.1
South Asia	1.8	6.8
Sub-Saharan Africa	6.1	0.2

Source: World Bank (1992)

Table 3.1 shows that the Asian NICs enjoyed strong GDP growth throughout the period from 1960 to 1990, even as growth rates in most of the rest of the South have slowed considerably in recent years. Among the sixty countries covered in a study by Summers and Heston (1988) of change in GDP per capita between 1960 and 1985, Taiwan was placed second, Hong Kong fourth, Singapore fifth, and South Korea sixth (figure 3.1). Much of this growth has been the result of increasing exports. The share of exports in gross domestic product for the Asian NICs climbed rapidly from 1960 to 1990, while it remained constant or declined in most other areas of the South (table 3.2).

The growth rate of exports in the Asian NICs has consistently remained well above the average in all of the South's major regions in both the periods 1965–80 and 1980–90 (table 3.3).

The Asian NICs have also greatly increased their share of total world exports and Third World exports, particularly of manufactures (table 3.4). The Asian NICs increased their share of total world exports from 1.5 percent to 6.7 percent and their share of total manufacturing exports from 1.5 percent to 7.9 percent during the 1965–90 period. If the New Southeast Asian NICs (Indonesia, Malaysia, and Thailand) are included, the share of world exports increases to 9.1 percent and of manufacturing exports to 9.4 percent in 1990. NIC growth in terms of share of Third World exports has been even more spectacular. The Asian NICs increased their share of total Third World exports from 6.0 percent to 33.9 percent and of Third World manufacturing exports from 13.2 percent to 61.5 percent between 1965

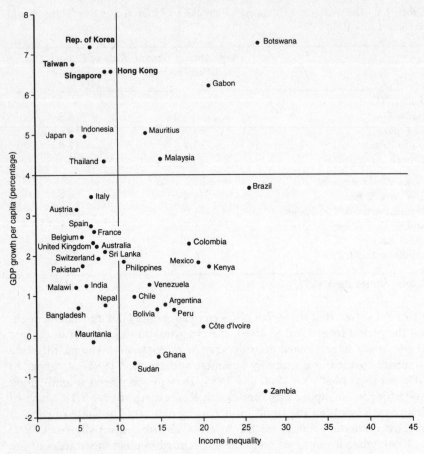

Note: Income inequality is measured by the ratio of the income shares of the richest 20 per cent and the poorest 20 per cent of the population.

Source: World Bank (1993b) figure 1.3, p. 31

Figure 3.2   Income inequality and growth of GDP for the Asian NICs and other selected countries, 1965–1989

and 1990. With the Southeast Asian NICs included, the share of Third World exports climbs to 46.3 percent and of Third World manufacturing exports to 73.5 percent in 1990. If the four original Asian NICs are considered collectively, they now rival the world's leading exporters of manufactures (table 3.5). By 1990, their share of global manufacturing exports was 9.6 percent, which compares favorably with Great Britain (6.0 percent) and France (6.6 percent) and is not far behind the world

Table 3.4 Export penetration of the Asian NICs and Third World countries, 1965–90

	Share in World Exports			Share in Third World Exports		
	1965	1980	1990	1965	1980	1990
Total Exports						
Asian NICs	1.5	3.8	6.7	6.0	13.3	33.9
New S.E. Asian NICs[1]	1.5	2.2	2.4	6.2	7.8	12.4
Total Asian NICs[2]	3.0	6.0	9.1	12.2	22.1	46.3
All Third World	24.2	28.7	19.8	100.0	100.0	100.0
Exports of Manufactures						
Asian NICs	1.5	5.3	7.9	13.2	44.9	61.5
New S.E. Asian NICs[1]	0.1	0.4	1.5	1.1	3.8	12.0
Total Asian NICs[2]	1.6	5.7	9.4	14.3	48.7	73.5
All Third World	11.1	11.8	12.9	100.0	100.0	100.0

1 New Southeast Asian NICs are Indonesia, Malaysia, and Thailand.
2 Total Asian NICs are the original Asian NICs (Hong Kong, Singapore, South Korea, Taiwan) and the New Southeast Asian NICs.

Source: World Bank (1993b) table 1.5, p. 38

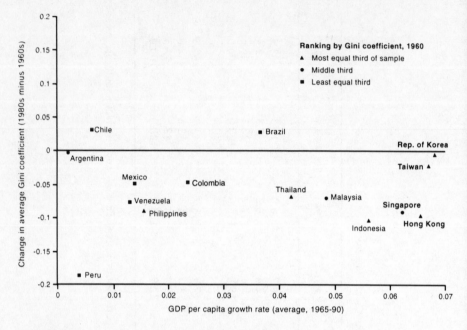

Note: Figure 3.3 plots the relationship between average per capita income growth and changes in the decade average of the Gini coefficient from the 1960s to the 1980s; a negative number indicates that income became less concentrated. The decade average is used because data are available for different years in different economies; the decade average for 1960s begins with data from 1965.

Source: World Bank (1993b) figure 3, p. 4

Figure 3.3   Change in inequity and the GDP per capita growth rate for the Asian NICs and other selected Third World countries, 1965–1990

leaders of Germany (14.5 percent), the US (11.9 percent), and Japan (11.2 percent).[2]

A relatively egalitarian income distribution has also accompanied rapid export-led growth among the NICs. Figure 3.2 measures income inequality and GDP growth during the 1965–89 period. Among the 39 countries considered in the graph, the four Asian NICs are clustered together in the upper-left quartile, which indicates the highest GDP growth and lowest income inequality. Moreover, rapid growth from the 1960s through to the 1980s has not increased income inequalities in any of the NICs. Along

---

[2]   It should be noted that, because they are taken from different sources, the figures for share of global manufacturing exports differ somewhat between tables 3.4 and 3.5. However, this does not alter the conclusions that can be drawn from the data because the trends are unmistakable.

Table 3.5   Leading exporters of manufactures, 1973–90 (% share of world)

	1973	1980	1990
Germany	17.0	14.8	14.5
United States	12.6	13.0	11.9
Japan	10.0	11.2	11.2
France	7.3	7.4	6.6
Great Britain	7.0	7.4	6.0
Hong Kong	1.3	1.6	3.1
Singapore	0.5	0.7	1.5
South Korea	0.8	1.4	2.5
Taiwan	1.1	1.6	2.5
Total Asian NICs	3.7	5.3	9.6

Source: GATT (1985–1986 and 1990–1991)

with the GDP per capita growth rate, figure 3.3 measures changes in the decade average of income inequality (Gini coefficient) from the 1960s to the 1980s. Although income inequality was reduced somewhat more in Hong Kong and Singapore than in South Korea and Taiwan during this period, all of the NICs show an improved income distribution. While it should be remembered that such aggregate figures may mask growing inequalities between specific groups within a society, the NIC record of generating strong growth with relative equity, especially when compared with other Third World countries during the same period, must be seen as remarkable.

Objections to Neoliberal Interpretations

Given this spectacular export-led development performance, it should come as no surprise that neoliberals have attempted to buttress their arguments for a new outward-oriented, market-driven development model by contrasting the recent accomplishments of the Asian NICs with the inferior performance of other Third World areas such as Latin America. However, in recent years a number of objections have been raised to the 'spin' that the neoliberals have given to the development performances of these two Third World areas. Banuri (1991: 9) states: 'The identification of "successful" Asia with openness and "successless" Latin America with illiberalism is little better than a crude caricature.' Fishlow (1991) contends that the selectivity of neoliberal comparisons which contrast all of Latin America with only the best Asian performers has exaggerated differences between the development performances of the two regions. He claims that a larger sample of countries portrays no significant correlation between economic growth and outward-oriented development strategies. Moreover, the direction of causality between exports and growth has not been firmly established. This means that no positive correlation can confidently be made between export levels and growth rates. Many intervening factors which have not generally been considered in neoliberal analyses may have contributed to both exports and overall growth. Ocampo (1990) adds that the neoliberal notion of a uniform Latin American development model of inward-oriented state interventionism is false. In reality, postwar development strategies differed substantially throughout the region – many small- and medium-sized countries followed traditional agroexport strategies of *desarrollo hacia afuera* (outward-oriented development), while many larger countries adopted industrially based strategies of *desarrollo hacia dentro* (inward-oriented development). By the end of the 1960s, however, both of these older development strategies had shown clear signs of exhaustion and many countries began turning to a 'mixed model' that incorporated elements of both an outward and inward orientation.

Moreover, it is also claimed that the strategies followed by many of the Asian NICs diverged substantially from the neoliberal ideal of laissez faire (e.g., Appelbaum and Henderson 1992; Bradford 1987; Eshag 1991; Vogel 1991; Wade 1992, 1993), while outward-oriented policies that have increased financial openness and deepened dependence on global financial and commodity markets have aggravated many of Latin America's macroeconomic problems. It is argued that this relatively high degree of financial openness and external dependency made Latin American economies particularly vulnerable to fluctuations in global markets (especially for primary commodities) and to capital market shocks (i.e., interest-rate escalation, capital flight, debt strangulation), both of which have recently contributed to macroeconomic imbalances (e.g., Dietz 1992; Hughes and Singh 1991). At the same time, the Asian NICs established strict controls over their external sectors to maximize benefits from trade and reduce their vulnerability to fluctuations in global financial and commodity markets.

It is further pointed out that particularities in Latin American socioeconomic and political structures have also made it difficult to replicate the East Asian model of export-oriented industrialization (EOI) based on labor-intensive manufactures. While primary-export development has historically been largely insignificant to the Asian NICs, much of Latin America was inserted into the world economy as exporters of agricultural goods and other primary commodities. A class alliance tied to the agroexport model has traditionally commanded the majority of the productive base, consumption share, and political apparatus of many Latin American countries. In some of the larger countries this dominant class alliance, which is commonly linked to powerful transnational capitals, has recently also diversified into import-substitution industrialization (ISI). A shift in development away from either the agroexport model or ISI toward an East Asian-style EOI strategy would have substantially shifted economic and political power in many Latin American countries and was therefore rejected by the dominant class alliance (Wade 1992).

A considerable income sacrifice would have also been required, at least initially, from many classes to achieve the low wage levels that would have been required to compete with East Asian industrial exporters (Mahon 1992). Because most Latin American countries have had a relatively long and successful history of labor mobilization, policies designed to support EOI by reducing real wages and standards of living would have encountered extensive opposition from well-organized and politically powerful labor organizations (Amadeo and Banuri 1991). By contrast, wage levels during the initial stages of East Asian industrialization were already at levels low enough to derive a comparative advantage on world markets. Moreover, labor mobilization in most of East Asia has had a relatively shorter history and labor organizations have commonly been too weak and fragmented to exert much political influence. These types of historical variations mean that elements of development models are only rarely directly transferable from one Third World region to another. The appropriateness and viability of specific development policies for individual countries depends in large part on the historical experience of those countries and the complex web of sociocultural, political, and economic structures that condition development in them.

Role of the State in NIC Development

As was indicated previously, many neoliberals contend that Asian NIC development has largely been based on the successful implementation of a laissez-faire growth strategy that has permitted the efficient operation of free markets. Indeed, the free-market and outward-oriented policy recommendations of structural adjustment programs are often supported by reference to the example of the Asian NICs. High growth rates among these NICs are attributed to the supposed absence of state economic intervention and the ability of key markets (e.g., external sector, labor, capital markets) to operate smoothly without undue regulation (e.g., Balassa 1988, 1991; Hughes 1988; Riedel 1988). However, this free-market explanation of NIC development has recently been subjected to increasing criticism, especially from specialists in the development of Taiwan and South Korea (e.g., Appelbaum and Henderson 1992; Haggard 1986; Hughes 1988; Kearney 1990; Vogel 1991; Wade 1990, 1992, 1993). Many analysts contend that neoliberals, in their haste to fit NIC development into an ideologically driven model of free-market growth, have ignored considerable evidence that contradicts their argument.[3] Wade (1992: 283–4), for example, states:

> My own evidence . . . suggests that neoliberal economists have been pioneer-ing a whole new principle of causal inference – that to explain superior economic performance one may either simply ignore everything that is not in line with neoliberal prescriptions or assert that it hindered what would otherwise have been an even better performance . . . the result is an aversion to serious investigation of the role of the state in economic development.

Criticism of the Neoliberal Version of Events

Kearney (1990: 209) notes that, although there have been some variations, NIC development has generally been 'more characterized by the Long Arm of state intervention than . . . the Invisible Hand of the free market.' Because of this, Vogel (1991: 111) contends that the neoliberal explanation of rapid NIC growth 'is fraying at the edges . . . there are signs that the kinds of cooperation and strategy coming out of East Asia are rapidly overtaking Westerners who believe that decisions should be made entirely by the market and the quarterly balance sheet.' Among the NICs, only Hong Kong could be said to have followed a laissez-faire type of development strategy – and even there, the government's 'positive non-intervention' policies are heavily involved in a broad range of activities (e.g., public housing, public services and social welfare, export promotion, economic diversification and technological change). Beyond Hong Kong, state intervention in the other NICs has played a key role in stimulating growth and facilitating structural change to an advanced industrial society. In both South Korea and Taiwan the state has used its ownership of all major commercial banks and a comprehensive system of trade controls and industrial licencing to shape decisions concerning investment and production. In Taiwan some analysts claim that about half of all assets are either directly owned by the state or

---

3   Given evidence of extensive state intervention into many aspects of NIC development, neoliberals have responded with two arguments. The first may be termed the theory of the 'virtual free trade regime,' which argues that various measures of state intervention canceled one another out to produce a neutral, market-led incentive structure (e.g., Lal 1983, World Bank 1987). The second is the theory of 'prescriptive state intervention,' which contends that state intervention did not hinder growth because it left room for 'private initiatives' (e.g., Bhagwati 1988). A recent World Bank (1993b) report on NIC development contains major elements of both of these theories. In effect, it is argued that, whatever state intervention there may have been, it did not affect the workings of the market mechanism because it was either self-canceling (virtual free trade) or porous (prescriptive state intervention) (see Chang 1993: 134). Therefore, despite evidence of substantial state intervention in NIC development, neoliberals continue to argue that such development conformed to laissez-faire principles.

controlled by the ruling Kuomintang political party (Bello and Rosenfeld 1990). In Singapore the state is deeply involved in public monopolies and parastatals (Grice and Drakakis-Smith 1985, Rodan 1989), and in South Korea the state has encouraged the growth of huge family-owned conglomerates (chaebol) vis-à-vis foreign-owned industries (Liang 1992).

If anything stands out about state economic intervention in the NICs, it is probably the highly selective and strategic nature of such intervention (Hughes 1988; World Bank 1993b). Governments have been very careful both in selecting specific areas for intervention and in carrying out their policies efficiently. The results of interventionist policies have been monitored closely and, if changes were needed, they were generally accomplished quickly and effectively. Moreover, state economic interventions, especially in areas related to export promotion and protection of infant industries, have largely been proactive and future-oriented rather than reactive and tradition-bound. In close consultation with leading capitals, labor organizations, and development scholars, governments have set policies designed to exploit promising niches within dynamic world markets rather than to prop up failing industries, as is so often the case in other countries. Governments have also taken great pains to make the scope and types of economic interventions appropriate to their particular institutional frameworks which, in turn, are dependent on a complex web of sociopolitical factors in each country (Eshag 1991). An important lesson of this experience is that neoliberal and other general theoretical arguments are of little value in indicating the role that a particular government can play in promoting economic growth and development.

Given differences in institutional frameworks and state–society relations, forms of state intervention that may have succeeded in the context of a particular time and place may be quite inappropriate to the historical conditions found in other countries.

Late Industrialization and Economic Nationalism in the NICs

Nevertheless, the experience of the Asian NICs offers strong support to those who claim that an activist state may spur growth and development, particularly for many Third World 'late industrializers.' The conventional neoliberal view of NIC export-led growth derived from open economies with competitive market prices responding to global demand is not supported by the evidence. Instead, the evidence points to a 'supply-push' development model in which the state has played a key role in stimulating capital formation and accelerating structural change (Bradford 1987: 314). Rather than laissez faire, the Asian NICs provide examples of 'guided market economies' in which state intervention is focused on 'strategic industries' based on criteria such as their high demand elasticity in world markets and their potential for technological progress and labor productivity growth (Chang 1993; Oman and Wignaraja 1991; Onis 1991). While states in the NICs have not pursued policies of generalized import protection, they have frequently implemented policies designed to protect strategic industrial sectors, especially infant industries associated with substantial learning economies.[4] However, protectionist measures have often lasted for a limited time, after which these industries are expected to become internationally competitive. This strategy seeks to ensure the initial survival of strategic industrial sectors without either forfeiting the overall gains from trade, or subsidizing for prolonged periods industries that cannot compete on world markets.

Rather than fitting into the neoliberal orthodoxy of free trade and laissez faire, the development strategies followed by the Asian NICs seem to conform more closely to classical Listian mercantilism (Burmeister 1990; Hoogvelt 1990; White 1988). Arguing against liberal economists such as David Ricardo, the nineteenth-century German economist Friedrich List (1844) claimed that the theory of comparative advantage represented a doctrine of the dominant; the dominated could expect to derive little advantage from it. Instead of allowing their markets to be dominated by established industrial powers through free-trade policies, List counseled late industrializing countries to protect strategic infant industries in order to strengthen and deepen their productive forces for future development. In fact, a detailed examination of the early period of European industrialization reveals that most countries pursued Listian policies of economic nationalism rather than neoclassical strategies of free trade (Senghaas 1984). The European experience shows that, during the initial stages of industrialization, free trade was a luxury that only the first and leading developer, Britain, could afford. Other countries followed policies of economic nationalism which have striking parallels to the contemporary strategies of the Asian NICs, including strong state economic intervention, protection for infant industries, and 'temporary dissociation' of their economies from international competition during the initial industrialization phase (Hoogvelt 1990: 354–5).

[4]   Oman and Wignaraja (1991: 86–7) provide a theoretical justification for state intervention to support infant industries and other domestic firms in international markets. Subsidies, import protection, and other forms of state intervention can tilt the terms of oligopolistic competition in many global industrial sectors so as to shift monopolistic rents or the benefits of positive externalities (e.g., moving rapidly down the 'learning curve') from foreign to domestic firms. State intervention can, under certain circumstances, play a role analogous to strategic moves by oligopolistic firms to increase market share or capture future markets (e.g., investment in excess capacity, research and development into new product lines). Moves which sometimes may appear inefficient from a short-term static point of view may make good sense from a longer-term dynamic perspective.

Policies of economic nationalism in the Asian NICs strongly resemble interventionist strategies to support early industrialization not only in most European countries but also in Japan. During the early postwar period, the Japanese state created the powerful Ministry of International Trade and Industry (MITI) to coordinate industrial development and protect strategic sectors (e.g., steel, oil refining and petrochemicals, automobiles, industrial machinery, electronics) from foreign competition during their infancy. Pervasive state intervention in early industrialization enabled Japan to rapidly escape the trap of static comparative advantage and to incorporate itself into the global markets for more technologically advanced products. In effect, the MITI and other arms of the Japanese state deliberately created comparative advantages for certain industrial sectors. Sectors selected by the state for interventionist measures typically enjoyed elastic foreign demand and offered opportunities for gains in labor productivity and technological advance (Lawrence 1993) – the same characteristics of industrial sectors that became the focus of state intervention later among the Asian NICs. It is now widely acknowledged that this Japanese model served as an example for later state-directed industrialization drives among the Asian NICs. As Abegglen (1980: 11) notes:

> This 'Gang of Four' . . . [presents] evidence of Japan's revolutionary impact . . . Japan serves as a model . . . [and] is a basic source of training for the leaders of many of these countries . . . The NICs use government even more explicitly than does Japan in economic planning and guidance. One would have to say that all except Hong Kong have a more centrally directed pattern of economic growth than Japan, although they would tend to see themselves as basically market, rather than planned, economies.

State Intervention in the NICs

In order to achieve their goals of rapid industrialization and economic diversification, the states in the NICs used a broad range of policy instruments. Use of these policy instruments, which, taken as a whole, represents a widespread interference in the operation of market forces, was especially intended to provide profit incentives for strategic sectors in order to meet production, trade, and other targets set out in state economic plans. Policy instruments ranged from direct controls over investments, imports, wages, and some product prices to indirect regulation of investment, production, and trade through measures such as subsidies, tariffs, and tax rebates (Eshag 1991: 629). State economic interventions focused on both organizational and financial aspects of industrial development (Irwan 1987: 396). Organizational interventions were intended to organize and coordinate financial and industrial capital under the guidance of the state. In effect, large private capitals were 'disciplined' by the state to pursue national economic goals. In addition, financial interventions provided favorable conditions for increased private investment in strategic sectors. Common areas of financial intervention among the NICs have included credit allocations, interest-rate subsidies, and export financing in South Korea; credit allocations, fiscal subsidies, and exchange-rate adjustments in Taiwan; and fiscal subsidies, tax concessions, and regulatory incentives in Singapore (Gibson and Ward 1992; Lewis and Kallab 1986; Rodan 1989; Tak-wing 1993; Wade 1993; World Bank 1993b).

Various forms of state financial intervention in the NICs have deliberately distorted relative prices to attain desired levels of private investment in strategic industrial sectors. The systematic 'underpricing' of investment goods relative to private and public consumption goods encouraged capital formation and investment (Bradford 1987). Some interventions took the form of domestic monetary policies affecting interest rates and credit allocations to industrial investors. Others took the form of direct subsidies affecting the price of domestically produced investment goods. These monetary and fiscal policies acted to increase the supply and demand of investment goods, which, in turn, facilitated capital accumulation, industrialization, and structural change. Moreover, strong measures to promote macroeconomic stability (e.g., strict curbs on fiscal deficits, restrictions on the growth of money supply, exchange-rate controls) encouraged a positive overall investment climate. Taken together, these policies allowed the states in the NICs to generate industrial structures that were radically different than those that unguided private capitals would have produced. Much higher levels of investment and growth in key industries were obtained than would have occurred in the absence of state intervention (Onis 1991; Wade 1992).

These policy instruments have allowed the state to 'guide' or 'govern' the process of resource allocation in order to produce an investment and production profile that serves national development goals and differs substantially from that which would have resulted under a free market system. Political, institutional, and organizational arrangements were put in place to coordinate the economic activities of the state apparatus and large private capitals as well as their mutual interaction. Various incentives, controls, and mechanisms were established to increase the profitability of strategic sectors and spread the risks of investment in those sectors. In effect, state policies have acted to 'socialize' many of the investment risks associated with new industrial ventures. This has allowed domestic capitals to 'externalize' such risks, much as the early industrial capitalists did in the developed world (Petras and Hui 1991:

185–6). However, in return for socializing investment risks, state policies have also acted to 'discipline' the behavior of domestic capitals in strategic sectors. Investment incentives and subsidies have been closely tied to stringent performance requirements. Those firms meeting performance targets were rewarded, while support was quickly withdrawn from those that performed poorly. Typically, the state refrained from bailing out badly managed firms in otherwise profitable industrial sectors. This allowed the state in the NICs to avoid much of the resource waste that has often characterized efforts by other middle-income states to prop up declining industries or firms experiencing protracted financial difficulties (Onis 1991).

'Disciplining' Capital to Serve Development

State policies in the NICs have attempted to discipline the behavior of not only domestic capitals but also foreign capitals involved in strategic industrial sectors.

Policies concerning outside investment were designed to attract foreign capitals only on terms and conditions that would permit their activities to be integrated with national development goals. State intervention limited and directed the impact of foreign capitals on local economies and regulated external sectors in terms of both trade and capital flows. Strict controls were commonly established over foreign loans and direct investments, exchange rates, and financial flows. Domestic industries were typically encouraged to compete in external markets and were protected, at least in their infancy, from foreign competition in internal markets. The ability of the strong states of the NICs to subordinate the behavior of foreign capitals to a strategic industrial strategy may be contrasted with the dependent relationships that have allowed transnationals to dominate industrialization in many other Third World countries. In much of Latin America, for example, relatively weak states have allowed the industrialization process to be shaped in the interests of powerful transnational capitals rather than national development goals (Dietz 1992; Ellison and Gereffi 1990; Stallings 1990). The industrialization experiences of the Asian NICs and Latin America, then, give strong support to the view that directive state intervention is necessary if foreign capital is to play a constructive role in national development.

Empirical evidence from the NICs demonstrates that foreign capitals, in fact, played a relatively minor role in industrialization and economic growth. A study by Jenkins (1991) shows that between 1951 and 1967 direct foreign investment as a percentage of total long-term foreign capital flows was only 1 percent in South Korea and 8 percent in Taiwan, whereas official transfers (bilateral and multilateral aid and loans) represented 86 percent and 74 percent, respectively.[5] Further research on all of the Asian NICs indicates that, apart from Singapore, foreign direct investment contributed little to domestic economic growth (Petras and Hui 1991: tables 2 and 3).[6] In fact, rather than being destinations for foreign investment, many of the NICs in recent years have actually become sources of investment for other less developed countries, especially in Asia.

Importance of the Institutional Framework

The ability of the NICs to discipline the behavior of domestic and foreign capitals to serve national development objectives is closely tied to institutional and organizational changes that the strong states of these countries have orchestrated. Changes within the state apparatus itself and in state–society relations have been sensitive to the initial conditions prevailing in NIC societies and have focused on creating mechanisms for public–private cooperation to further national interests under the guidance of the state (Ranis 1989). Institutional/organizational changes have been particularly important in four areas. First, changes in the fiscal, monetary, and taxation systems have provided incentives for private investment in strategic industrial sectors. Direct state ownership has been of secondary importance in industrialization relative to the institutional capability of the state to manage and direct industrial capitals. Second, institutional changes in the educational system and other key elements of social infrastructure have facilitated economic restructuring by providing education, training, and research. In South Korea and Taiwan, infrastructural change was complemented by an agrarian reform which further broadened economic participation. Third, the state has performed a central role in the promotion of cooperative labor–management relations. In particular, the activities of trade unions have been strictly controlled and wage rates have been allowed to rise only at rates lower than productivity growth so that international competitiveness is not adversely affected. Fourth, and probably most sig-

5   These proportions are often reversed for countries in other areas of the South such as Latin America. During the same period, offical transfers of long-term capital flows comprised 11 percent in Argentina, 31 percent in Brazil, and 8 percent in Mexico, while direct foreign investment was 53 percent, 51 percent, and 57 percent, respectively (Jenkins 1991).

6   According to Petras and Hui, annual foreign direct investment as a share of gross domestic capital formation fluctuated between: 0.8 and 4.3 percent in Hong Kong from 1972 to 1978; 0.4 and 2.8 percent in South Korea from 1966 to 1980, 0.5 and 2.8 percent in Taiwan from 1960 to 1978; and 1.6 and 23.2 percent in Singapore from 1965 to 1976. The share of foreign-invested firms' exports in total exports was: 11 percent in 1974 and 17.8 percent in 1984 in Hong Kong; 31.4 percent in 1974 and 18.3 percent in 1978 in South Korea; 30 percent in 1975 in Taiwan; and 66.5 percent in 1970 and 92.9 percent in 1980 in Singapore.

nificant, the state has established an institutional framework that allows it to create comparative advantages via economic restructuring. The state has systematically managed the market as a means to create the conditions for long-term economic transformation and sustained growth.

Significantly, the NICs have practiced a highly selective form of state intervention that requires neither a large public sector nor a large public-enterprise sector. The effectiveness of state intervention has been based on the existence of a coherent institutional framework that has strengthened administrative capacities and created opportunities for public–private cooperation in national development planning. As Onis (1991: 124) notes, the state apparatus in the NICs is characterized by 'tightly organized, relatively small-scale bureaucratic structures with the Weberian characteristics of highly selective, meritocratic recruitment patterns and long-term career rewards, which enhance the solidarity and the corporate identity of the bureaucratic elite.' Successful state intervention has been based not on the absolute size but on the coherence of the state's institutional framework; it has been based not on the quantity but on the quality and selectiveness of interventionist policies. Sector-specific forms of indicative planning have been a key component of state-directed, but market-oriented development. Relatively small but powerful state agencies, such as the EPB in South Korea (which is patterned on the MITI in Japan), have avoided much of the unwieldiness of larger bureaucratic structures and have permitted a highly trained, select group of experts to provide timely and imaginative strategic guidance to key economic sectors. Because of their cohesiveness and potency, these state agencies have been able to act, sometimes in the face of opposition from special interests, with the persistence and forcefulness necessary to maintain a stable environment of policy continuity. At the same time, however, their relatively small size and high quality has permitted them to exhibit the type of flexibility, pragmatism, and quick responsiveness that rapidly changing economic conditions and development priorities often require (Dietz, 1992).

Relative State Autonomy and Public–Private Cooperation

Underlying the institutional framework that has supported effective state intervention in the NICs are two important conditions often associated with 'developmental' states: a high degree of state-relative autonomy coupled with close public–private cooperation (Douglass 1993; Jenkins 1991; Onis 1991). These two conditions have allowed key state agencies to develop independent national goals and to translate these goals, within the broader polity, into effective policy action. Moreover, the coexistence of these conditions is critical; each of them is necessary for developmental states to operate effectively. States with a high degree of relative autonomy are able to pursue policies that conform to broad national interests, even if they sometimes conflict with the interests of powerful fractions of capital. Mechanisms facilitating public–private cooperation enhance the ability of the state to generate consensual support for its development goals and to carry out its policies more effectively within the larger society. Conversely, states that lack relative autonomy often find their development goals reduced to narrow special interests, while states without adequate mechanisms of public–private cooperation commonly cannot carry out their development policies effectively within broader society.

Although states may take concrete actions to increase their relative autonomy and improve public–private cooperation, these conditions are also historically determined by many factors outside of immediate state control. This means that general theoretical arguments concerning the role that a particular state should play in promoting development are of little value. Policies that succeed in the NICs may prove unsuitable for states in other Third World countries under different historically determined conditions.[7] Development strategies therefore need to take account of the complex and historically changing web of internal and external factors that act to structure state–society relations in each country. Within the NICs, for example, state-relative autonomy has been furthered both by their specific histories of class formation and class struggle and by the international (especially geopolitical) context of postwar East Asia (Jenkins 1991). The fragmentation and disorganization of the working class meant that NIC economic policies could ignore short-term labor interests to prioritize investment over consumption expenditures much more than might have been possible in other countries with stronger traditions of working-class militancy. In addition, geopolitical factors in East Asia led to a massive influx of US aid and permitted the NICs to implement policies that the US and other Western powers would probably have vigorously opposed in other less strategically important countries.

This particular mix of internal and external conditions, many of which are missing in most Third World areas, made it possible for the NICs to strengthen the state apparatus, increase state autonomy, and improve public–private cooperation under the auspices of a strong state. Given these initial advantages, the state was able to unite the bureaucratic and business elites behind a coherent, nationalist development strategy which adhered to the popular consensus that economic growth was paramount.

[7]   Indeed, there is considerable variation among the NICs themselves in the institutional mechanisms that have been created to foster public–private cooperation and guide development. Douglass (1993: 154) reports that the state in South Korea and Singapore has promoted the rise of large-scale domestic enterprises, whereas in Hong Kong and Taiwan it has fostered a process of international subcontracting by indigenous, small-scale firms.

Even though this consensus never included all social groups and classes and has recently shown signs of fragmentation, especially in South Korea and Taiwan, it did provide the states of the NICs with a unique opportunity to pursue broadly based nationalist interests. This is undoubtedly one of the most striking features of the NICs, contrasting them with most other developing countries, in which the ruling elites are culturally, ideologically, and institutionally fragmented and integrated into the international bourgeoisie (Bienefeld 1988: 24). In much of Latin America, for example, systems of public–private cooperation have arisen within the neocolonial context of weak, dependent states that lack autonomy either from international capitals or from powerful fractions of the domestic elite (Dietz 1992; Jenkins 1991; Stallings 1990). Under such conditions, public–private cooperation has often degenerated to the point that state goals, rather than reflecting any real popular consensus, are directly reducible to the narrow interests of dominant classes and social groups.

By contrast, the widely shared perception of a severe external threat coupled with the cohesive internal structure of state–society relations in the NICs encouraged a sense of collective social responsiblity. This, in turn, provided the basis for a long-term, nationalist strategy of economic growth that required relatively few concessions to the demands of special interest groups. Under state direction, strong nationalist sentiments were transformed into the single-minded pursuit of industrially based growth at the expense of other objectives. A consensus was created in favor of rapid industrialization as the best means to achieve national economic independence and eradicate poverty – goals that have eluded almost all other developing countries. However, as Wade (1992: 314) notes, what emerged in the NICs was an 'unattractive kind of regime' that suppressed individual freedoms and promoted a type of 'puritanical nationalism' enforced by authoritarian rule. The transferability of this type of political system to other Third World countries under different historical conditions is highly questionable. In particular, the unusual degree of state autonomy in the NICs is closely related to their geostrategic position and the relative weakness of their internal organizations representing both capital and labor – conditions that are rare in other countries. Moreover, attempts to sustain both state autonomy and public–private cooperation over the long term may ultimately prove contradictory, as is evidenced by the growing power and autonomy of large-scale capitals in many of the NICs (e.g., the Chaebol in South Korea). And finally, state autonomy and authoritarianism are not particularly compatible with broadening political participation. Growing social unrest and demands for greater democratization in many NICs call into question whether their type of political economy can coexist with more liberal and democratic polities.

The Influence of Internal Conditions on NIC Development

There are a number of recent studies that stress how historically determined factors within NIC societies have shaped development, especially the role that the state has played in generating growth (e.g., Amsden 1989; Chang 1993; Lie 1991; Petras and Hui 1991; Winckler and Greenhalgh 1988). These studies caution against overly hasty generalizations and the uncritical acceptance of any formalistic theory, especially one derived simply from abstract economic principles, to explain NIC development. An understanding of the recent development experience of the NICs cannot be gained by simply reading off from theory, no matter how elegant or comprehensive that theory may seem. As in all countries, development in the NICs has unfolded within a specific historical and sociocultural context which is too involved to be addressed by general economic models and abstract principles.

Explanations of NIC development point to the messier realm of interdisciplinary research in which sociocultural and historical factors are interwoven with the political economy. Moreover, attempts to reduce the NIC development experience to generalized economic principles which can be applied to other Third World countries are dangerously ahistorical. Given the tremendous diversity of the South, no approach to development can successfully claim universal validity. In the end, each country must devise strategies that are appropriate to its own historically changing conditions.

As we saw in the last section, state-centered studies of NIC development have largely succeeded in debunking neoliberal analyses that seek to link the successful economic performance of the NICs to market-led growth with little state intervention. However, these state-centered approaches are also partial and incomplete because they have generally ignored the realm of state–society relations. The social composition of the state strongly influences both the content of state policies and the manner in which those policies are carried out. As Petras and Hui (1991: 191) note, state-centered approaches typically treat the state as a 'black box' structure and neglect the matrix of class and other social relations in which the 'box' is inserted. In order to explain the efficacy of state institutions and policies in the NICs, studies need to theorize the state within its broader context of social relations and structures. In particular, the rise of specific state forms and actions in the NICs is closely interrelated with the historical course of processes of capital accumulation and social reproduction in those societies.

Both capital accumulation and social reproduction also necessarily involve classes and social groups in complex processes of conflict and accommodation. These, in turn, are related to the broader social structures and have evolved under particular historical conditions in each of the NICs.

The Colonial Legacy

Japanese occupation and colonialism heavily affected the early twentieth-century development of both South Korea and Taiwan. Japanese colonial rule led to the rise of modern state and industrial structures, Japanese foreign investment and technology supported early industrialization, and Japanese marketing firms managed most exports (Bello and Rosenfeld 1990). In addition, the Japanese carried out extensive agrarian reforms in both countries, creating relatively egalitarian rural structures and permitting a more broadly based and internally articulated pattern of growth to arise which spurred domestic demand and supported early industrialization (e.g., Lin 1989; Nolan 1990).[8] Externally imposed agrarian reforms also effectively destroyed the feudal rural oligarchy as a dominant class, thereby removing a potential obstacle to urban-based industrialization. By contrast, the prolonged domination of economic and political structures by the traditional rural oligarchy in much of Latin America significantly delayed industrialization and other economic diversification (Bagchi 1990; de Janvry 1981; Jenkins 1991).

Agrarian reforms in South Korea and Taiwan also generated a rapidly expanding agricultural surplus that the state could utilize to promote industrial growth. In South Korea the state extracted surplus from the peasantry by means of mandatory grain payments for rents and loans, as well as the occasional extension of a legislated state monopoly over grain purchases (Petras and Hui 1991: 186). Agricultural prices were also kept below world levels, thus lowering the costs of living and wage demands of industrial workers, by permitting abundant imports of grain and other agricultural products from the US. In Taiwan the state extracted agricultural surplus by means of land taxes, compulsory state purchases of rice at below-market prices, and a rice-fertilizer barter scheme (ibid.: 186). Rapidly rising agricultural productivity allowed the state to extract a share of rising peasant incomes to finance provisions of social infrastructure and the development of strategic industrial sectors. It is estimated that net capital outflow from agriculture during early Taiwanese industrialization financed approximately 34 percent of gross domestic investment (Grabowski 1988: 63).

Like South Korea and Taiwan, the development of Hong Kong and

---

[8]   It is notable that an extensive agrarian reform was also implemented in Japan itself by the Meiji administration during the nineteenth century that facilitated modernization by breaking the power of the feudal oligarchy and promoting more egalitarian rural structures.

Singapore has also been influenced by colonialism, albeit of a British rather than a Japanese variety. Because both Hong Kong and Singapore are essentially city-states, neither agriculture nor agrarian reforms have been important to their development. However, as important outposts in the British Empire, they rapidly established themselves as significant regional centers of entrepôt trade and were able to dominate development in their respective hinterlands (Kearney 1990; Nolan 1990). Under British colonialism, modern state institutions were established that could play an active role in directing and managing economic growth. The creation of advanced industrial, financial, and administrative/managerial structures provided important advantages to attract capitals seeking investment opportunities in East and Southeast Asia. The construction of a highly efficient transportation and communications infrastructure further distinguished these centers from most other large Asian cities. Provisions of basic social infrastructure created a relatively healthy and well-educated population, supplying an important base of human capital for industrially oriented growth. Taken together, these supply-side conditions offered Hong Kong and Singapore important advantages to pursue opportunities in global markets that have eluded many other Third World countries which inherited quite different conditions from their colonial legacies.

The Weakness of Capitalist and Working-Class Organization

The historical weakness and disorganization of both the working class and domestic capitals in the Asian NICs has also strengthened state autonomy to direct a strategy of rapid industrially-based growth. A variety of reasons have been offered for this weakness, including the relative lateness of industrialization and high degrees of cultural and ethnic heterogeneity in some of the NICs, especially Taiwan and Singapore (Banuri 1991: 191). In addition, a series of internal and/or international conflicts (e.g., Japanese occupation for varying lengths of time of all of the NICs; the Korean War and subsequent US military presence; the seizure of Taiwan by the Kuomintang) effectively destroyed many labor organizations and domestic capitals. Authoritarian state structures, which in many NICs were strengthened as a result of conflict and external threats, were also used to place strict controls on the actions of both labor and capital. State-corporatist institutions were established and laws (in some cases, emergency decrees) were enacted that are only quite recently beginning to be challenged in many of the NICs. Highly centralized state structures monopolized political and economic power, thereby preventing alternative organizations and movements from acquiring the type of political strength and national identity that they possess in many other Third World countries. Jenkins (1991), for example, contrasts this situation in the Asian NICs with the difficult position faced by many Latin American states due to the historical militancy of labor organizations and the hegemonic control exerted by powerful domestic and international capitals in the region. Because they enjoyed a high degree of autonomy, the states in the Asian NICs were able to direct structural economic change away from ISI (import-substitution industrialization) toward EOI (export-oriented industrialization) with relative ease. By contrast, the powerful entrenched interests of both the industrial bourgeoisie and organized labor in many Latin American countries have strongly opposed state attempts to de-emphasize ISI relative to EOI. While the autonomous states of the NICs have been able to largely ignore the special interests of capital and labor organizations in formulating and implementing economic policies, this has seldom been the case in other Third World areas such as Latin America. Critical differences in the historical evolution of key societal structures often present Third World countries with conditions that restrict possibilities for state autonomy.

Confucianism and other Sociocultural Factors

The early postwar consolidation in the Asian NICs of structural conditions that fostered state autonomy and rapid industrially-based economic growth was also facilitated by large-scale migrations of Mainland Chinese, especially to Hong Kong, Singapore, and Taiwan. Many of these migrants were well educated and possessed considerable entrepreneurial expertise and other skills important to industrialization. In addition, they brought with them traditional Confucian values and ideological beliefs which have been at the core of Chinese society for the last two thousand years. Until recently, many development theorists, especially those in the modernization camp familiar with Weber's (1951) classic work on *The Religion of China*, regarded Confucianism as an archaic religion inappropriate to the sociocultural and ideological requirements of industrial capitalism (e.g., individualism, social utilitarianism, liberal democratic principles). It was believed that Confucian traditions in China and other Asian societies would delay the adoption of modern, Western-style sociocultural attributes necessary for industrialization. However, since 1980s a number of scholars have turned this hypothesis on its head – they contend that instead of being hindered, the Asian drive to industrialize has been helped along by values, mores, and beliefs rooted in Confucian traditions (e.g., Gray 1988; Kuah 1990; Leung 1987; Morishima 1982; Nolan 1990).

Among these scholars, Morishima (1982) first related rapid postwar economic growth in Japan to an ethical value system rooted in Confucianism and related religions. From their two major religions (Confucianism and Taoism turned into Shintoism), the Japanese acquired 'an ideological driving force for solving problems which their society had confronted' (ibid.: 19).

Rather than focusing on individual achievements and rewards, the Japanese 'ethos' tended to stress the group effort needed to achieve collective economic development and social mobility. Likewise, a number of analysts have linked the more recent process of state-directed economic growth in the Asian NICs to the strong presence of neo-Confucian values in these societies. These values include the importance of social harmony based on familial and community obligations, a de-emphasis on self in favor of group cooperation, dedication to work and the need for achievement, the preeminence of education, and respect for authority (e.g., Gray 1988; Leung 1987; Kuah 1990). The Confucian 'work ethic,' loyalty to organization, and appreciation for the value of education are regarded as particularly important to rising industrial productivity and the lack of labor militancy in the NICs. Moreover, it is believed that Confucian respect for authority and social harmony have helped to legitimize highly centralized and authoritarian systems of governance. This, in turn, has facilitated 'social engineering' efforts by the 'strong' states in the region designed to hasten structural economic changes needed to spur industrially based growth.

Analyses which link Confucian traditions to the successful performance of state-directed industrialization strategies in the NICs underscore the important role that sociocultural factors can play in economic and political development. Confucian values facilitated the creation of human capital needed for industrialization and provided consensual support for a development strategy that subsumed the immediate interests of specific classes and social groups to long-term goals set by an authoritarian state to meet broad national development objectives. The sociocultural makeup of NIC societies proved to be particularly compatible with state efforts to mount this type of development strategy. However, it should be noted that a similar strategy might prove quite incompatible for other societies with different histories, cultural traditions, and social structures. Key elements of the NIC development strategy (e.g., state autonomy, authoritarianism, private–public cooperation, EOI) might, at least in similar form, produce disastrous results in other Third World countries. It follows that internal factors based on the histories and sociocultural traditions of individual countries ought to be given careful attention alongside economic and political factors in any considerations regarding the transferability of development strategies.

Internal Growth

Neoliberals commonly juxtapose the (successful) performance of outward-oriented growth strategies in the Asian NICs to the (failed) history of inward-oriented strategies in many other Third World areas. A uniform model of market-led, export-oriented industrialization is constructed for the NICs that is then contrasted to the strategy of state intervention and import-substitution industrialization which is said to characterize development in much of the rest of the South, particularly Latin America. However, just as it is false that state-directed ISI dominated postwar development in all Latin American countries to the exclusion of outward-oriented strategies, so too is it incorrect that internally oriented growth did not play a significant role in NIC development. No one denies that exports have provided an important stimulus to growth in the NICs; but rising domestic demand and the creation of internal linkages between key economic sectors and social groups also played a crucial role in their development (e.g., Bradford 1986; Gereffi and Wyman 1990; Gibson and Ward 1992; Irwan 1987; Liang 1992).

The NIC development performance demonstrates that selective policies supporting ISI and other domestically oriented sectors may not necessarily be incompatible with export promotion and other outward-oriented policies. Characteristically, state economic intervention in the NICs promoted selective market opening in competitive sectors (i.e., those in which domestic firms were expected to compete internationally), while protecting local markets for ISI and other non-competitive sectors. The parallel and interwoven existence of these two strategies may well become a trend for developing countries in the future (Dinh 1993). It enables them to make use of the rational core of the theory of comparative advantage to enlarge their participation in international markets, while simultaneously providing conditions for a more participatory, internally articulated form of development which can utilize a broader range of domestic resources.

The development of internal linkages proved especially important to the early stages of industrialization in many of the NICs, providing both domestic demand and supply-side conditions (e.g., technology, skills, human and physical capital) to foster economic diversification and the subsequent construction of an export platform. Moreover, even after the achievement of a more mature, specialized stage of industrialization, a complementary mix of inward- and outward-oriented policies was followed in order to broaden economic participation and provide for more socially, sectorally, and spatially balanced growth (Burmeister 1990; Luedde-Neurath 1986).[9] State policies were used to direct investment to strategic industries that not only showed a high growth potential but also had good possibilities for producing 'demonstration effects' (e.g., technological change, human skills development) and for developing forward/backward linkages with related

---

[9]   In fact, Douglass (1993: 163) argues that, following the experience of Japan, South Korea may now try to lessen its export dependence in favor of a more inward orientation that stresses domestic market growth as a key basis of accumulation.

local industries (Irwan 1987; Schive and Majumdar 1990).

Rural Development and Agricultural–Industrial Linkages

State policies were also instrumental in linking industrialization with rural development in a mutually complementary manner, especially during the early stages of growth immediately following the Second World War. In Taiwan, for example, the state encouraged rural industrialization by a comprehensive rural electrification program, the charging of equal energy rates in rural and urban areas, and the early establishment of rural industrial estates, export processing zones, and bonded factories (Ranis and Stewart 1993: 91). Such policies helped to alleviate the urban bias that has marked the initial phase of import-substitution industrialization in many Third World countries. Rural industries also contributed significantly to the rapid expansion of non-traditional exports (e.g., canned mushrooms, asparagus) in Taiwan. Much of the growth in both import-substitution and export industrialization that took place in rural areas was concentrated in relatively small industries and led to significant job provisions. As a result, the percentage of Taiwanese rural workers engaged in non-agricultural activities increased from 29.1 percent in 1956 to 47.0 percent in 1966 to 66.9 percent in 1980 (ibid.: 93). The increasing non-agricultural jobs in rural areas promoted a more egalitarian income distribution in socioeconomic and spatial terms; this helped to slow the tide of rural–urban migration and fostered rural development by providing farm families with needed additional income.

In all of the NICs, state policies were designed to spread income distribution and increase economic participation by different social sectors, raise domestic demand for ISI and other internally oriented sectors, and legitimate the overall role of the state in directing economic growth to serve the 'common good.' De Janvry (1981) provides a useful analysis of the role that domestic market expansion (for wage-goods) and broadened economic participation may play in creating more internally linked and 'articulated' patterns of growth in Third World countries. He argues that a 'market-widening' process frequently accompanies economic growth in the capitalist periphery, but that a 'market-deepening' process seldom occurs. The market-widening process results from the expansion of capitalist relations and structural economic transformation which, in many highly polarized countries, have undermined traditional local economies and converted peasants, artisans, and others into an impoverished working class. Within such economies, domestic demand is focused on luxury goods for the elite rather than wage-goods for the popular sectors, thereby providing incentives for luxury imports and disincentives for local wage-good production. By contrast, the market-deepening process results from real wage increases and more egalitarian income distribution, which gradually increases domestic demand for a continually expanding array of wage-goods, especially more technologically demanding durable goods produced by import-substitution industries. This leads away from the type of polarized development that dominates so much of the South to a broader, more participatory form of development in which social classes, economic sectors, and regions interact in a mutually reinforcing manner. In a study of South Korea, Irwan (1987) finds that, in contrast to most Third World countries, the process of market-widening was accompanied by market-deepening, enabling a relatively large proportion of the population to share in the benefits of rapid economic growth. His findings for South Korea are equally applicable to the other Asian NICs, all of which have managed to avoid widening inequalities by implementing policies which linked economic growth with rising real incomes.[10]

In the NICs that contain a substantial rural sector (South Korea, Taiwan), state policies directed toward agricultural and agroindustrial development were especially important in promoting broader economic participation and more egalitarian income distribution (e.g., Burmeister 1990; Lie 1991; Ranis 1992; Ranis and Stewart 1993; World Bank 1993b). State policies helped to incorporate the rural sector into the national economy in a way that simultaneously accelerated agricultural and industrial production, while generating widely distributed increases in income for both rural and urban households. This helped to avoid the problems of massive rural marginalization that have accompanied industrialization in other Third World countries such as Brazil and Mexico (Senghaas 1984).

Particularly in South Korea, the state designed a development strategy that articulated agricultural policies related to wage-goods linkages, agroinput industrial linkages, rural consumption linkages, and human-

[10]   However, following several decades of rising real incomes and diminishing inequalities, at least some of the NICs may be entering into a period of widening income inequalities. In his study of South Korea, for example, Irwan (1987) notes that real wages declined in the early 1980s as a result of the global economic recession and growing economic concentration by a few giant, family-owned conglomerates (chaebol). He comments that future income distribution will depend on economic and political struggles between the state, domestic firms, and foreign capital, on one side, and workers, students, and others working for the democratization of the state, on the other side. Irwan also contends that state policies appear to be producing differential effects on income distribution among the next tier of Asian NICs. In Thailand, policy changes designed to shift the Thai economy to a more outward orientation have been implemented in a gradual and timely manner in order to minimize harm to domestic sectors, maintain stable macroeconomic conditions, and prevent increasing inequalities in income distribution. In Indonesia, however, internally oriented economic sectors have not been well supported, the income share of the bottom 80 percent of the population has stagnated, and real per capita income has increased much more slowly than it did during a comparable period of early industrialization in, for example, South Korea.

capital investment linkages (Burmeister 1990). The state guaranteed producer markets and consumer distribution channels by creating parastatals. State intervention provided critical support for agroinput markets, such as the import-substitution fertilizer industry. State programs helped to provide human capital in some key sectors (e.g., engineers for the entire petrochemical industry via the state-supported fertilizer industry). State policies supported agricultural mechanization for both small/medium and large producers, thereby raising yields and output and spreading rural income distribution. Broadened income distribution also allowed rural households to allocate more disposable income to finance the education of their children beyond primary school. This enabled rural households to release an employable (i.e., literate and disciplined) pool of surplus labor for the industrial sector, while yield increases continued to raise agricultural output.

As was mentioned in the previous section, land reforms in both South Korea and Taiwan during the period of Japanese occupation also produced structural and organizational changes that promoted broadly based economic growth. These land reforms generated important economic fallout effects by raising rural productivity, redistributing income and stimulating domestic demand, increasing food production for urban areas, and releasing skilled and highly employable labor for industrialization (e.g., Eshag 1991; Lie 1991; Ranis 1989).[11] In addition, the reforms created decentralized farmers' organizations that provided a useful network for the allocation of rural credit, the diffusion of both agricultural and non-agricultural technology, the pooling of small savings, and the development of irrigation and other rural infrastructure (Ranis and Stewart 1993).

However, while the land reforms established initial conditions that facilitated the integration of the rural sector into an expanding national economy, subsequent state policies reinforced and extended these conditions. Strategies of rural development typically combined elements from both a 'development from above' approach, which stressed active intervention by the centralized state, and a 'development from below' approach, which emphasized local participation (Boyer and Ahn 1991). A large population of relatively well-off farmers was thereby created that made a significant contribution to the dynamic expansion of the domestic economy and provided critical support for the early industrialization process. Especially important to income and wealth distribution in rural areas were state programs aimed at raising productivity by accelerating and broadening

---

11   Khan (1987: 98) states that South Korea and Taiwan were also characterized by a highly egalitarian distribution of 'operational holdings' (i.e., actually working farms as opposed to ownership units) before the land reforms implemented by the Japanese. Because of this, he contends that the subsequent redistribution of land via the reforms created relatively little disorganization and met with only sparse resistance.

technological diffusion to all rural social sectors. In South Korea, this was largely accomplished by way of a 'Korean version of the Green Revolution' (Burmeister 1990; Lie 1991). In contrast to the widening inequalities that have accompanied the spread of Green Revolution technologies in most other Third World countries, the Korean Green Revolution succeeded in fostering equity with growth because it was largely scale neutral. Relatively equal access was created to many of the key factors required for the technological transformation of agriculture, including rural credit, farm inputs (especially machinery, chemical fertilizers, improved seeds), basic education, and specialized technical assistance programs.

Infrastructure Provision and the Stress on Human Resources

State programs to speed the process of technological advance and structural change in NIC societies were also complemented by generous provisions of basic economic and social infrastructure. The construction of transportation networks, communication systems, electrical power grids, irrigation systems, and other aspects of basic physical and economic infrastructure provided major 'preconditions' for the process of structural transformation to a modern industrial society. Equally important was the provision of a basic social infrastructure. Among others, Behrman (1990), Kuznets (1965), and Lebeau and Salomon (1990) have stressed vital interconnections between a country's ability to undergo structural economic change and the capacity of its institutions to support human-resource development. The quality of a country's human resources especially influences the ways in which it can absorb, adapt, and disseminate new technologies associated with structural change. The experience of the Asian NICs demonstrates that future Third World growth may be based not just in natural resources but on the development of 'created' comparative advantages through investments in human capital and social technology (Patel 1992; Sengupta 1993).[12] The growing importance of these determinants of long-term, sustained growth – not just for the North, but for the South also – were foreseen by Bernal (1965: 17) several decades ago: 'The real source of wealth lies no longer in raw materials, the labor force or machines, but in having a scientific,

12   The concept of social technology is explained by Patel (1992: 1872–3): '[It] refers to all advances in skills acquired by people individually and collectively . . . Social technology encompasses not only the individual's skills employed in carrying out his or her own economic activity. The collective influence of the working together of all components of society, including policies pursued by governments, and economic, social and political institutions, must also be included in social technology . . . Social technology has a dual character. It is needed as a means to raise the level of output of goods and services. But it is also by itself a goal, an end of development. For instance, better education, greater health, wider spread of social welfare facilities help not only raise productive capabilities but also satisfy basic needs and urges of the people.'

educated, technological manpower base. Education has become the real wealth of the new age.'

The states in the NICs have demonstrated a deep commitment to enhancing human resources, particularly via the expansion of education at the primary and secondary levels (Kearney 1990, World Bank 1993b),[13] but also through programs to develop scientific, engineering, and technical expertise needed to permit diversification into new high-technology growth sectors (Hon 1992; Kim et al. 1992; Yoon 1992). Strong cultural traditions in NIC societies that place a high value on education and achievement have facilitated state efforts to broaden both specialized technical expertise and general levels of education among the masses. As in Japan previously, the land-poor NICs stressed the role of humans as their greatest resource to propel modernization and development – as an 'ever-increasing basic resource' that should be nurtured by the state for the common good (Somjee 1991: 63). This placed an enormous emphasis on education, the spread of information, the learning of new skills, and, above all, on the enhancement of human capacities to participate in the structural changes needed to create a new technologically advanced, industrially based society. Because of this, public expenditures on education and other basic social infrastructure met with little resistance. As a result, the NICs are among the few Third World countries that have invested properly in human-resource development.

The Influence of External Factors on NIC Development

In addition to being shaped by internal conditions, the development performance of the NICs was also influenced by a series of external factors related to their geographical location and the historical period in which their export-led industrialization drives were carried out. Situating NIC development within two broader contexts helps to gain a better understanding of these external factors. The first concerns the geostrategic locations occupied by the NICs on the periphery of the Eurasian landmass, and American-led efforts to contain the spread of Communism in the Cold War era (Jenkins 1991; Petras and Hui 1991). The second is related to development opportunities afforded to the NICs by their advantageous position within the New

---

[13]   Kearney (1990) notes that the NICs focused on primary and secondary education in order to enhance the ability of all classes and social sectors to participate in the national development project. By contrast, the focus for education in many other Third World countries (especially in Latin America) has been on post-secondary education, which has created a well-educated elite but has not permitted the masses to acquire the type of practical literacy and other basic skills needed to broaden their economic participation. Chakravarty (1990) contends that this was also a problem that was not effectively grasped by development planners in India.

International Division of Labor during an unprecedented period of global economic growth (Browett 1985; Douglass 1993; Gereffi and Wyman 1990). Both of these considerations call into question the transferability of the NIC model to other Third World countries that are not presented with such fortuitous external conditions.

A number of authors contend that the integration of the Asian NICs into the global economy occurred in a 'moment of opportunity' in the structure of the world system, which was distinguished by the strategic concerns of OECD countries (led by the US) in containing the spread of Communism and by the interests of core capitalist countries (especially the US and Japan) in extending their economic influence in East and Southeast Asia (e.g., Irwan 1987; Petras and Hui 1991; Robison 1989). These authors claim that both neoliberals on the Right and dependency theorists on the Left have paid insufficient attention to the influence of empire and international security alliances on NIC development. NIC growth strategies were profoundly influenced by global geopolitical relationships and the transnational networks that they produce – from British and Japanese colonialism to the postwar system of US alliances.

The Cold War and Geostrategic Concerns

Geostrategic concerns during the Cold War era were critically important in influencing relationships between the US and the Asian NICs – particularly, but not exclusively, South Korea and Taiwan. Geostrategic interests conferred special advantages on the NICs (e.g., in terms of trade, exchange rates, state-to-state loans and aid, military expenditures, technology transfers) that few other developing countries have enjoyed (Gulati 1992). The US permitted the NICs to establish 'mercantilist' trading relations which coupled protectionist measures of import substitution with expansionary policies of export promotion aimed largely at American markets.[14]

The US also allowed the NICs to systematically undervalue their currencies *vis-à-vis* the dollar in order to facilitate access to American markets.[15] In addition, US state-to-state loans, aid, military expenditures, and other

[14]   Leamer (1990: 365) estimates that exports from Hong Kong, South Korea, and Taiwan were on average suppressed by trade barriers of 12–15 percent, which is substantially lower than for most Latin American countries.

[15]   Tang (1988) notes that similar policies were previously followed by the US to build up Japan as an East Asian bulwark to communist expansionism. The US occupation authorities created the MITI and put Japan on the firm footing of a mercantilist trade regime of import substitution and export promotion. The US also allowed Japan to fix its exchange rate at 360 yen to the dollar – a rate that stood for more than 20 years. Moreover, the US pressured its labor organizations (especially through George Meany at the AFL-CIO) to cooperate in the opening of American markets to the products of Japan's budding industries.

forms of largesse transferred large amounts of capital to the NICs on quite favorable terms. Between 1952 and 1962 (during the key initial phase of import substitution in the NICs), US loans and aid to South Korea and Taiwan funded 70 percent and 85 percent of imports and 80 percent and 38 percent of domestic capital formation, respectively (Robison 1989: 373). US military expenditures and the stationing of large numbers of American troops in these countries also brought capital and other benefits that proved especially important to state development.

As a result of this special relationship with the US, the NICs were able to increase their state autonomy, strengthen state institutions and mechanisms for public–private cooperation, and accelerate the progress of their state-directed development projects. US aid helped to increase levels of public and private consumption without the usually associated fiscal and monetary problems. Moreover, US aid allowed the state to create conditions for real wages in the industrial sector to rise by giving the NICs, particularly South Korea and Taiwan, the advantage of not having to encourage direct foreign investment to initiate labor-intensive industrialization. In contrast to the massive US aid effort in the NICs, the main source of capital inflows to most other developing countries has been via direct foreign investment by TNCs – which tends to reduce rather than increase state autonomy and limits the ability of the state to direct development strategies to serve broadly based national interests (Jenkins 1991: 212–13). Commenting on the relationship between US geostrategic interests and accelerated NIC development, Hamilton (1983: 53–4) concludes:

> For the larger part of the '50s and early '60s fully one half of Korean Government revenue came from the USA . . . Over the 1951–65 period US aid to Taiwan contributed about 34% of total gross investment . . . [Aid] more than doubled the annual rate of growth of GNP [in Taiwan], quadrupled per capita GNP and cut 30 years from the time needed to attain 1964 living standards.

Geographical Location and the Asian Regional Division of Labor

The geographic location of the NICs in East and Southeast Asia also gave them special development advantages beyond those gained from geostrategic concerns. Hong Kong occupies a pivotal position astride trading routes between Northeast and Southeast Asia. It has also benefited enormously as the main link to the outside world for its surrounding region of southeast China. Singapore continued to act as the principal port for Malaysia after independence and is strategically situated at the southern end of the Strait of Malacca which funnels trade flows between the Pacific and Indian Oceans. Singapore is also centrally located relative to the rest of Southeast Asia, facilitating its rise as a financial, commercial, and administrative/managerial center for the region (Parsonage 1992). Moreover, all of the NICs (but especially South Korea and Taiwan) are ideally located to take advantage of growing trade and other ties with Japan. There are important complementary factors between the economies of Japan, the NICs, and other surrounding Asian countries that have fostered a regional division of labor that has been profitable, up to now at least, for all concerned (Emmerij 1987; Kim 1993). In many ways the NICs have followed in the footsteps of postwar Japanese growth. The NICs have consciously emulated many aspects of the Japanese development model (e.g., mercantilism, export-led industrialization, strong state autonomy). Japan has also acted as a major regional 'growth pole' for the NICs, providing them with important benefits such as favored trading relations, direct investment and other capital inflows, subcontracting by Japanese capital, and technology transfers (Edgington 1993).

In recent years, as their industrialization processes have matured, the NICs have begun to occupy a more intermediate position in the regional division of labor between Japan and other less developed countries in East and Southeast Asia. Asian economic growth is sometimes described as the 'flying geese pattern' of development, with Japan at the head followed by the NICs and then the new NICs of Southeast Asia (Kim 1993: 29). The evolution of this regional division of labor has increased opportunities for some of the NICs, especially Hong Kong and Singapore, to develop as mid-level centers for administrative/managerial, financial, and commercial functions. In some cases, NIC domestic capitals have recently opened up branch plants in surrounding Asian countries with lower labor costs – presenting new capital accumulation opportunities, but also posing new challenges to the NICs to develop ever more sophisticated economic sectors to employ their own workers at higher wage levels. In other cases, the NICS have begun to play the role of regional 'command and control' centers for foreign capitals with production facilities in neighboring lower-wage countries. Singapore, for example, has supported the creation of a 'Golden Triangle' of regional economic cooperation, in which it acts as the administrative and financial center for TNCs that have set up branch plants in low-wage areas of the adjacent states of Johor in Malaysia and Riau in Indonesia (Parsonage 1992). Likewise, Hong Kong has developed as the major center through which productive investments by TNCs in the New Economic Zones of Kuangdung and other provinces in southeast China are managed.

NIC Integration into the Global Economy

Over and above the development advantages of their geographical location, the NICs' rise from peripheral to semiperipheral status within the world economy was also facilitated by broader global conditions that may be fast disappearing. Typically, a large part of any country's development story can be attributed to external circumstances and events beyond its control. This was true of Western Europe during the Industrial Revolution, the New World during the nineteenth and twentieth centuries, and the NICs in the 1960s and 1970s (Kearney 1990: 198). Rapid NIC development based on export-led industrialization was encouraged during this period by a number of fortuitous external circumstances: dramatic reductions in transport costs and trade barriers for many industrial products entering the US market; intensified competition within many industrial sectors in the US market; unparalleled growth in the world economy, particularly in the US and other OECD countries; and enhanced comparative advantages for labor-intensive products in the NICs relative to the capitalist core. These factors combined to prompt global capital to initiate an unprecedented horizontal expansion into Third World areas that offered good accumulation opportunities based on lower production costs.

The growing concentration and centralization of capital at an international scale, the restructuring of labor-intensive production processes, the emergence of globally efficient transportation and communications networks, and the rise of international financial circuits, combined under the broad stimulus of rapid global economic growth to produce a 'New International Division of Labor' (see, e.g., Fröbel et al. 1980; Palloix 1977). The rise of this NIDL was closely linked to the internationalization of productive capital in search of global accumulation opportunities under the more flexible labor-supply conditions offered by some developing countries which had relatively productive, inexpensive and less militant labor forces (Browett 1985).

The Changing Global Conditions Facing Aspiring NICs

For a number of authors, however, a series of new conditions confronting Third World manufacturers underscore the fragility of much of the South's outward-oriented industrialization and the difficulty that new industrializers will face trying to replicate the NICs' development experience (e.g., Gereffi and Wyman 1990; Harris 1986; Wade 1992). These new conditions include: the collapse of global financial circuits as a result of rising Third World indebtedness in the 1980s; increasing contradictions and crises within the market-widening strategies of many countries; the rise of new productive technologies permitting the return of some previously exported manufacturing to First World countries; the global economic slowdown and the increasing unevenness of growth, both within and between countries; and the spread of protectionist sentiments, especially within the capitalist core, versus Third World products. For Bello and

Rosenfeld (1990: 57), these new conditions mean that the NIC model of export-led growth may be running out of steam just as neoliberals have enshrined it as the new development orthodoxy: 'The troublesome truth is that the external conditions that made the NICs' export successes possible are fast disappearing, while the long-suppressed costs of high-speed growth are catching up with these economies.'

Just as in Japan previously (Tang 1988), ready access to an expanding US market supplied much of the demand for the export-led industrialization drive of the NICs. By 1964 the US market was absorbing about one-half of the manufactured exports of the largest NIC exporters (Alger 1991: 885).[16] By 1984 this percentage had risen to more than two-thirds, representing approximately one-third of the total manufacturing output of these NICs (Harris 1991: 120). By comparison, about two-fifths of manufactured exports from all developing countries in 1984 were destined for the US (Alger 1991: 885). This underscores the necessity for other Third World countries seeking to replicate the export-led growth strategies of the NICs to find ways to penetrate the US market (or some other equivalent).

However, the protracted slowdown of growth in the US and other OECD countries, the rise of neoprotectionist trade policies in much of the North, and the division of the world into regional trading blocs (e.g., the EEC, NAFTA) all present formidable demand-side limitations to new industrial exporters seeking to mimic the NIC model. Recent studies by Bhagwati (1991) and the Commonwealth Secretariat (1990) show that protection-ism among Northern countries is growing and that bilateral agreements and rising use of non-tariff measures are steadily undermining the GATT system of fixed, universal trading rules. Many analysts contend that the erection of non-tariff barriers (e.g., anti-dumping duties, import licencing, technical specifications, voluntary export restraints) by Northern countries has become one of the most serious obstacles to Third World exports (e.g., Roarty 1993; Watkins 1992). Whereas the GATT has enjoyed some success in reducing overt tariffs between countries, less conspicuous non-tariff bar-riers have proliferated and have become a major impediment to freer trade. Moreover, a disproportionate share of these non-tariff barriers in Northern countries have been aimed at Southern products. According to the World Bank, some 31 percent of the South's manufacturing exports are subject to non-tariff barriers, compared with the North's 18 percent (Watkins 1992: 35). Given rising protectionism in the US and other Northern countries against Southern exports, it may be that the 'moment of opportunity' in the global economic system that the NICs exploited so successfully in the 1960s and 1970s has largely passed for other countries.

In addition to demand-side constraints, new Third World industrial

---

[16]   Alger's figures include Brazil and Mexico, as well as the four Asian NICS.

exporters are also facing difficult supply-side conditions associated with increased competition from the existing NICs, aspiring NICs seeking to export similar products, and older industrialized countries trying to maintain their manufacturing base in order to limit rising unemployment and restore trade balance. The original NICs were able to benefit from the relative lack of concern in OECD countries for protecting their manufacturing sectors during a period of rapid growth and job creation, as well as from the lack of other Third World competitors, many of whom were still pursuing ISI and other inward-looking strategies. However, it is apparent that both of these conditions have largely passed. While the capitalist core has become preoccupied by rising unemployment resulting from industrial restructuring, scores of peripheral countries are turning to outward-oriented strategies (often under the aegis of structural adjustment programs, or SAPs) to supply the motor for future growth. Fierce competition from a multitude of aspiring NICs is crowding many export sectors at the same time that many developed countries, given their own economic woes, are proving unable and unwilling to absorb increasing imports from the South.

There is also growing evidence that the development of new production technologies and marketing techniques is beginning to allow the re-importation of some industries to the capitalist core that had previously been located in peripheral areas (see, e.g., Ariff and Hill 1986; Jenkins 1985; Harris 1991). In the productive sphere, the rise of labor-displacing microelectronics technologies (e.g., computer-aided process controls) is altering factor proportions in some activities, which disadvantages labor-abundant and capital-poor Third World locations. Something of this sort is occurring, for example, in one of the most notoriously mobile sectors of manufacturing, the textile and garment industry (Griffith 1987; Harris 1991). American-based firms have recently introduced automated fabric-cutting processes that are cheaper per unit of output than the labor-intensive alternatives.[17] In many industries, new organizational and marketing techniques (e.g., 'just-in-time' inventories) have also been altering locational requirements. Many of these new techniques require high quality control, increased flexibility within production processes, and rapid decision-making to respond to sudden market changes – all of which enhance the locational advantages of First World rather than Third World sites. As technological change has gradually eroded the low-wage comparative advantage of many NIC export sectors, possibilities have increased for mutually destructive bidding wars among Third World countries desperate to attract capital

[17]   This has created a new pattern of global specialization within the garment industry in which fabric cutting is shifting to the US and sewing remains in Third World areas such as East and Southeast Asia. The paradoxical result is that, among categories of goods imported to the US, garments now have one of the highest proprotions of value added produced in the US itself (Harris 1991: 118).

to the reduced range of sectors in which they remain globally competitive (Alger 1991).

Many new Third World exporters are thus faced with stiff competition within limited markets from both aspiring and already established NICs in a fiercely contested succession process (Athukorala 1989). The ability of aspiring NICs to take the place of the original NICs especially depends on the capacity of the latter to shift production into higher-value, more technologically advanced sectors, thereby leaving the more labor-intensive sectors at the lower end of the export market to the new arrivals. There is some evidence that concerted efforts by the NICs to shift export production toward more technology- and skill-intensive goods has shown some success (Ranis 1992; Sengupta 1993; World Bank 1993b). Table 3.6 shows that by 1986, Hong Kong, South Korea, and Taiwan had shifted almost 30 percent and Singapore some 78 percent of exports to the US into more sophisticated industrial sectors. Nevertheless, traditional less sophisticated products still composed 62.2 percent in Hong Kong, 52.7 percent in South Korea, 49.1 percent in Taiwan, and 13.9 percent in Singapore of all exports to the US.

It has been noted that technologically sophisticated industrial sectors are particularly dominated by well-established transnational intra-firm trade (Helleiner 1979: 306). This means that if the original NICs are to continue to broaden their process of technological transformation into more advanced sectors, they will need to penetrate export categories currently dominated by transnational corporations (TNCs) based in the capitalist core. In recent years, efforts by NIC capitals to break into more sophisticated industrial sectors have been limited by factors such as insufficiently developed research and development infrastructures, dependence on special licencing arrangements with established TNCs, the reluctance of these TNCs to share advanced technologies, and selective trade barriers and other forms of market control by OECD countries designed to protect their high-technology sectors from outside competition (Lin 1989).

In effect, this has subjected the NICs to a 'structural squeeze' in which they are able to graduate into only a limited number of more advanced capital-intensive sectors and are priced out of their older labor-intensive sectors by rising wage levels (Bello and Rosenfeld 1990; Clark and Kim 1993). The succession process by which Third World countries are supposed to gain upward mobility has largely been blocked. One of the principal constraints to the succession process has been the increasing use of new forms of monopolistic market control by TNCs and their political allies in core capitalist countries. While the original NICs have been struggling, against mounting odds, to break into more sophisticated export sectors, very few other developing countries have been able to make the initial transition from basic primary exports to labor-intensive industrial sectors (Tan 1993). Because they are effectively excluded from

Table 3.6 Percentage distribution of NIC exports to the US of selected product groups, 1966 and 1986

Product Groups	Hong Kong 1966	Hong Kong 1986	South Korea 1966	South Korea 1986	Taiwan 1966	Taiwan 1986	Singapore 1966	Singapore 1986
Traditional	67.9	62.2	56.5	52.7	44.6	49.1	73.6	13.9
R&D intensive (general)	9.8	23.8	2.0	19.2	15.8	22.3	0.0	58.2
R&D intensive (sophisticateds)	17.5	29.5	3.9	29.6	20.3	29.2	0.2	78.1

Among the product groups, traditional represents low-range goods made using cheap, unskilled labor with little research and development; R&D intensive (general) represents mid-range goods made using semi-skilled labor and globally generalized research and development procedures; and R&D intensive (more sophisticateds) represents high-range goods made using highly skilled labor, technologically sophisticated production processes, and specialized research and development procedures that have not been globally generalized.

Source: Kellman and Chow (1989) table 5, p. 271

participating in more capital-intensive, high value-added sectors, developing countries find themselves locked in a desperate and mutually destructive struggle with other Third World exporters similarly confined to traditional, low value-added sectors within a rigid international division of labor.

Consequences of the NIC Model of Development

While most studies of the NICs have concentrated on explaining the causes of their growth, the broader consequences of NIC development also deserve serious scrutiny – especially in debates over the appropriateness of the NIC model for other Third World countries. Among areas deserving more attention are: the democratization process, respect for personal liberties and basic human rights, freedom of association, distribution of income and wealth, equality of opportunity among classes and social groups, working and living conditions, and environmental sustainability. A growing number of authors (e.g., Amirahmadi 1989; Amsden 1989; Bello and Rosenfeld 1990; Douglass 1993; Ogle 1990; Petras and Hui 1991) claim that the NICs, despite their rapid growth, have serious shortcomings in many of these areas that may detract from their usefulness as models of development for the rest of the South. It is asserted that progress in these areas, all of which must be included in any broadly based definition of development, has been sacrificed by the NICs in their all-out pursuit of rapid export-oriented growth. This calls into question the appropriateness of constructing a new development orthodoxy on the basis of a model that many analysts contend gives precedence to exports over domestic needs, economic growth over environmental sustainability, and the accumulation interests of a few over the basic human rights and democratic interests of the many.

Authoritarianism and Repression in the NICs

The role that the state has played in directing NIC development has not been confined to direct economic planning or exerting strict controls over economic institutions. Authoritarianism, repression, the exercise of strict social control, and the disciplining of the working class and other popular sectors to serve the accumulation interests of capital have also been central elements of the national development projects of the NICs. Although some variation exists, none of the NICs have made much progress in creating democratic structures that would facilitate meaningful political participation by the majority. South Korea and Taiwan spent much of the postwar period under military and/or one-party rule. Only recently have there been some signs of a halting democratization process beginning to emerge in either of these countries (A. Lee 1993; S. Lee 1993; Takwing 1993). Moreover, any tendencies toward democratization continue to be strictly circumscribed by the authoritarian nature of broader state structures.[18]

Hong Kong and Singapore hold multi-party elections and appear, on the surface at least, to enjoy many elements of a functioning democracy. However, closer scrutiny of their political systems reveals a pervasive authoritarianism and frequent use of coercion by the state to limit and direct political participation. In some respects, democratic rights among the NICs have been most restricted in Singapore, despite its holding of regular elections, by the ubiquitous presence of the state engaged in massive social control and social engineering (Bello and Rosenfeld 1990). Singapore is effectively controlled by a tightly knit ruling elite of civil service technocrats and politicians from the dominant People's Action Party (PAP), aligned with powerful domestic and transnational capitals. Williams (1992) describes the ideology of the PAP as an 'ideology of survival' in which all considerations are subservient to economic and political survival. He notes that the 'effect of this ideology [is] to legitimate the existing social order, justifying questions of social control and the distribution of resources on rational and scientific grounds (p. 368). Paul (1993: 298) also draws links between the imposition of authoritarian social control in Singapore and the ideology of its ruling elite:

> The political culture of Singapore's ruling elite is authoritarian in character and includes ideological elements familiar to the totalitarian ethos of the right. There is the strong belief in the genetic differentiation of society into the have and the have-not. This is reflected in their preference of a hierarchically organized and patriarchally led society as the most successful model of political territorial organization ... They also argue that democracy is not suitable for the country and leads to moral decadence and the economic impoverishment of society; and that the world at large is a jungle where only the morally and militarily strong survive.

In order to restrict dissent and ensure strict compliance from all social sectors to state-directed development goals, all of the NICs have created

18   Petras and Morley (1992) offer a useful analysis, based on the experience of Chile and other Latin American countries, of the limited nature of the so-called democratization processes taking place in much of the Third World. In this study, they make the important distinction between state and regime. While offering a formalistic democratic facade, many recently elected regimes operate only within the strict authoritarian parameters set by longstanding state structures.

large police forces and internal security apparatuses. But it is in South Korea and Taiwan that concern for security has become most extreme. From 1961 to 1987 South Korea was ruled almost continuously by a military dictatorship, while Taiwan endured one of the longest periods (from 1949 to 1987) of martial law in modern history. During the 1950s, South Korea and Taiwan had among the highest military/civilian ratios in the world – with about 600,000 soldiers in each army (Petras and Hui 1991: 187). For much of the postwar era, both of these countries achieved international notoriety for the extreme repression carried out by their internal security forces against labor, farmers, students, and other popular organizations.

In Taiwan, the omnipresence of its highly-developed secret police force has, until quite recently, stifled any meaningful dissent against the dictates of the Nationalist regime (Petras and Hui 1991). Kuomintang (KMT) party organizations were established to represent virtually every sector of society, including labor, farmers, commerce and industry, occupational and professional groups, schools and universities, women, and Buddhist religious associations (Tak-wing 1993: 5). Until the 1980s, competing organizations outside of KMT control were banned. The first opposition party, the Democratic Progressive Party (DPP), was inaugurated only in 1986. In South Korea, the creation of an enormous paramilitary police force (estimated to number some 150,000 men) and the all-pervasive Korean Central Intelligence Agency (KCIA) has allowed the state to extend its control into virtually every arena of Korean economic and social life (Amsden 1989; Ogle 1990; Petras and Hui 1991). For most of the postwar period, the KCIA and the police, often aided by company 'goon squads,' have intimidated and harassed union organizers and labor leaders (Choi 1989).

Although restrictions on union organization have been relaxed somewhat in recent years, the violent repression of a strike at the Hyundai shipyard in 1990 by thousands of riot police shows that coercive labor control is not entirely a phenomenon of the past (Douglass 1993: 162). As in Taiwan, independent social movements have only become a legitimate force in Korean society within the last few years. S. Lee (1993) reports that many of these social movements arose following a nationwide popular uprising against the Chun regime in 1987 and have drawn large numbers of urban professionals (the 'new middle strata') into their ranks alongside traditionally more militant sectors of Korean society, such as students and organized labor.

The Mixed Record on Distributional Issues

While the record of the NICs concerning respect for personal liberties, basic human rights, and democratization has generally been unfavorable, their record is more mixed in areas such as employment, poverty reduction, wage and income levels, and working conditions. There is little doubt that the majority of workers in the NICs have benefited greatly from the improved job prospects that have accompanied export-led industrialization. Poverty has been significantly reduced and real wage and aggregate income levels have risen dramatically (see Addison and Demery 1988; Chakravarty 1990; Wade 1992). The Asian NICs, especially South Korea and Taiwan, stand out among virtually all other Third World countries for having reduced the income gap with the core capitalist countries of Western Europe and North America over the past two decades.

For the most part, the NICs have also succeeded in improving income distribution. Not only have aggregate income levels risen, but the benefits of NIC growth have been much more equally distributed in comparison with other prominent Third World industrializers such as Brazil and Mexico (e.g., Eshag 1991; Irwan 1987; Wade 1992). In the two NICs with a significant rural population (South Korea and Taiwan) state programs have created forward/backward linkages between rural producers and agroindustries that have fostered the rise of middle-class farmers and a more egalitarian rural income distribution. Based on escalating real wage and income levels, the rise of a substantial urban middle class has also formed a key element of NIC development. In recent years, the urban middle class has not only begun to shape the dominant pattern of domestic consumption and urban lifestyles, but it has also begun to make political demands (Koo 1991). In both South Korea and Taiwan, for example, the growing politicization of the urban middle class has been a critical element in the recent transition from dictatorship to limited democracy (S. Lee 1993; Tak-wing 1993).

However, the generally positive performance of the NICs in the areas of employment, poverty, real wage and income levels, and income distribution is somewhat offset by their poorer record concerning unionization and freedom of association, working conditions, and the disproportionate costs that some social sectors, particularly young women, have paid to fuel economic growth. Notwithstanding the benefits that NIC development has conferred on workers, they have also had to pay a high price for the export success of the NICs. State policies in the NICs have been designed to heighten capital accumulation opportunities in key sectors by ensuring corporations the cheapest, most productive, and least militant workers possible. The state has played an active role in disciplining the working class to accept these conditions via a number of means, including state control of labor organizations, restrictions on freedom of association and other repressive labor laws, state-directed violence against labor activists, and weak or non-enforced legislation concerning work hours and workplace conditions (Addison and Demery 1988; Amsden 1989; Bello and Rosenfeld 1990; Ogle 1990).

In many NIC export sectors, hazardous and unhealthy working conditions and extremely long work hours, often compounded by shift work, extract particularly high costs from the workers. Despite relatively high wages in these sectors, rates of labor turnover are especially high as workers quickly 'burn out'.[19] Up to now, these workers have been relatively well compensated because increasingly tight labor markets in the NICs have produced real wage growth. However, if the global markets upon which the NICs depend take a turn for the worse, as they did in the early 1980s, workers may be left on their own to cope with the ensuing austerity. It is in such times that basic labor rights such as freedom of association and collective bargaining become critical. It remains to be seen if denial of basic labor rights will remain a permanent feature of NIC industrialization or will gradually disappear as the NICs assume more economic maturity and the working class struggles to become more organized and assertive.

The Disproportionate Burden of Female Workers

Although other social groups have also been systematically exploited,[20] it appears that young women in particular have borne a disproportionate burden to accelerate export-led growth in the NICs. Much of the labor-force in export-oriented industrial sectors has been composed of a youthful female 'temporary' proletariat that work during the transition between school and marriage (Lin 1989; Park 1993). These women usually either commute from nearby homes, in which they live with their parents, or are housed in barrack-like dormitories in large factory compounds. They tend to be concentrated in entry-level, shop-floor jobs in industrial sectors with low pay and long hours – jobs that are left vacant

---

[19]   Worker burn-out from excessive work hours, shift work, and unhealthy working conditions seems to have become a generalized problem for export-oriented industrial sectors throughout the South. Evidence shows that free trade zones and other export-oriented industrial concentrations are experiencing especially high rates of labor turnover and that most workers cite burn-out as their reason for leaving.

[20]   Bello and Rosenfeld (1990), for example, report that South Korean and Taiwanese farmers are increasingly being driven into debt and squeezed off their land by low producer prices and rising imports of US agricultural goods. It seems that farmers, after making a significant contribution to the initial 'take-off' stage of growth in these countries, are now being sacrificed. Bello and Rosenfeld also note a growing tendency to use foreign workers for unskilled jobs in some of the NICs, especially Singapore. These workers (e.g., from Bangladesh, India, Indonesia, Philippines, Sri Lanka) are strictly controlled and heavily exploited. They tend to occupy jobs that most nationals will no longer take. In Singapore foreign workers now represent about 12.5 percent of the overall workforce and 25 percent of manufacturing labor.

by older workers because the wages are too low to support an entire household.[21]

In Taiwan, women comprised 59 percent of the workers in the food-processing industry, 79 percent in the textile industry, 85 percent in the apparel industry, and 65 percent in the electrical equipment and supplies industry in 1978 (Kung 1984: 109). Most were young and often had been recruited directly out of rural schools by factory representatives. In South Korea, the number of women workers increased fourteen times between 1963 and 1980. 'Female manufacturing industries' (e.g., textiles and clothing, rubber and plastics, electronic goods, shoes, china and pottery), in which women account for more than half of all workers, were responsible for 70 percent of total national export earnings in 1975 (Park 1993: 132). Women in these industries typically work extremely long hours for relatively low wages. According to data from the South Korean ministry of labor, the ratio of female to male wages was only 52.8 percent in 1989. Manufacturing, in which the majority of women are employed, had the lowest wage level among all industries and was the only sector in which wages had always been below the average (ibid.: 133). Moreover, in 1988 South Korea was the only country in which women's working hours were longer than men's among the fifteen countries that released data to the International Labor Organization (ILO). In manufacturing, women worked for an average of 245 hours per month, or 9.7 hours a day, in 1984 (ibid.: 134).

The economic rationale behind the use of 'temporary' young women in export manufacturing is readily apparent – because the workforce is female, transitional between generations, and does not generally have to support a family, a true 'living wage' does not have to be paid that would be commensurate with familial social reproduction requirements (Lin 1989). In many cases, the social reproduction of the families of these young female workers rests on the mobilization of all members of extended families. However, traditional patriarchal structures often lead to the creation of an intra-familial sexual hierarchy in which parents have different expectations of their daughters *vis-à-vis* their sons. While sons are normally educated to enter into higher-status occupations to eventually become the principal breadwinners for their families, daughters are often asked to sacrifice their education to take up dead-end, low-paying jobs until they get married and assume the household reproductive tasks for their own families.

For Greenhalgh (1985: 303–4) this situation illustrates the interlocking

[21]   A similar pattern of female employment exists in many of the export-oriented industrial concentrations of Third World countries such as Bangladesh, Brazil, India, Mexico, Sri Lanka, and Thailand. Moreover, many of the mill-towns of the nineteenth and early twentieth centuries in countries such as Great Britain and the US were also characterized by extensive use of this type of female labor.

and mutually supportive nature of capitalist industrial institutions and traditional patriarchal structures in the NICs. Industrial capitalism and the state provided new means (jobs and education) for parents to use old tools (sexually differentiated inter-generational expectations) to recreate and extend traditional hierarchies (sexual inequalities). Industrial capitalism, in turn, took advantage of the sexual hierarchies created within families by using women's lower skill levels, familial obligations, greater docility, and temporary laborforce status to offer them dead-end, low-paying jobs that no other social group would fill. These discriminatory features of the industrial labor market also acted to reinforce the subordinate status of women in the family, providing justification for parents to continue treating their daughters as tools for the advancement of others, particularly sons.

While the structures of capitalism and patriarchy were interlocked and served to reinforce one another, capitalism by itself cannot be held responsible for creating these sexual hierarchies – it simply used and extended gender differences which have been rooted in traditional Asian societies (as well as many others) for generations (ibid.: 304). Nevertheless, as Park (1993: 142) notes, 'both the domestic and international capitalist systems [were] structured to maximize profit by using the culturally marginal members of a society, thus making them also economically and politically marginal.' As a result, the role of women in NIC development has been neither recognized nor rewarded in any way commensurate with their actual contribution. While in material terms the situation of some women may have improved, in relative terms it generally has not. Indeed, gender discrimination has allowed the NICs to join the ranks of the industrialized world much more quickly than would have otherwise been possible. From this perspective, rapid export-led growth in the NICs has been made possible only by the unfair use of female labor, i.e., exploitation (Pettman 1992: 53).

Widespread Environmental Degradation

A further consequence of NIC development has been widespread environmental damage. Although this problem is by no means unique to the NICs, their single-minded pursuit of rapid economic growth at all costs has caused particularly severe environmental consequences that, given the increasing wealth of the NICs, could have been largely avoided by giving more priority to goals of more balanced and sustainable development. All-out growth has left much of the countryside in both South Korea and Taiwan severely and perhaps irreparably damaged. South Korean rural areas suffer from extensive deforestation, which has also caused associated problems of soil erosion and flooding. In many areas, there is evidence of serious chemical contamination of groundwater, rice fields, and rice crops from excessive levels of chemical fertilizer applications (Wade 1992). Similarly, many rural areas in Taiwan have suffered severe environmental damage; industrial waste water has polluted approximately 20 percent of all farmland and 30 percent of the annual rice crop is contaminated with heavy metals (Bello and Rosenfeld 1990). In addition, both South Korea and Taiwan have become heavily committed to nuclear power generation. Storage problems for nuclear waste and poor quality-control in the components and construction of nuclear power plants raise the possibility of a major disaster in the making (ibid.).

Although Singapore enjoys a somewhat better record, the other NICs are also confronting massive environmental problems in their major urban areas. Rapidly rising urban congestion coupled with the lack of enforcement of environmental regulations have produced escalating costs in terms of severe air and water pollution. Air pollution and acid rain have become major health hazards for those living in urban areas. The air over Seoul, for example, has one of the highest concentrations of sulfur dioxide in the world. Ranis (1992: 239) reports that 'traffic jams and accompanying problems such as noise and air pollution have made [Taipei] one of the worst places to live in the world.' Much of the urban tap water in the NICs is said to be unfit for drinking as a result a worsening water pollution (Wade 1992).

Rising environmental costs in both urban and rural areas in the NICs are materializing in poor health, physical damage, loss of amenities, and other problems that will call for extensive remedial spending in the near future (Winpenny 1991). This will show up in future NIC development as the negative counterpart of earlier growth-first strategies that failed to properly consider environmental consequences. In order to stimulate rapid growth, the NICs have used up significant environmental capital that can only be restored, if at all, at considerable cost to future generations.

As these consequences of the NIC development model become more apparent, they may prompt many neoliberals and other development analysts to reconsider their image of the NICs as paragons of Third World development. While it is undoubtedly true that the NICs have made great strides in some areas of development that are the envy of much of the rest of the South, there are also serious shortcomings to their development model that ought to be given attention alongside its successes. In any case, given the important role that geographical and historical particularities have played in NIC development, analysts should carefully assess the applicability of many elements of the NIC development model for other Third World countries. To neglect these issues would be to risk replicating many of the problems of formalism and universalism that have accompanied the imposition of inappropriate development models on Third World countries.

Such models, from modernization to structural adjustment, have not only produced an intellectual impasse in mainstream development theory, but have also extracted particularly high costs from those in Third World countries who can least afford them.

4

The South (1): Neoliberal Policy and Strategy

While the previous chapter focused on the Asian NICs, this chapter turns to the recent development experience of the rest of the South. The rapid spread of neoliberalism throughout the South in recent years, particularly with the rise of structural adjustment programs (SAPs), now gives us the opportunity to assess the performance of this development strategy. Although it is recognized that many factors beyond the immediate control of Third World states have profoundly affected the outcome of SAPs, emphasis will be placed on the influence of variations in state policies. The role that ideological considerations have played in the framing of neoliberal policies, especially within the IMF and World Bank, will also be analyzed. Many of the shortcomings of outward-oriented policies and other liberalization measures will be revealed and alternative development strategies suggested. Particular attention will be given to problems of increasing polarization as well as to the social costs of SAPs.

The Spread of Neoliberalism and Structural Adjustment Programs

The origins of neoliberalism in the South can be traced back to a few experimental programs initiated in a small number of countries during the 1970s. In Latin America, various elements of neoliberal development strategy were first implemented in Chile under the Pinochet regime and soon thereafter by a few other countries such as Bolivia and Mexico. In Africa, Ghana was an early testing ground for neoliberal policies, which then were emulated in a handful of other countries (e.g., Kenya, Nigeria, Gambia). In Asia, a few countries (e.g., Turkey, Indonesia) embraced broad neoliberal development programs relatively quickly, but most others (e.g.,

India, Pakistan, Thailand) moved rather slowly and hesitantly to adopt neoliberal policies.

From these meager beginnings, neoliberalism has quickly spread throughout the South, so that today there are very few countries that have not adopted major neoliberal elements into their development strategies. Perhaps the strongest moves toward neoliberalism have taken place in the external sectors of many countries. Currencies have been regularly devalued or realigned with convertible monetary systems, and many restrictions governing trade flows and external financial movements have been reduced. However, neoliberal policy instruments have also been directed at the internal economies of many countries. Internal markets have been deregulated, often involving the abolition of agricultural marketing boards and the removal of price subsidies for basic foodstuffs and other wage-goods. In many cases, internal deregulation has also been extended to labor markets through de-unionization and the abolition of minimum wage laws and other labor regulations. Many of these efforts have been aimed at reducing private consumption so that an increasing proportion of the national economy may be diverted toward private investment, thereby allowing trickle-down mechanisms to function. Complementary measures have also usually been directed toward reducing public consumption, especially by privatizing state-owned enterprises and cutting the size of many government bureaucracies. In addition, government spending on social and economic infrastructure (e.g., education, health care, social welfare, transportation and communication systems) has commonly been curtailed.

The Role of the IMF and World Bank

Although a few countries initiated neoliberal measures during the 1970s, and some countries have subsequently implemented such policies on their own, the rise of neoliberalism in the South has particularly coincided with the spread of IMF/World Bank structural adjustment programs (SAPs) among indebted countries. The origins of structural adjustment lending can be traced back to the creation, in 1974, of the Extended Fund Facility (EFF) by the IMF to supervise economic stabilization programs in some financially troubled countries. For most Third World countries, however, structural adjustments began during the next decade, following the introduction of Sectoral Adjustment Loans (SECALs) in 1979 and Structural Adjustment Loans (SALs) in 1980 by the World Bank. The mutual focus of the IMF and World Bank on structural adjustment lending was further formalized in 1985 with the establishment of the Structural Adjustment Facility (SAF), jointly managed by the Fund and the Bank. Although the IMF had traditionally concentrated on short-term stabilization measures, while the World Bank had focused on longer-term adjustments and project lending, the roles of these two preeminent international financial institutions converged in the 1980s in support of structural adjustment programs.

In the wake of the international debt crisis, which had shaken many of the world's largest banks and financial institutions, SAPs quickly became the accepted vehicle by which Third World countries would regain financial solvency and begin repaying their foreign debts. Future IMF–World Bank lending to indebted countries (which comprise virtually all of the South) was made conditional on their submission to officially supervised structural adjustment programs. Moreover, other multilateral financial institutions, private banks, and international development agencies commonly began to insist on an IMF–World Bank 'seal of approval' as an indispensable condition for further loans and/or aid. In effect, the submission to SAPs had become the decisive factor in restoring the international creditworthiness of most Third World countries, without which their access to foreign capital would be withdrawn.

Given the historical dependence of most Third World economies on external sources of capital, very few countries have been able to withstand IMF–World Bank pressure to submit to structural adjustments. A handful of countries (notably Argentina, Brazil, Israel, Peru, Zimbabwe) chose to include some heterodox elements in their adjustment programs, but the overwhelming majority of countries submitted to an orthodox package of SAPs under IMF–World Bank supervision.[1] By 1983, three-quarters of Latin American countries were operating under IMF-supervised SAPs (i.e., 'upper credit tranche arrangements with a high degree of conditionality under the Stand-by Arrangement or Extended Fund Facility', Pastor 1989: 90). As the decade continued, most other Latin American countries also fell under IMF control, and the few countries that avoided direct IMF intervention were often under indirect IMF supervision (ibid.). Likewise, two-thirds of African countries had submitted to some form of IMF-supervised structural adjustment by the mid-1980s and many others were under different types of indirect IMF regulation (Landell-Mills et al. 1989).

Factors Affecting the Performance of SAPs and Neoliberal Strategies

As might be expected, the ability of Third World countries to sustain structural adjustment programs has been quite variable. Much of this

---

[1]   In addition to the usual fiscal and monetary instruments of orthodox SAPs, some countries chose to add a number of heterodox elements (e.g., wage and price freezes, exchange rate pegging, deindexation measures) aimed especially at producing drastic and immediate reductions in inflation. In recent years, Russia and many Eastern European countries have also carried out heterodox SAPs.

variability can be attributed to both internal and external conditions over which many of these countries have only limited control. Price movements in international commodity markets have exerted a determinant effect on the outward-oriented adjustment programs of many poorer, smaller countries that have export sectors concentrated in a few traditional primary products. The capacity of countries to attract investment capital has in large part been determined by internal socioeconomic structures (e.g., levels of human resource development, the efficiency of transportation and communications infrastructure) that are inadequate in much of the South and can only be changed very slowly through concerted state intervention. Likewise, because the creditworthiness of countries is largely the result of past fiscal policies, many current administrations have had to contend with the consequences of huge foreign debts accumulated by previous governments. In some cases, unexpected events (e.g., droughts, floods) have also affected the ability of governments to sustain programs of expenditure reduction and economic stabilization. Such catastrophic natural occurrences have had a particularly devastating effect on many of the poorer, rural countries of Africa and Latin America. Many analysts now contend that outside economic experts who monitor the performance of SAPs have generally paid insufficient attention to special structural problems and uncontrollable events that have particularly hampered the efforts of many severely underdeveloped countries to sustain adjustment programs (see, e.g., Banuri 1991; Cheru 1992; Colclough and Green 1988; Green 1985; Helleiner 1992; Riddell 1992; Streeten 1993).

The Influence of the State and Policy Framework

At the same time, however, considerable evidence has accumulated that a few key elements of state policy have had a strong influence on the performance of SAPs and neoliberal development strategies in general. First, effective development strategies require the fusion of specific policies aimed at immediate problems with a broader structurally oriented focus on long-term development needs. The multifaceted elements of the policy framework need to be integrated into a coherent overall strategy which eschews one-dimensional solutions. Abstract, idealistic models ought to be rejected in favor of realistic, achievable strategies based in the diverse empirical realities of the development experiences of different countries. Pragmatic solutions based in real-world development processes should replace one-dimensional, dogmatic worldviews. In the rather messy and highly changeable field of Third World development, adherence to rigid orthodoxies almost always produces poor results.

Second, a consistent and well-conceived policy framework should be established that is not subject to frequent or sharp reversals. Credible and predictable economic conditions, which are widely expected to be sustained into the indefinite future, are particularly important to stimulate long-term investments associated with structural economic change. Coordinated fiscal, monetary, and exchange rate policies play a vital role in creating a stable macroeconomic environment for investment decisions. The weight of current opinion is that needed liberalization measures should be introduced gradually, but ought not to be drawn out over too many years (Michaely et al. 1991). For countries in which massive economic imbalances exist, liberalization policies should probably start with a strong step to break with past conditions and heighten the credibility of the new program. Under circumstances of rapid inflation, for example, strong measures aimed at economic stabilization (e.g., restrictive fiscal and monetary policies, currency devaluation) need to precede other liberalization policies. Particularly if widespread liberalization is envisioned, attention should be focused on policy coordination and sequencing in order to avoid problems of policies negating one another or operating at cross-purposes.[2]

At the same time, however, it should be remembered that a sound policy framework comprises many elements and, even in cases of extensive liberalization, should not be simplistically equated with the free operation of market forces. In the rural sector, for instance, levels of real producer prices strongly influence agricultural production, but so do a range of other factors (e.g., rural credit, agricultural extension programs, transport and marketing systems, access to consumer goods and agro-inputs). Raising producer prices without complementary policies designed to address the special needs of small/medium farmers may generate perverse results, especially in the highly polarized rural sectors of many countries. Many of these farmers, who have traditionally dominated domestic food production in most countries, may be driven off their land because they are unable to meet the new conditions of heightened competition with transnational agribusinesses and other larger producers. Rather than just focusing on prices, the countries that have succeeded in stimulating equitable agricultural growth have created and maintained a reasonable balance and efficiency in the entire policy package affecting rural development (Ghai 1987; Jaeger and Humphreys 1988).

Third, the appropriateness of a country's domestic policies strongly influences its capacity to expand exports. Moreover, export expansion may provide a strong stimulus to growth, especially among smaller Third World countries with relatively underdeveloped internal economies. Export performance is affected not only directly by trade measures themselves, but also indirectly by a host of other supporting policies. These include pricing

2   For example, the experience of Chile, Mexico, and other countries demonstrates that liberalization of the capital market should be attempted only after initial adjustments to the goods market have been completed. A more lengthy analysis of the important issue of policy coordination and sequencing appears later in this chapter.

policies for internal products and factors of production; fiscal and monetary policies affecting exchange rates and domestic inflation; and investment policies to build up social overhead capital needed to realize potential comparative advantages (Myint 1987: 115–16). The positive association between exports and economic performance is often attributed to increasing returns to scale, the attraction of capital for investment and imports, the dynamic spillover effects of export growth on the remainder of the economy, and other externalities (e.g., technological diffusion, 'demonstration' effects on human capital) related to global competition (Esfahani 1991; Sengupta 1991). A growing body of statistical evidence indicates that an outward orientation is positively related to rates of growth, particularly in the industrial sector (e.g., Chow 1987; Dollar 1992; Cox Edwards and Edwards 1992; Michaely et al. 1991). However, it should be cautioned that the direction of causality has not been well established in this relationship (Helleiner 1986; Toye 1987). The effectiveness of trade liberalization and other outward-oriented policies may largely depend on the structure of exports and the general level of economic development (Dodaro 1991). It may be only when a relatively advanced level of economic development has been achieved that extensive trade liberalization becomes feasible.

Fourth, outward-oriented policies should not sacrifice economic sectors and social groups linked to the domestic market in favor of those tied to export production. If it is properly planned, export-led growth can stimulate broadly based development by a number of means, including direct job provision, which generates a 'ripple' effect on the rest of the economy; indirect job provision through backward linkages with other economic activities; technological diffusion and other externalities; and increased net foreign resource inflows, which are especially important for enhancing import capacity (Colclough and Green 1988; Esfahani 1991; Myint 1987). The experience of the Asian NICs demonstrates that outward-oriented development need not produce the type of severe socioeconomic and spatial polarization that has characterized export-led growth in most other Third World countries. However, care must be taken to avoid policies that stimulate growth in some sectors at the expense of others. For example, trade liberalization often needs to be accompanied by policies supporting small/medium rural producers and other domestically oriented groups if they are to survive new conditions of greatly increased foreign competition. Likewise, technological diffusion may decrease rather than increase job opportunities in areas of high unemployment if policies and institutional arrangements are not put in place which facilitate technological adaptation among small-scale, labor-intensive operations. Generally, such policies should expand the focus of outward-oriented development beyond just increasing exports, creating conditions for export-led growth that will promote broadly based structural change.

Fifth, fiscal and monetary policies should aim to create stable macroeconomic conditions without causing undue hardship through drastic economic contraction. Many analysts note that improved fiscal and monetary management is a key element of macroeconomic reform among highly indebted countries, both to restore balance in domestic accounts and to contain inflationary pressures (e.g., de Gregorio 1992; Khan 1990; Moran 1989; Myint 1987). The creation of a stable macroeconomic environment is also considered to be a crucial first step for longer-term economic restructuring. The major objective of fiscal policies under SAPs and other neoliberal programs has been to reduce government budget deficits, usually by restraining expenditure. Reductions in government spending and deficits may, in turn, help to decrease government borrowing, both domestically and externally, which is one of the major goals of neoliberal monetary policies. However, fiscal and monetary policies also strongly affect the overall production and domestic expenditure levels of an economy through various multipliers and indirect effects. Excessively restrictive measures, especially under conditions of stagnant growth, may tip an economy into a deflationary spiral. This may cause irreparable harm to fragile economic sectors and social groups. Perversely, it may also actually increase government deficits, as revenues shrink through economic contraction and expenditures rise to meet growing social welfare needs. To avoid this type of no-win situation, restrictive fiscal and monetary measures need to be closely coordinated with other more expansionary policies designed to stimulate growth in specific sectors according to national development goals (e.g., broadening economic participation, increasing economic diversification, promoting structural change).

Sixth, SAPs and other neoliberal programs should be carefully crafted to suit the institutional and organizational structure of both the state itself and state–society relations in different countries. Given wide variations in Third World political structures, programs which enjoy success in some countries may prove disastrous in others. Moreover, the success of reform programs is highly dependent on the support of national decision-makers, who have the capacity both to subvert otherwise sound policies and to control the response of influential economic actors and social groups. One of the potentially most beneficial aspects of the neoliberal policy agenda is its focus on reducing waste and inefficiencies within the state apparatus. It is healthy to avoid the old assumption that the state can do anything and everything, which unfortunately has marred many neo-Keynesian development strategies. However, it should not be assumed that the market by itself can automatically meet the broad development goals of all countries under all conditions. Properly conceived policy instruments should work to improve the effectiveness of both the state and the market in a mutually supportive manner. Given the structural constraints and underdeveloped markets that characterize most Third World economies, development strategies normally need to cautiously combine market and prudent administrative policy instruments.

Seventh, liberalization measures and other policies should consider important variations within the socioeconomic and spatial structures of Third World countries. The development experience of the South certainly supports the contention that prices do matter. Most analysts have accepted the often quoted observation of Timmer (1973: 76) that '"getting the prices right" is not the end of economic development. But "getting the prices wrong" frequently is.' Nevertheless, the effects of price movements depend strongly on both country and product contexts (Colclough and Green 1988: 2). Internal liberalization measures in many Third World countries, particularly in Africa, have concentrated on the privatization of state-owned enterprises (SOEs) and the dismantlement of state/parastatal marketing boards. However, while both of these measures may normally be economically logical under competitive market conditions, competition within many economic sectors in Africa and other parts of the South is severely limited.

In a study of privatizations of SOEs, Prager (1992) contends that there is a strong bias in favor of private ownership when a competitive market exists and the privatization program is met with little resistance. However, under conditions of imperfect competition, public enterprises may often prove less inefficient than private sector firms. Moreover, SOEs can often be made to operate more efficiently if the general economic climate is favorable, political interference is eliminated or substantially reduced, and proper incentives are installed. Likewise, Maddock (1987) notes that the scrapping of state agricultural marketing boards may reduce waste, inefficiencies, and corruption that have provided serious disincentives to rural producers in many countries. However, in some cases (e.g., Sri Lanka), governments may want to retain some market controls, such as the maintenance of buffer stocks for foodstuffs or regulations over export quality. In other cases (e.g., Malawi), there may be no viable alternative to state marketing boards. In these instances, it may be better to strengthen state institutions and encourage them to adopt a more market-oriented perspective.

Eighth, in order to be sustainable, policies need to gain consensual support and must foster political and social stability. These factors may be partially dependent on prior conditions and other circumstances beyond state control. However, they are also largely dependent on the methods by which policies are implemented and the relative distribution of the costs and benefits of such policies across economic sectors and social groups. Through most of the 1980s, structural adjustment efforts concentrated almost exclusively on stabilizing macroeconomic conditions and liberalizing markets to improve the efficiency of resource allocations. Neither the social costs nor the political feasibility of SAPs were given much attention.

However, as many countries began to experience increasing instability and unrest under the pressures of adjustment, questions began to be raised over the social and political sustainability of SAPs and other neoliberal policies. The mounting social costs of neoliberal programs, especially on traditionally disadvantaged classes and social groups in many countries, have become a source of rising concern, as has the incapacity of democratic governments to carry out unpopular policies without resorting to repressive measures. Both of these areas of concern imply that the focus of SAPs and neoliberal strategies in general should be broadened to include real-world social and political considerations alongside abstract economic factors.

Ideological Biases of Structural Adjustment Programs

As the IMF and World Bank have applied SAPs throughout the South, increasing objections have been raised over the ideological biases of the programs themselves and of the financial institutions that are imposing them. Many analysts contend that the neoliberal policies around which SAPs are structured are based more on an ideological commitment to the 'virtues of the market' than on a logical and well-tested body of theory (e.g., Bernstein 1990; Helleiner 1990; Stewart 1987). George (1988: 56) comments: 'The Fund lives in a never-never land of perfect competition and perfect trading opportunities, where dwell no monopolies, no transnational corporations with captive markets, no protectionism, no powerful nations getting their own first.'

The Ideological Thrust of Neoliberalism

It has been asserted that the neoliberal counterrevolution led by the IMF and World Bank has a hidden agenda: 'its attempt to depoliticize its own political intentions even as it refuses all other political economies' (Corbridge 1989: 250). Similar to neoclassical theory in general, neoliberalism presents itself as a positive, value- and ideology-free science. On the surface, the technical language and modeling procedures of the neoliberal framework appear to be purely objective and scientific, stripped of all values and ideological content. Moreover, the discourse of neoliberalism is especially seductive because it combines the seemingly objective language of neoclassical economics with policy proposals that serve dominant global power structures (Levitt 1990: 1594). However, as Amin (1990: 39) notes: ['The neoliberals'] language does not conform to the basic criteria of scientific analysis. It is a language of ideology in the worst sense of the term.' According to Levitt (1990: 1594), 'In reality it is an instrument whereby the rich and powerful impose a set of values and rules of the game which reinforce inequality and injustice.' This conforms to the longstanding role of neoclassical economics (and mainstream frameworks in the social sciences in general) as an instrument of social control, a point that is recognized by at least some economists themselves:

> [E]conomics serves as social control. Social control institutions organize, proscribe, prescribe, structure, channel, and integrate behavior and choice . . . Those who desire a particular system of economic organization and control, or specific policies and performance, will favor a congenial and supportive definition of economic reality. They also will actively work to establish that definition of economic reality as the basis for or means to the achievement of their normative end . . . The creation and (re)creation of economic theory is part of the process of the creation and (re)creation of public opinion, and the manipulation of public opinion is part of the process by which the masses, various classes, and the state both control and are controlled. All these manifestations of economics as social control are important aspects of the sociology of economics as an institution. (Samuels 1988: 350–1)

Bias Toward the Interests of TNCs and Core Capitalist Countries

From this perspective, the imposition of SAPs on Third World countries by the IMF, World Bank, and other multinational financial organizations plays a vital role in the establishment of new conditions facilitating the expansion and deepening of global capitalism in the South (Biggs 1987; Foxley 1982; Pastor 1987). As well as directly acting on behalf of the international banking system and its investors, the IMF and World Bank indirectly serve the broader interests of Northern-based transnationals in penetrating Southern markets (George 1988; Kreye and Schubert 1988; Stein 1992; Wade 1992). A critical part of the outward-oriented 'trickle down' strategy that is a centerpiece of SAPs is the provision of a hospitable environment in the South for trade and foreign investment by TNCs. Harris (1989: 21) states: 'The main role of the IMF and World Bank is the construction, regulation and support of a world system where multinational corporations trade and move capital without restrictions from national states.' For Bernstein (1990: 23), this means that we cannot understand the real significance of SAPs without first 'locating the distinctive place and global role of the World Bank [and IMF] within imperialism, and within its postwar nexus of international financial and regulatory institutions.'

Because they serve the interests of transnational capital, the IMF and World Bank also necessarily serve the interests of the corporate elite in the core capitalist countries of the North. Indeed, the structure of these international financial institutions ensures continuing core capitalist domination. The managing director of the IMF has always been a West European, while the president of the World Bank has always been an American. The key decision-making body of the IMF, the Executive Board, is dominated by core capitalist countries and their clients. In 1985, although more than 150 countries were official members of the IMF, the following six countries controlled 44.71 percent of all votes: United States (19.29 percent), United Kingdom (6.69 percent), Federal Republic of Germany (5.84 percent), France (4.85 percent), Japan (4.57 percent), and Saudi Arabia (3.47 percent) (Bradshaw and Wahl 1991: 254). By contrast, the 41 sub-Saharan African countries controlled just 4.91 percent of total IMF votes (ibid.). A completely united Third World bloc, which would represent about three-quarters of the total population of IMF member countries, could control no more than one-third of total votes (Schoenholtz 1987: 405).

Given this structure, decisions endorsed as official IMF policies are invariably made by the Group of Five, representing the permanent members of the Executive Board (US, UK, Germany, France, and Japan) (ibid.: 405–6).[3] Moreover, the constitutional Articles of Agreement of the IMF provide the US with an effective veto, because any major changes, such as the allocation of votes, requires an 85 percent majority (ibid.: 405). Similarly, the US and other core capitalist countries dominate the key decision-making bodies in the World Bank and its regional development banks (Inter-American, Asian, and African). For example, the US controlled 34.54 percent of total votes in the Inter-American Development Bank (IDB) in 1985, giving it virtual veto power over all IDB loan allocations and substantial influence in the direction of bank policy (Dewitt 1987: 284).

Many observers contend that domination of the IMF and the World Bank by core capitalist countries has permitted the manipulation of policies not only to serve the interests of Northern-based transnational capitals, but also to discriminate in favor of or against selected Third World countries for geostrategic, ideological, or other reasons (e.g., Bienen and Gersovitz 1985; Biggs 1987; Black 1991; Loxley 1987). In particular, it is contended that the US has used its power to reduce or deny Fund/Bank assistance to a series of countries at odds with American foreign policy, while assistance has increased to a number of US client states despite problems of pervasive corruption and human-rights violations. In Latin America, for example, assistance was curtailed to the leftist Allende administration in Chile but was immediately returned upon the ascendancy of the Pinochet dictatorship in a US-backed military coup (Bienen and Gersovitz 1985). In Nicaragua, assistance was denied to the leftist Sandinista government but was resumed as

---

3   Some analysts also refer to a 'Group of Ten' in the IMF, composed of the five permanent members of the Executive Board, as well as Canada, Italy, the Netherlands, Belgium, and Sweden, which were joined by Switzerland in 1984 (Schoenholtz 1987: 406).

soon as the pro-US Chamorro administration assumed power. Moreover, at the same time that funding was denied to Nicaragua under the Sandinistas, assistance was increased to the right-wing governments in El Salvador and Guatemala (Black 1991; Schoenholtz 1987).

Similarly, in Africa, assistance in the 1980s was extended under favorable terms to American client states (e.g., Morocco, Sudan, Zaire), while other countries were treated much less leniently (e.g., Sierra Leone, Tanzania) or were denied funding completely (e.g., Angola, Mozambique) (Haynes et al. 1987; Loxley 1987; Schoenholtz 1987). Such inconsistencies have not escaped the notice of Third World countries, which have often protested strongly over the lack of objectivity in IMF/World Bank decision-making. For example, the so-called Arusha Initiative, which was signed by members of the Organization of African Unity (OAU) in 1980, states:

> [T]he IMF is not objective in the application of its own criteria. Double standards have been applied to similar situations. Examples show that certain countries, because of their geographical situation, international weight or political orientation, receive more lenient treatment than others. (in Schoenholtz 1987: 409)

The Anti-Third Worldist Posture of Neoliberalism and SAPs

There is a widespread perception in the South not only that Fund/Bank policies have unfairly treated many countries because of ideological and geostrategic considerations, but also that SAPs and neoliberalism in general are part of a concerted ideological offensive by the capitalist core to reassert its global domination and prevent the rise of alternative, more autonomous development projects from the South. A vigorous ideological challenge from the South in the 1970s confronted mainstream development strategies of both the North in general and the IMF/World Bank in particular. During this period, a 'Third Worldist' argument gained favor throughout the South that placed much of the blame for Third World underdevelopment on Northern governments, transnational capital, and international financial institutions. It was argued that many of the structural problems causing Third World underdevelopment were the direct result of the historical domination of Third World countries by the capitalist core and its transnational corporations. Instead of being allowed to develop according to its own needs, the South had been systematically underdeveloped by a global capitalist system designed to serve Northern interests. Moreover, current policies by the IMF and other financial institutions not only failed to meet the South's structural requirements for overcoming its legacy of underdevelopment, but they punished Third World countries for problems (e.g., balance-of-payments shortfalls, foreign indebtedness) that were fundamentally externally caused and beyond the South's capacity to control.

This 'Third Worldist' position was advanced intellectually in the 1970s by a variety of alternative development frameworks (e.g., dependency and world systems theory, structuralist economics) and was politically supported by many influential Third World leaders (e.g., Allende in Chile, Castro in Cuba, Kuanda in Zambia, Manley in Jamaica, Nyerere in Tanzania, Sukarno in Indonesia). By the late 1970s, it had united Third World governments, both authoritarian and democratic, capitalist and socialist, behind demands for a New International Economic Order (NIEO) in general and more flexible IMF/World Bank policies in particular (Dietz and James 1990; Pastor 1989; Toye 1987). At the same time, the widespread availability of private capital resulting from the glut of 'petro-dollars' in global financial markets and the economic slowdown in the North had substantially reduced the influence of the World Bank and other international financial institutions. Faced with an unprecedented surplus of capital and a declining demand for lending in the capitalist core, many of the transnational banks offered enormous loans to developing countries with few if any conditions attached. Much of the South used this source of seemingly unlimited private credit to avoid both structural adjustment in general and IMF/World Bank conditions in particular. In order to regain their eroding influence, the IMF and World Bank were forced to lower the conditionality of their loans and back away from harsh adjustment demands. In addition, they had to become more responsive to Third World demands that lending programs take account of the unstable political conditions and structural development needs facing many countries in the South.

However, the international debt crisis at the turn of the 1980s dramatically reversed this favorable lending situation for many Third World countries. Most sources of private international credit were abruptly cut off, as the specter of widespread Third World defaults caused a panic in the international financial community. Suddenly, the power of the IMF and World Bank was ascendant in a capital-scarce world in which the private banks had reversed their profligate lending practices and were looking for international leadership to guide them out of the debt quagmire. The IMF and World Bank, which had been the 'lenders of last resort' for much of the South in the 1970s, were quickly transformed into the 'lenders of first resort,' as the only institutions capable of carrying out debt and lending negotiations between the Northern banks and Southern governments. From this omnipotent position, the international financial institutions succeeded in organizing a 'creditors cartel,' which both dictated macroeconomic policy to Southern debtors through SAPs and forced individual Northern banks to continue 'involuntary lending' to avert the possibility of systemic collapse due to widespread defaults (Pastor 1989). As a result, since the early 1980s there has been a concerted attempt, spearheaded by the core capitalist countries and international financial institutions, to 'put the genie of the South back into the bottle' (Cypher 1990: 43). Most Third World countries, already overdependent on international capital and severely weakened by a protracted economic crisis, have proved suitably docile. This Northern initiative to impose new conditions of development on the South has taken two interrelated forms. First, Northern governments and international financial institutions have used the tremendous leverage afforded to them by the debt crisis to dismantle alternative, more autonomous development projects in the South in favor of mainstream strategies that stress global integration, austerity, and 'trickle down' economics. Secondly, an ideological offensive has been mounted that both supports mainstream development thinking (i.e., neoliberalism) in defense of the existing international economic order and discredits alternative Third Worldist frameworks which seek radical structural change in favor of a new, more equitable global development agenda.

Much of this ideological offensive has been directed at creating a coherent explanation for Third World economic woes that is compatible with core capitalist interests. During the 1970s, alternative development frameworks largely placed the blame for continuing Third World underdevelopment on external factors, such as the legacy of (neo)colonial domination and the inferior position of many developing countries within an inequitable and rigid international division of labor. While it was generally acknowledged that some internal policies might need correction, the economic crisis afflicting Third World countries was regarded as fundamentally global in nature. Accordingly, there would be little possibility for progressive development in the South in the absence of global structural change. However, the neoliberal counter-revolution in development thinking responded with a stance that turned Third Worldist explanations of underdevelopment on their head. Rather than being caused by external factors, Third World underdevelopment was basically attributable to inappropriate internal policies. In particular, introverted state-led development strategies, profligate government spending, and poorly conceived interventionist policies had prevented market forces from operating efficiently, thereby inevitably generating macroeconomic imbalances and stagnant growth. Therefore, the way out of the crisis for the South is to reject the failed inward-oriented and state-interventionist policies of the past in favor of Northern guidance to create a new, economically sound development model. The key components of this model are global economic integration according to principles of comparative advantage and the reduction of the role of the state in development so that market forces can create the macroeconomic conditions necessary for future growth.

The Loss of IMF/World Bank Legitimacy in the South

The perceived bias of these policies and their sponsors in the international financial community has caused widespread resentment in the South. In the eyes of the popular sectors, the IMF and World Bank have become irrevocably associated with harmful austerity measures, economic stagnation, widening inequalities, and the outward transfer of capital to wealthy Northern bankers. Bienefeld (1985: 77) comments that, in Africa, the popular consensus is that 'the main thrust of [neoliberalism is] . . . to explain why the destitute and starving people of Africa should accept the payment of extortionate interest rates to overfed and wealthy people, as an overriding economic priority.' Pastor (1989: 110) ends an article on the debt crisis in Latin America with a political cartoon from a leading Mexican daily, *El Excelsior*, which, he contends, accurately summarizes the popular attitude in Latin America toward the international financial institutions. The cartoon depicts a working-class Mexican hanging from a scaffold while a well-dressed man with a briefcase stamped 'IMF' is reaching into the dying man's pocket to take the last of his money. In India, Sarkar (1991: 2309) states that the popular image of relations between Third World countries and the international financial institutions 'is similar to that between poor peasants and the village moneylenders – under difficult circumstances the poor peasants (here, mainly the debtor LDCs) are forced to accept the bondage of the cruel moneylenders (the IMF and World Bank).'

Many political leaders in both North and South have also criticized the bias of IMF/World Bank policies. Following negotiations with the IMF, Tanzanian President Nyerere remarked:

> The IMF always lays down conditions for using any of its facilities. We therefore expected that there would be certain conditions imposed should we desire to use the IMF Extended Fund Facility. But we expected these conditions to be non-ideological, and related to ensuring that money lent to us is not wasted, pocketed by political leaders or bureaucrats, used to build private villas at home or abroad, or deposited in private Swiss bank accounts . . . The IMF . . . needs to be made really international, and really an instrument of all its members, rather than a device by which powerful economic forces in some rich countries increase their power over the poor nations of the world. (in Schoenholtz 1987: 418)

In an article in the *Washington Post*, an influential member of the US Senate Finance Committee, Senator Bill Bradley, commented on the link between the effects of structural adjustment programs and rising impoverishment in Latin America:

> Obsessed with debt collection, the administration endorsed austerity pro-

grams that offered a trickle of emergency lending if debtors cut consumption
and investment to the bone. Growth in Latin America was already faltering.
Austerity threw the region into recession. Latin countries could no longer feed
their poor or invest in their future . . . The sucess may have been, in relative
terms, a windfall for the banks. But it proved disastrous for US farmers,
factory workers, and exporters. (in Kreye and Schubert 1988: 268)

The widespread perception of bias within IMF/World Bank programs has
seriously eroded the legitimacy of these international financial institutions,
especially as representatives of the common interest in development between
the North and the South, the rich and the poor. Indeed, as recent elections in
both the Third World and Eastern Europe have shown, identification with
IMF/World Bank SAPs has become a serious political liability for many
governments. Increasingly, the IMF and World Bank have come to be
viewed 'as the fiscal vanguard of a heartless system' that serves transnational
corporate interests at the expense of all others (Horowitz 1985: 38). Given
the outward transfer of capital that has accompanied the imposition of SAPs
in much of the South, it is contended that adjustment lending represents aid
not to indebted Third World countries but to the largest transnational banks
headquartered in the North (Streeten 1993: 1294).

If they are to maintain any semblance of global legitimacy, it is asserted
that the IMF and World Bank, at a bare minimum, 'should . . . be acting
so as not to generate a net transfer of resources from countries that are at
present in desperate circumstances in consequence of terms-of-trade deterioration, heavy levels of external debt, and other factors' (Helleiner 1992:
790). Reducing the external cash-flow obligations and payments on debt
account probably represent 'the most cost-effective form of official external
resource transfer' that is currently available to assist development among
poor, indebted countries (ibid.: 781). A viable debt 'workout strategy' is
urgently required for these countries, for their own economic well-being
as well as the stability of the world as a whole (Culpeper 1988: 136).
Ways must be found to delink policies aimed at resolving the international
debt crisis from those that are designed to increase domestic savings and
investment in Third World countries for development projects (Emmerij
1987: 15). Until this is done, efforts to restrain consumption, heighten
efficiency, and increase output will do little to improve the well-being of
Third World countries. They will merely help to service part of a seemingly
ever-growing debt at the expense of the popular majority.

Common Shortcomings of Liberalization Policies

Price incentives and 'getting the prices right' is a major emphasis of
IMF/World Bank SAPs and other neoliberal strategies. The underlying

assumption is that, even in a world of pervasive imperfections, unrestricted markets can normally sustain economic growth better than government intervention. As Helleiner (1989: 110) notes: 'Even in the world of the second-best, [the neoliberal's] approach is consistently to liberalize that which can be liberalized.' Consequently, liberalization of trade, domestic markets, and the financial sector commonly form principal components of neoliberal programs. The message from the international financial institutions and their sponsors among Northern governments is that the South should do the following: allow market forces to determine patterns of resource allocation; remove state intervention in both external and internal markets; provide incentives to foreign capital for investment and job creation; accept outward-oriented growth according to principles of comparative advantage as the basic engine of development; and rely heavily on foreign experts to guide development and ensure efficient project selection.

Inadequacies of the Neoliberal Focus on Export-led Growth

Much of the literature promoting liberalization appears to be guilty of a basic ecological fallacy (i.e., countries X, Y, and Z (such as Singapore, South Korea, and Taiwan) have developed rapidly as a consequence of outward-oriented liberalization; therefore, this strategy must cause development and should be emulated elsewhere). As we saw in the last chapter, many of the factors that propelled growth in the Asian NICs are largely absent in other countries. Moreover, liberalization has hardly characterized the development strategies of these NICs.

Given the relatively undeveloped industrial structures and narrow internal markets of most developing countries, the production of primary commodities for export is viewed as the main engine of future economic growth for much of the South. Conventional comparative advantage theory links the ability to compete in world markets with the interaction between commodity production characteristics (i.e., the technical requirements of production as represented by factor combinations and national attributes) (Dodaro 1991: 1156). However, as the experience of the Asian NICs shows, many important comparative advantages for global markets do not exist naturally but are socially constructed, often with the assistance of an interventionist state. In addition, economic growth in the NICs has focused on highly elastic industrial exports rather than on primary commodities, which have suffered from demand restrictions and falling prices in recent years. Global competitiveness generally entails both a price and product quality dimension, with a tendency for the latter to increase in importance as products become more sophisticated or move closer to their final consumption stage. Product quality, in turn, largely depends on human capital and other created comparative advantages, which have generally received little attention from neoliberal strategies focused on prices.

Many analysts fear that outward-oriented neoliberal policies will confine many developing countries to a 'nineteenth-century' niche as primary commodity producers in the international division of labor (e.g., Bitar 1988; Corbridge 1988; Saha 1991). Taken to the extreme, such policies would reduce much of the South to a mere source of supply of primary commodities for the North. Export-led growth would be focused on low value-added agroexports and raw materials with few forward/backward linkages and little potential to contribute to needed structural change. The continuing exploitation of cheap labor and land would provide the sole source of comparative advantage on world markets. At the same time, the South would become increasingly dependent on Northern imports of food, clothing, manufactures, and virtually everything else.

The excessive concentration of developing economies on a few primary exports has long been a source of concern for many development theorists. International commodity markets have traditionally been characterized by wide fluctuations in demand and prices, which are often aggravated by oligopsonistic market controls exercised by Northern-based transnationals. Without other sources of growth, highly dependent Third World economies are extremely vulnerable to global market conditions over which they exercise little if any influence. Sudden downturns may be transmitted and amplified throughout dependent economies – causing not only a precipitous decline in export sectors, but a generalized economic contraction as well. An old adage among economists in the South is that when Northern economies catch a cold, Southern economies catch pneumonia.

Continuing dependence on a few primary exports may lock developing economies into relatively low-wage, low-skill, and low-productivity sectors that show few prospects for sustained growth. Neoliberal policies focused on the exploitation of static comparative advantages of cheap labor and land may block private and social investments that, over time, could create more dynamic comparative advantages with positive implications for stable economic growth, structural change, and income distribution. A recent study by Firebaugh and Bullock (1987) concludes that concentration on a few primary exports retards growth in developing economies because it blocks structural changes associated with increasing forward linkages and export upgrading. Research by Maizels (1987) finds that global commodity markets appear likely to remain unstable due to a combination of factors, including low elasticities on both the supply and demand sides of the markets, continuing low levels of stocks held by risk-adverse private traders, and the effects of fluctuating exchange rates of the major currencies and of intermittent rounds of destabilizing speculation. For Levitt (1990), current problems in world commodity markets have a deeper cause within long-term structural change taking place in North–South trading relations – a prolonged decline in the North's relative need for the South's primary commodities.

As a result, he concludes:

> Prebisch and Schumpeter were right; Malthus, Ricardo and the Club of Rome were wrong; there are no scarcity rents accruing to natural resources. Rents accrue to those who innovate, and can collect monopolistic quasi-rents on their innovation . . . [C]oncentration on the export of primary commodities cannot be a long-term strategy for development; at best it can serve only as a temporary means to access foreign exchange at a high opportunity cost in terms of getting locked into a trap of export dependence. (p. 1586)

From Prebisch (1950) onwards, many analysts have found a long-term tendency of declining terms of trade for Third World primary exporters. According to Levitt (1990: 1586), a recent study of 33 major non-oil commodities showed a decline in terms of trade from 1900 to 1988 at the overall rate of 0.57 percent per annum; for the basket of commodities most important to developing countries, the rate of decline was even faster at 0.67 percent per annum. Lele (1984: 677) reports that international prices for many of sub-Saharan Africa's primary exports have been falling since 1977–78. Levitt (1990: 1590) states that export prices in Latin America and the Caribbean deteriorated approximately 20 percent during the 1980s. For the South as a whole, a study by UNCTAD (1985) reveals a $55 billion loss of foreign-exchange earnings between 1980 and 1984 due to falling prices of major commodity exports, representing 63 percent of the total value of these exports in 1980.

Both the domination of global commodity markets by a few Northern-based TNCs and an oversupply in many commodity sectors due to excessive production by Third World countries have exacerbated the problem of declining terms of trade for Southern exports. Global commodity markets have become increasingly concentrated in recent years. For example, five transnational agribusinesses control 90 percent the global market in foodgrains, six TNCs market 60 percent of the world's coffee, and three preside over 75 percent of the world's bananas (Kolko 1988). The supply of commodities to key Northern markets is thus strictly controlled by transnational oligopsonies based in the capitalist core. In most commodity sectors, Third World exporters have little recourse but to accept the prices and marketing conditions dictated by these TNCs.

Problems of oversupply in global commodity markets have often adversely affected the bargaining power of individual Third World exporters *vis-à-vis* transnational agribusinesses. Moreover, this situation seems to be worsening as outward-oriented adjustment programs are being imposed on exporters of similar primary commodities throughout the South. Neoliberal strategies of export promotion tend to assume that recent price trends for primary commodities will not be adversely affected by additional supply from one country because it produces only a small share of the aggregate global product. However, this argument appears to suffer from a 'fallacy of composition' (Sarkar 1991: 2309). Each country is expected to implement a more-or-less fixed set of policies (e.g., real currency devaluation, wage cuts) to increase its exports. But no account is taken of the impact of export growth in one country on the export performance of other countries producing similar goods. At the same time, other countries are also advised to increase their exports using broadly identical measures (Sarkar and Singer 1991). This may often exacerbate problems of oversupply and declining terms of trade in global commodity markets.

Dell (1982: 607) offers an example of IMF-sponsored export promotion in a group of Third World countries which illustrates this problem. In 1975, Chile, Peru, Zaire, and Zambia, facing balance-of-payments problems due to a price slump in their major export (copper), requested assistance from the IMF. As part of its export promotion strategy, the IMF called for currency devaluations to take place in all four countries. The result was overproduction, a further price crash, and declining export revenues despite increasing volume. This example suggests that the international financial institutions should take into account the impact of their policy recommendations on all the countries affected by such policies, rather than merely the specific country to which a policy is addressed (Bhaskar 1991).

The responsibility of the North to ease access to its markets for Southern exports should also be stressed, especially since the core capitalist countries have strongly supported IMF/World Bank export promotion strategies throughout the South. In the latest 'Uruguay Round' of GATT negotiations, many of the subsidies and other market distortions that have proliferated over the last forty years in the North, particularly in the agricultural sector, were eliminated or greatly reduced. However, many Third World representatives left the GATT negotiations profoundly dissatisfied that Northern countries had done relatively little to increase access to their markets for Southern products, despite the considerable progress that was made in facilitating North–North trade. From a Southern point of view, the unevenness of the GATT agenda in favor of Northern interests only fueled resentment that international institutions appear to discriminate against poorer countries (Helleiner 1990). The GATT seemed willing to tolerate abuses of its fundamental principles by industrialized countries so long as the effects were felt only in the developing world. Little was done, for example, to dismantle the growing array of tariff and nontariff barriers that Northern countries have recently erected against Southern products

(Leamer 1990; Maizels 1987).[4] At the same time, a series of new GATT rules (e.g., over patents, and intellectual property rights) were applied strictly and rigidly to Third World countries.

Alternatives of Regional Cooperation and Strategic Trade Policies

Given the apparent unwillingness of the North to further open its markets to Southern products, it is imperative for developing countries to explore new methods for trading among themselves. The outward-oriented policies being adopted by developing countries have generally not encouraged forms of South–South cooperation. On the contrary, many of the initiatives being implemented tend to be competitive (e.g., devaluations, wage cuts, relaxation of labor regulations, creation of tax-free export zones). Especially for cases in which their exports have low elasticities, countries should make efforts to restrict supply through collective agreements and/or to encourage diversification into other products. International and regional commodity agreements can be appropriate under unusually severe circumstances, which certainly describes the recent state of many of the world's commodity markets (Helleiner 1990).

In many parts of the South, measures to increase intra-regional trade could stimulate export diversification. Previous economic integration schemes in regions such as Latin America may have generally been too ambitious (Urrutia 1987), but this does not preclude countries from exploring new methods to cooperate in areas such as intra-regional trade, financial relations, technological research, and industrialization. In the case of Latin America, Urrutia (1987: 64) notes:

> The recent agreement between Argentina and Brazil in capital goods is an innovation that may have interesting possibilities. Credit schemes for intraregional trade must also be developed, as well as a revitalization and

---

[4]   As was mentioned in the previous chapter, the proliferation of nontariff barriers in the North has presented a particularly serious problem for Southern exporters. Many of these nontariff barriers have been directed at products in which developing countries have a comparative advantage, notably agricultural goods, textiles, and clothing (Nolan 1990: 53). In a study using UNCTAD data to examine Latin American trade with the North in 1983, Leamer (1990: 337) finds that nontariff barriers were applied by 14 major industrialized countries against 19 percent of Latin American exports. As a result, he estimates that Latin American exports to these countries were reduced by a total of 34 percent, varying from 5 percent for Mexico to 75 percent for Argentina. Many Latin American countries faced extremely high barriers. For example, nontariff barriers were applied to 38 percent of Brazil's exports, 73 percent of Cuba's exports, and 62 percent of Paraguay's exports. On the other hand, certain favored countries were exempted from many of these barriers. Exports from Hong Kong, South Korea, and Taiwan were estimated to have been suppressed by trade barriers between 12 percent and 15 percent, which is considerably lower than for most of the Latin American countries.

stabilization of some of the trade preferences developed as part of the existing integration schemes.

Many analysts have recently concluded that adoption of a strategic trade policy may be preferable to a free trade stance, especially for less developed countries (e.g., Dodaro 1991; Furtado 1987; Helleiner 1990; Hirschman 1987; Krugman 1986). Free trade may often be desirable for countries that have already achieved a relatively high degree of economic development and internal productive efficiency which enables them to compete successfully in growing global markets. For less developed countries, however, a free trade stance may permanently confine them to a 'trap of static comparative advantage' in which they are unable to diversify away from primary commodities and other low-wage goods into more technologically sophisticated export sectors with higher demand elasticities and prospects for growth. As the experience of the Asian NICs demonstrates, a strategic trade stance permits export promotion policies to be situated within the broader context of national development goals. The efficiency of trade liberalization cannot be established a priori for individual countries with different needs and priorities. Any strategy that does not address wider aspects of development but focuses solely on liberalization measures risks generating unforeseen and destabilizing results, as well as missing new potential sources for future development.

IMF/World Bank liberalization policies have generally focused on short-term macroeconomic management, reducing the role of the state in development, and encouraging low-wage, labor-intensive export production according to principles of static comparative advantage. Little attention has been paid to possibilities for technological innovation, increasing labor skills and productivity, and improving infrastructure capabilities – all of which are critical to promoting economic diversification in most new vibrant export sectors. As Krugman (1986: 9) points out, 'A good deal of trade now seems to arise because of advantages of large-scale production, the advantages of cumulative experience, and transitory advantages resulting from innovation.' These factors may already be present in industrialized countries, but in most developing countries they need to be created through strategic government intervention. Instead, short-term liberalization measures lead in the opposite direction by curtailing needed expenditure on social and economic infrastructure, reducing support for indigenous research and development projects, and providing disincentives to economic diversification that could accelerate technological innovation and structural change. Because they neglect the overall context within which development is generated, liberalization measures may be sacrificing opportunities for dynamic future growth in favor of marginal and transitory gains derived from static comparative advantage. In a recent study of manufactured exports in Argentina and Brazil, Paus (1989: 178) emphasizes interconnections between export growth, processes of technological change, and the broader context of development strategy:

> I have argued that in order to understand the forces behind export-linked growth, one has to analyze manufactured exports in the context of the overall development strategy, because the general economic framework has to be consistent with the promotion of manufactured exports . . . And one has to analyze the development of technological change and productivity growth – conditioned by the very continuity or discontinuity of the accumulation process – because they are vital for the achievement and maintenance of competitiveness on the international market.

Much of the argument in the development literature for strategic trade policy rests on the advantages of protecting selected infant industries, at least during their formative period, from competition by well-established foreign TNCs (e.g., Dietz and James 1990; Dodaro 1991; Hirschman 1987; Urrutia 1987). In addition, certain small/medium producers, especially in the rural sector, may require support either to enter global markets or to meet the demands of increasing competition in their traditional domestic markets (Barham et al. 1992; Watkins 1992). Although most analysts agree that the old-style 'umbrella' approach to protectionism should be avoided, this does not preclude supporting specific sectors according to particular development goals (e.g., fostering structural change, creating employment opportunities, avoiding peasant impoverishment and rural polarization). Within the manufacturing sector, it has often been pointed out that infant industries may require initial state support to gain a foothold against foreign competition in domestic and international markets. This may allow domestic firms to capture economic rents from foreign competitors, thereby increasing national welfare. It may also create opportunities for significant 'spin-offs' in terms of technological diffusion, demonstration effects, and skills development that may extend well beyond the infant industries themselves.

Within the rural sector, many small/medium farmers may need help in pursuing promising new opportunities in global markets or to avoid being swamped by foreign competition in their established domestic markets. Well-targeted state programs (e.g., crop insurance, technical assistance, improved credit access, creation of diversified processing and distribution channels that offer competitive outlets) may substantially reduce many of the risks that have prevented peasants and other smaller producers from entering new potentially lucrative export markets (Barham et al. 1992: 48). In addition, various forms of state support may be used to protect peasant food producers from 'dumping' by highly subsidized Northern exporters. Policy-makers in the North have traditionally used subsidized exports to create outlets for surplus agricultural production within Southern markets. However, as Watkins (1992: 32) notes, the implementation of trade liberalization measures in many Third World countries has further assisted this form of export dumping:

> In Costa Rica, a World Bank structural adjustment package introduced in 1985 left domestic food staple producers exposed to competition from heavily subsidized wheat and maize exports from the US. The result was a 10% a year increase in imports, and a sharp decline in the area under bean and maize cultivation. The liberalization of agricultural imports in the Philippines, again under the auspices of a World Bank adjustment program, had similar effects, with domestic rice and course grain prices being depressed by subsidized imports. From a position of near self-sufficiency in the mid-1980s, by 1990 the Philippines was importing some 600,000 tons of rice annually, equivalent to some 16% of national consumption.

While it is recognized that many trading practices of the North with the South are fundamentally unfair, most analysts agree that any state intervention to protect domestic producers in developing countries must be pursued carefully and selectively. Once more, the successful experience of the Asian NICs in this area may offer lessons. The NICs used various incentives, controls, and mechanisms to generate an investment and production profile that served national development goals and differed substantially from that which would have resulted under a free market system. However, investment incentives and subsidies were closely tied to stringent performance requirements. This allowed the NICs to avoid much of the resource waste that has often characterized efforts by other states to prop up domestic industries. The NICs also succeeded in overcoming a number of dichotomies (e.g., import substitution versus export promotion, planning versus the market, rural versus urban development) that have fragmented development efforts in many other countries. The NICs showed that the different sides of each of these dichotomies need not necessarily be mutually exclusive. In fact, they could be mutually reinforcing, given appropriate and properly coordinated policies with regard to exchange rates, pricing, investment, and trade (import–export) regimes.

Shortcomings of Financial Liberalization

IMF/World Bank SAPs have typically applied liberalization measures not only to the trade sector but also to the financial markets of developing countries (e.g., increasing financial openness and liberalizing foreign exchanges, removing interest rate ceilings, liberalizing the capital account of the balance of payments). However, as the experience of many Third World countries, particularly in the Southern Cone of Latin America (Argentina, Chile, Uruguay) and in Africa (e.g., Ghana, Kenya, Malawi, Tanzania) demonstrates, financial liberalization, if not properly designed, may cause instability in the financial system which, in turn, may aggravate macroeconomic instability and choke off investment (Cho and Khatkhate 1989; Diaz-Alejandro 1985; Helleiner 1989, 1992; Rodrik 1990; Stewart 1991; Toye 1987). Poorly coordinated and inappropriate financial liberalization measures have often been implemented with little regard for their consequences in terms of overall economic stability and sustainability.

The following areas of financial liberalization have proved particularly problematic. First, the removal of interest rate ceilings has frequently put financial sectors in a frenzy and ultimately caused them to crash. Diaz-Alejandro (1985: 1) summarizes the case of the Southern Cone countries in Latin America as 'good-bye financial repression, hello financial crash.' Moreover, given the presence of many structural constraints, private consumption and savings have not responded to real interest rate changes in many low-income countries in the same way as might be expected in higher-income countries. In fact, rising interest rates in many African countries have not led to an increase in domestic savings and seem to have choked off borrowing for investment (Helleiner 1992; Stewart 1991). Second, capital-account liberalization has increased the cost of financing the deficits of many countries because it has reduced the private sector's demand for government liabilities (Rodrik 1990). In addition, as real exchange rates have constantly been devalued for reasons of competitiveness, a premium has been built into domestic real interest rates relative to foreign rates, thereby adversely affecting domestic investment. Third, increasing financial openness and the liberalization of foreign exchanges have aggravated problems of capital flight in many countries (Banuri 1991; Eshag 1989). Instead of alleviating instabilities created by sudden trade fluctuations, financial openness has increased the vulnerability of many economies to such external 'shocks' by opening up new avenues for capital flight. Fourth, financial liberalization has endangered the broader structural reforms initiated in many countries (Rodrik 1990). Increased interest rates have driven up costs for firms struggling to adjust to sharply altered prices. Firms in difficulty have had to refinance their loans at ever-increasing interest rates, non-performing loans have multiplied on the balance sheets of the banks, and much of the domestic banking sector has ended up insolvent (ibid.: 942). This story underlies many financial crashes that have accompanied liberalization measures, especially in the Southern Cone countries of Latin America.

The often disastrous experiences with financial liberalization emphasize the need for proper coordination and sequencing of policies. In countries with unstable macroeconomic environments or imperfect markets, liberalization policies must be carefully coordinated and implemented to avoid creating imbalances and instability (Killick and Stevens 1991). Financial liberalization does not generally improve the allocation of resources in countries with distorted price structures. Therefore, major structural reforms should be completed before the introduction of financial liberalization measures (Coats and Khatkhate 1991). Current opinion among economists supports a sequencing of liberalization: 'The goods market should be liberalized first, and liberalization of the capital market should be added only when much of the adjustment to the former has been completed' (Michaely et al. 1991: 277). Proper sequencing must be developed not only with respect to overall liberalization measures, but also for financial liberalization policies themselves. Bajpai (1993: 993) contends that 'If the capital account [of the balance of payments] is opened when the domestic capital market is still repressed and interest rates are fixed at artificially low levels, massive capital outflows will take place.' Therefore, he reasons that the capital account should be opened only after the domestic capital market has been liberalized and domestic interest rates have been raised. The experience of countries in all parts of the South shows that premature or poorly coordinated liberalization policies may produce little positive effect on savings and investment and may cause many adverse side-effects. The risks appear to be particularly severe in countries with unstable macroeconomic environments, high levels of indebtedness, and markets which function imperfectly – all of which are common characteristics of the overwhelming majority of Third World countries undergoing SAPs.

Inadequacies of Internal-Market Liberalization

Complementary to their focus on liberalizing trade and financial markets, SAPs and other neoliberal strategies also frequently apply liberalization measures to the domestic markets of developing countries. However, many development analysts contend that neoliberals have a rather naive view of Third World markets. Neoliberals suggest that market failures due to state intervention are the primary cause of the economic crisis currently afflicting most countries. Hence, market restoration is seen as the solution. But this neoliberal solution characteristically contains little or no analysis of the ways in which real-world markets operate in the South. As Toye (1987: 86) points out, a recent World Bank study found that some two-thirds of the economic performance of Third World countries could not be accounted for by policy-induced price distortions. It must be concluded, therefore, that we need to know much more about other factors affecting development before embracing liberalization measures.

Neoliberal policies commonly assume that the market permeates everywhere in Third World countries. However, as Riddell (1992: 61) notes, this assumption ignores the fact that in many countries significant sectors of the population are only partially integrated into the market. For example, in many of the rural areas of less developed countries, capitalist relations of market exchange are geographically concentrated in 'enclaves' formed around agroexport or mining activities, while the bulk of the peasantry operates chiefly according to traditional relations of social exchange, such as reciprocity and redistribution. Neoliberal policies are also rooted in abstract theories that are suited for 'benevolent,' if not perfect, market environments (Schoenholtz 1987: 428). Consideration is rarely given to the processes and relationships which define the social context within which production and exchange take place. Given patterns of severe socioeconomic polarization and political repression in many Third World countries, it should be evident that the environment in which many people live and work is hardly benevolent. This realization throws into question many of the neoliberal assumptions based on 'trickle-down' theory.

Even a cursory familiarity with Third World markets ought to uncover many structural constraints that prevent the bulk of the population from responding to price signals as prescribed by neoliberal policies. Corbridge (1989: 234) comments: 'To promote as panacea an abstract "market" is to conceal the necessary imperfections and inequalities of particular economic systems.' Real-world markets exist within diverse structural contexts and are constituted according to varying principles and power relations. Market failures may be common, even pervasive, in the context of developing economies. Structural constraints to development frequently exist in areas such as transportation and communications networks; education, health care, and other social infrastructure; credit and financial systems; and the productive sphere itself. Frequently, productive sectors are fragmented into many parts, each of which may have a different market orientation and be subject to a different set of policies (Lele 1990). In many cases, differential access to key factors of production (e.g., land, credit, technology) may profoundly affect the supply response of various sectors to price changes. Simple price liberalization, without complementary measures designed to address the structural constraints facing many disadvantaged classes and social groups in the South, has little potential to generate the type of supply response called for in the neoliberal models. Indeed, it may lead to deepening polarization and impoverishment, as privileged producers with greater access to the resources necessary to expand production take advantage of new policies to drive other less fortunate producers out of competition.

These types of structural considerations underscore the point that 'the correctness of prices must be decided by reference to a comprehensive development strategy, not independently of it' (Fishlow 1985: 141). Rather than being simplistically equated with the free operation of market forces, a sound policy framework should comprise many diverse elements. Focusing on rural development, for instance, the maintenance of real producer prices at reasonable levels is important, but so are a range of other factors such as well-functioning transport, marketing, credit, and agricultural extension systems, as well as ready access to a wide range of consumer goods and agricultural inputs (Ghai 1987: 123).

Given extreme levels of polarization within rural development in many countries, Reusse (1987: 315) notes that measures to enhance market transparency and competition are especially necessary. He recommends policies designed to facilitate market entry by small/medium producers, to remove physical and institutional obstacles to the establishment of a fully competitive system, to improve producer and consumer knowledge of seasonal price developments, and to increase popular consultation in policy decision-making. In a study examining the effects of SAPs on the fragmented rural sector of Malawi, Lele (1990: 1207) states: 'Broad-based growth in such a sector requires the adoption of an entire gamut of policies toward prices, taxes, subsidies, markets, and asset distribution involving all factors of production, and requiring a long time period to obtain a strong and sustained supply response.'

As Ghai (1987: 123) notes: 'It is countries which have succeeded in establishing and maintaining reasonable balance and efficiency in the entire policy package that have attained sustained expansion of agricultural output.'

As we have seen, however, the adjustment programs of many Third World countries have been characterized by inappropriate and poorly coordinated policies, as well as a narrow focus on liberalization measures to the exclusion of other factors vital for development. Directly contrary to the central thrust of structural adjustment, macroeconomic imbalances (e.g., fiscal deficits, foreign debt, inflationary pressures) have often been aggravated. Resulting uncertainties and instability have adversely affected growth and investment in many key economic sectors – thereby negating possibilities for trickle-down effects, which are another key component of the neoliberal programs.

Neither rapidly developing Third World countries, such as the Asian NICs, nor the core industrialized countries have ever practiced the rigid liberalization measures that are being imposed through SAPs. It must be seen as curious, therefore, that the current obsession with liberalizing markets has not given way to more flexible policies which are capable of addressing the broader concerns of development in the South.

Increasing Polarization and Social Costs under SAPs

Many analysts contend that SAPs and other neoliberal programs have not only neglected many of the broader structural concerns of Third World development, but have also produced widening polarization and rising social costs in many countries (e.g., Colclough and Green 1988; Cornia et al. 1987; Helleiner 1989; Jolly and van der Hoeven 1991; Singer 1989). Neoliberal strategies have subordinated important development issues concerning equity and income distribution, poverty alleviation, and access to basic needs to the exigencies of an abstract 'free market.' The technical focus on improving market efficiency and macroeconomic conditions has all but ignored the human dimension of development. Until quite recently, specific targets for improving human conditions were not even included within most SAPs.

Even now, there is a feeling that only lip service is being paid to the human and social concerns of development, while the central thrust of SAPs on abstract macroeconomic factors remains unchanged. As a result, neoliberal policies (e.g., cuts in real wages, food subsidies, and health care and education expenditures) continue to generate high social costs, especially for the poor and other disadvantaged groups. In addition, because they ignore many of the structural constraints to development in Third World countries, neoliberal policies can offer, at best, only palliative recommendations concerning the poor as target groups, rather than attack the basic forces that make them poor in the first place.

Falling Investment Levels and Economic Contraction

Basically, neoliberals argue that their focus on liberalizing markets is consistent with the long-term needs of the poor and will avoid the inefficiencies and anomalies of previous state efforts to assist the poor through non-market means. Freeing markets and creating more favorable macroeconomic conditions should spur investment and growth, as well as improving overall economic efficiency and productivity. If markets are allowed to allocate goods, capital, and labor rationally without interference, the poor and others will inexorably reap higher incomes derived from increased efficiency and productivity through the operation of 'trickle-down' forces. Moreover, in most Third World countries, rational resource allocation would produce a labor-intensive bias for development projects that would inevitably favor the poor majority over time. Non-intervention, then, represents the best way to help the poor in the long run. Past state interventions to assist the poor and redistribute income (e.g., social service spending, subsidized credit, price controls, agrarian reforms) have caused excessive government spending and have detracted from market efficiency, thereby reducing overall output and job creation. In many cases, state intervention has also produced perverse, regressive effects because programs designed for the poor have been manipulated to support wealthy government client groups. Thus, such programs should be avoided in favor of a macroeconomic approach which stimulates private investment to create real long-term jobs according to principles of comparative advantage and trickle down.

However, considerable evidence has accumulated from various developing countries that SAPs have generally failed to stimulate investment and growth and have produced increasing socioeconomic and spatial polarization, with particularly devastating results for the poor and other disadvantaged groups. A key element of SAPs has been adjustment of excess demand over domestic supply in many Third World economies, which was being met by an unsustainable volume of external resources, generating increasing debt. The intent of SAPs has been to administer a dose of deflation to these economies, which would lower external and fiscal deficits and provide a stable macroeconomic foundation upon which to stimulate the supply side. The general consequence of SAPs, however, has been severe economic contraction, particularly in production for the domestic market. A deflationary cycle has been created in many countries in which falling demand lowers production levels, which further contracts demand, and so on.

Discouraged by falling utilization of productive capacity and by the general recessionary economic climate, investment has not only failed to increase, but has declined precipitously in many countries. For the South as a whole, the investment share in GDP fell in the 1980s by about 20 percent for non-fuel-exporting countries and by 30 percent for fuel exporters (Bourguignon et al. 1991: 1496). Private capital flows to developing countries declined from annual levels of $60–80 billion prior to 1982 to $12–15 billion per annum at the end of the decade (Levitt 1990: 1590). By the late 1980s, direct private investment in the South had been reduced to its lowest level in the postwar era: a mere $5–10 billion annually. The decline in investment was most severe in poorer regions such as sub-Saharan Africa, where overall private investment has fallen nearly 25 percent since 1980 (ibid.). Foreign direct investment (FDI), a vital component of neoliberal strategies, declined even further. Cheru (1992: 505) reports that, from a level of $1.5 billion in 1981, FDI in sub-Saharan Africa declined to about $400 million annually at the start of the 1990s and was distributed among only a handful of countries.

It appears that investment has also declined more rapidly in countries undergoing SAPs than in other similarly indebted Third World countries (Faini et al. 1991; Mosley et al. 1991). Rather than stimulating growth through higher investment according to trickle-down principles, 'the results [of SAPs] show much foregone growth because of lower aggregate (public and private) investment levels during the period of adjustment' (Faini et al. 1991: 966). It appears that any stimulus SAPs have been able to impart to the supply side has been confined to the export sectors of a few countries; however, the deflationary blow suffered by the remainder of their economies has more than offset this (Mosley et al. 1991: 229). SAPs seem to have had a negative or, at best, neutral effect on already low rates of Third World economic growth, while they have aggravated problems of capital flight and slumping investment (Eshag 1989; Faini et al. 1991; Greenaway and Morrissey 1993; Helleiner 1992; Kreye and Schubert 1988; Mosely et al. 1991; Pastor 1989; Rodrik 1990; Stein 1992). This situation augurs particularly poorly for future economic growth in the South. The prospect of continuing, and perhaps catastrophic, economic decline appears only too real for many countries. As Taylor (1988: 168) notes: 'The risk of economic collapse under liberalization seems to be non-trivial, if the recent history [of countries undergoing SAPs] provides a guide.'

Regressive Income Redistribution

At the same time that SAPs have generally failed to increase growth and investment in the South, they have also had a profoundly regressive effect on income distribution in many countries (Bourguignon et al. 1991; Eshag 1989; Minocha 1991; PREALC 1988; Senses 1991). According to Pastor (1987: 258), 'The single most consistent effect [of SAPs] . . . is the redistribution of income away from workers.' In a 1985 study of Latin American development, the Inter-American Development Bank (the regional branch of the World Bank) concludes that there is evidence that a disproportionate part of losses in real incomes has 'been concentrated in the lower income strata' (in Pinstrup-Andersen 1988: 39–40). The Bank further suggests that 'to the extent that real wage containment remains a necessary element of the adjustment process, mechanisms will have to be found to shift some of the burden to the higher income groups in the interest of social justice and domestic peace' (ibid.).

Even in organizations such as the World Bank, there is widespread recognition that the working class and other popular sectors in developing countries have borne a disproportionate share of the social costs generated by SAPs. The brunt of structural adjustment has consistently fallen on the popular sectors for a number of reasons. First, liberalization measures have caused widespread job losses, especially in many labor-intensive, domestically oriented economic sectors. Unemployment has risen rapidly in many countries through job losses in many formal sectors and the failure of informal sectors to provide additional sources of steady employment (Bourguignon et al. 1991; PREALC 1988; Riveros 1990). Second, levels of both real wages and minimum wages have decreased as unemployment has risen and neoliberal policies have removed labor regulations. Studies of urban labor markets in Africa (Ghai and Hewitt de Alcántara 1990; Stein and Nafziger 1991) and in Latin America (PREALC 1988; Riveros 1990), for example, show a significant deterioration in real wages under the impact of SAPs. Third, prices for food and other basic goods have risen dramatically as liberalization measures have cut state subsidies designed to hold down prices for the urban poor and other popular sectors. Because these groups spend a proportionally larger share of their income on basic consumption goods, such measures have had a profoundly regressive effect on purchasing power. Fourth, access by the popular sectors to many basic social services has been reduced following cutbacks and/or privatization. In many cases, higher user fees accompanying privatization have significantly affected the ability of poorer groups to utilize basic services such as health care and education. Fifth, government cutbacks have eliminated many programs targeting particular groups for special forms of assistance. Such programs range from those designed to provide basic consumption and social reproduction needs (e.g., food banks, prenatal and infant care for poor mothers, shelters for the homeless) to others that offer assistance in production to disadvantaged groups (e.g., provisions of credit, production inputs, marketing assistance for peasants and small artisans).

As the brunt of the social costs of adjustment has fallen on labor and the popular sectors, SAPs have systematically redistributed income toward the more affluent and propertied classes (Barkin 1990; Kreye and Schubert 1988; Pastor 1987). This has had a profoundly regressive effect on the already polarized structures of many Third World societies. While new opportunities for accumulation and enrichment have been offered to the privileged few, the popular majority has suffered and many have slipped into deeper impoverishment. The central thrust of SAPs on increasing profitability and surplus generation in order to attract investment necessarily favors certain classes and social groups over others, especially capital over labor. Indeed, research consistently concludes that SAPs have increased the capital share of income at the expense of the labor share (e.g., Bernstein 1990; Black 1991; Ghai and Hewitt de Alcántara 1990; Pastor 1987; Ruccio 1991). In the ten largest countries of Latin America, for example, Ghai and Hewitt de Alcántara (1990: table 6) find that during the 1980–85 period per capita consumption by business (owners of capital) increased by 15.8 percent, while that of labor decreased by 25.7 percent.

SAPs have played a key role in the neoliberal strategy to impose new economic conditions on the South which both create new accumulation opportunities for capital and roll back gains achieved by labor through previous struggles. As Black (1991: 98) notes: 'The strategy [of neoliberalism] seeks not merely to freeze socioeconomic relationships and maintain the status quo but rather to promote accumulation or reconcentration – that is, to redistribute assets and income from the bottom up.' Ruccio (1991: 1326) states that, from the perspective of the capitalist class, many of the widely acknowledged failures of SAPs may actually be transformed into successes:

> Neoclassical and structuralist policies [orthodox and heterodox SAPs] may be seen as successes once their effects on the rate of exploitation and other class features of capitalism are taken into account. Each policy package, in its own way and under different circumstances, may participate in strengthening important conditions within which surplus value is appropriated from the direct producers. Thus, what may be a failure from the standpoint of achieving full employment, price stability, and balance-of-payments equilibrium can be considered successful in terms of promoting the widening and deepening of capitalist class processes.

Increasing Societal Polarization

Many authors have also noted that SAPs have had a polarizing effect not only between capital and labor, but within and between various other classes, class fractions, economic sectors, and social groups (e.g., Barkin 1990; Ghai and Hewitt de Alcántara 1990; Hugon 1991; Timossi Dolinsky 1990). The macroeconomic thrust of SAPs has tended to favor outward-oriented over inward-oriented sectors, speculative and commercial activities over production, and informal over formal sectors. The outward-orientation of SAPs has especially favored export sectors dominated by transnational capitals and their local allies. Similarly, capitals involved in the importation and commercialization of foreign goods have prospered as a result of decreased trade restrictions. At the same time, local capitals oriented toward the domestic market have faced increasing hardship due to rising foreign competition and contraction of the internal economy. The relaxation of financial regulations and other controls under SAPs has also generally favored speculative and commercial activities over domestic production. Moreover, rising unemployment and the removal of many labor regulations have promoted informal over formal activities. In many countries, large sections of the working class have become steadily informalized. Job layoffs and pay cuts have forced many elements of the middle class to find sources of income in the informal sector. As a result, the already indistinct lines between the middle and working classes of many countries have been further blurred, and the concept of the 'working class' itself has become increasingly fuzzy (Ghai and Hewitt de Alcántara 1990: 410–11).

SAPs have introduced important changes not only in class relations, but also in the broader range of social relations in many countries (e.g., these based on gender, ethnicity, age). Generally, the position of more privileged social groups has improved, while that of traditionally disadvantaged groups has deteriorated further. While more privileged groups have used their greater access to key resources and contacts to take advantage of new outward-oriented economic opportunities, disadvantaged groups have suffered through the contraction of the domestic economy; falling wages and the removal of labor regulations; rising prices for basic consumption goods; and cutbacks in many social assistance programs. Privatization and government spending cutbacks have adversely affected access to many basic social services (e.g., education, health care) in general, as well as curtailing programs designed to offer special assistance to particularly disadvantaged groups (e.g., poor women and children, the elderly, ethnic minorities).

According to many analysts, the effects of SAPs have particularly harmed poor women and children (e.g., Elson 1989; Geisler 1992; Ibrahim 1989; Sollis and Moser 1991; Standing 1989). Declining wages and the deregulation of labor markets have led to heightened rates of exploitation and the 'feminization' of much of the lower end of the job market. Many single mothers or women whose partners have become unemployed have been forced into dangerous, unregulated work at abysmally low pay to meet the social reproduction needs of their families. The 'double burden' of production and reproduction that such women must bear goes unrecognized within SAPs. The underlying economic assumptions of SAPs tend to treat society as an undifferentiated whole, thereby neglecting the special needs of particularly disadvantaged groups such as poor working women and their children. Moreover, programs targeted to address these special needs have been curtailed by government spending cutbacks designed to meet the profitability requirements of capital. However, the unfair burden that poor working women are subjected to by the exigencies of SAPs may have far-reaching, long-term consequences, especially in the realm of social reproduction. As Elson (1989: 58) notes, 'Women's unpaid labor is not infinitely elastic – a breaking point may be reached, and women's capacity to reproduce and maintain human resources may collapse.' Falling social indicators in many Third World countries may signal that many poor women and their families have already reached this point.

Many analysts also note that SAPs have had a devastating effect on small-scale producers and their families, particularly those whose production is oriented toward the domestic market (e.g., Allison and Green 1985; Ghai and Hewitt de Alcántara 1990; Geisler 1992; Hugon 1991; Kay 1985; Mengisteab and Logan 1990; Stewart 1991; Stein 1992). Peasants and other small/medium rural producers appear to be especially vulnerable to the harmful effects of SAPs. In most rural Third World countries, SAPs have concentrated on increasing agroexport production by transnational capitals and other large-scale producers. Domestic food production, which is dominated by peasants and other small/medium farmers, has largely been neglected. While trade liberalization has often increased agricultural exports, it has also allowed highly subsidized Northern producers to flood Southern markets with cheap foodstuffs. Increasing foreign competition and falling internal demand (resulting from rising unemployment and economic contraction) have combined to force down prices. Moreover, many state programs that previously gave small/medium peasants access to vital resources (e.g., rural credit, extension services, agricultural inputs, marketing assistance) have been drastically reduced or eliminated. Consequently, many peasants have been driven out of competition and displaced from their traditional plots into a destitute and insecure existence in the teeming informal sectors of the large cities. Others have been pauperized and forced into an equally desperate situation as mere subsistence producers or seasonal laborers for the large agroexport estates.

Privileged groups such as large-scale agroexport producers, and disadvantaged groups such as small peasants, tend to be located in distinct spatial concentrations in most developing countries. Because of this, tendencies toward socioeconomic polarization under SAPs have also been manifested in widening spatial and regional polarization (Amirahmadi 1989; Kay 1985; Riddell 1992). Spatial polarization is widely acknowledged as a severe impediment to development in much of the South. In many outlying rural areas, the bulk of the peasantry lives in abject poverty, deprived of essential facilities and services, and barely integrated into the national economy. On the other hand, core urban areas and other modern enclaves are often better linked to the outside world than to their surrounding rural hinterlands. Polarized development within and between regions has generated an internally disarticulated pattern of growth, which has blocked the rise of social, economic, and spatial linkages vital for broadly based development.

The market-led, outward-oriented focus of SAPs has further aggravated historical problems of polarized and internally disarticulated growth in many Third World countries. Investment and development projects have been concentrated in core locations and modern export enclaves with superior physical and social infrastructure, while underserviced peripheral areas have been further marginalized. With government spending cutbacks, dwindling public expenditures on basic infrastructure have been directed at economically and politically important core locations rather than poorer, more remote areas. In many outlying regions in which peasants are concentrated, roads and other basic physical infrastructure have fallen into disrepair. Farm inputs (e.g., seeds, fertilizer, pesticides, tools, machinery) have become increasingly expensive and scarce. Basic social services (e.g., health care, education, water and electricity) as well as more targeted rural development programs (e.g., agricultural extension, rural credit, marketing assistance) have declined or become nonexistent. As vital forward/backward linkages have been severed and basic infrastructure has crumbled, many peasant areas have become increasingly isolated and unable to compete in traditional markets. In this situation, price liberalization and other policies designed by SAPs to stimulate rural production can have little positive impact on peasant producers. In fact, because only larger producers have the means to take advantage of liberalization measures, widening rural differentiation and polarization inevitably occurs.

Deteriorating Living Conditions and Rising Social Costs

Considerable evidence has accumulated that the polarizing effects of SAPs have had severe consequences on the standard of living of the popular majority in most Third World countries (e.g., Bourguignon et al. 1991; Cornia 1984; Cornia et al. 1987; Geisler 1992; Pinstrup-Andersen 1988; Singer 1989; Stein and Nafziger 1991). The contraction of the domestic economy and the removal of labor regulations have caused unemployment to rise and wages to fall. Increasing foreign competition and the elimination of state assistance programs have forced many small/medium producers into bankruptcy. Liberalization measures have driven up prices for many foods and other basic goods. Government spending cutbacks have worsened problems of unemployment and the deterioration of basic infrastructure and social services. The cumulative impact of these factors has lowered income levels, diminished purchasing power, and reduced access to essential social services and other basic needs for the popular sectors. As always, it appears that the most vulnerable and disadvantaged sectors of Third World societies have been the most negatively impacted. As levels of absolute poverty have increased and social programs have been cut, rising hunger and malnutrition have put severe pressures on many poor families. Such pressures have begun to show up in various behavioral indicators (e.g., number of abandoned children, incidence of family violence, youth crime and delinquency). These problems show every sign of becoming chronic in many developing countries unless specific countermeasures are put in place and the macroeconomic focus of SAPs is fundamentally altered.

As the negative impact of SAPs on the poor and disadvantaged has become more apparent, criticism has mounted from various sources. Critics from the academic community and many international organizations contend that issues such as basic needs provisions, poverty alleviation, and sustainable development have been ignored in the macroeconomic, growth-oriented agenda of SAPs. Within the United Nations, the UNDP (United Nations Development Program) and UNICEF (United Nations Children's Emergency Fund) assert that the lack of a 'human dimension' in SAPs has caused particular hardship for vulnerable groups such as poor women and children (see Cornia et al. 1987; Helleiner 1987; Jolly and van der Hoeven 1991; Shaw 1991). Similarly, the ILO (International Labor Organization) maintains that adjustment policies have generally increased unemployment and income inequalities, and have adversely affected basic needs provisions for the poor (see García et al. 1989; Helleiner 1987; Pinstrup-Andersen 1988). Governments in both the South and North have also voiced concerns over rising impoverishment under SAPs. At the 40th Anniversary of the General Assembly of the United Nations, many Third World Heads of State focused their speeches on the human consequences of SAPs in mounting poverty and malnutrition, which, it was feared, would inevitably lead to social and political instability (Jolly 1988: 75). In the US, the Congress adopted legislation in 1987 which sought to encourage the IMF and World Bank to give poverty alleviation a higher priority within SAPs. As Sanford (1988: 267) relates:

> The House Appropriations, Senate Foreign Relations, and House Banking Committees all expressed concern about the [poverty] issue. The final authorization law directed that the US executive directors (EDs) at the multilateral development banks should encourage the multilateral agencies to undertake programs that help the poor, particularly the rural poor . . . The law directed the US EDs to urge the multilateral banks to do studies assessing whether their loan operations help or hurt the poor. The EDs were also required to recommend that the multilateral banks adopt formal guidelines which would be designed to identify and minimize any such negative impact on the poor.

Rising concerns over the social costs of SAPs have also made an impression on the IMF and World Bank themselves. In 1987, the incoming Managing Director of the IMF, M. M. Camdessus, made protection of the vulnerable during the course of adjustment a major part of his first speech to the UN Economic and Social Council (ECOSOC). According to Jolly (1988: 75): 'That speech marked the first time a Managing Director of the IMF had spoken out on the desirability of adjustment policy paying explicit attention to issues of income distribution, health, nutrition and poverty.' In the late 1980s, the President of the World Bank, Barber Conable, also began to refer to the Bank's 'reemerging concerns about poverty alleviation' in many of his speeches (Singer 1989: 1314). Indeed, a special 'poverty task force' of senior Bank staff members submitted a report to Conable in 1987 which found that in spite of 'encouraging activity on poverty in many countries – and creative innovation in a few – overall the Bank's efforts were considered insufficient' (in Singer 1989: 1315). Following this report, the Bank has issued a number of major studies (including its 1990 World Development Report and the 1989 Sub-Saharan Africa, from Crisis to Sustainable Growth) which openly acknowledge its neglect of poverty and other issues of human development in the 1980s, but suggest that this is being remedied by a change in the priorities of SAPs and other Bank development projects.

The Need for Alternative Policies

Many critics of SAPs, however, contend that changes in IMF/World Bank priorities have been largely rhetorical and have had little impact on actual policies in the South (e.g., Bernstein 1990; Cobbe 1990; Singer 1989). For example, in an article analyzing the World Bank's *Sub-Saharan Africa*, popularly known as the 'Berg Report,' Saha (1991: 2755) concludes: 'The contents of the new program (World Bank 1989) do not appear to be much different than the earlier one. It seems that Berg's agenda of action has simply been repackaged and represented in a more user friendly language.' Some of these critics have called for compensatory policies to reduce the social costs of adjustment for the poor and other disadvantaged groups (e.g., Killick and Stevens 1991; Pinstrup-Andersen 1988; Shaw and Singer 1988). Such policies might be designed, for instance, to provide food aid or other forms of income transfer to the poor; create jobs in the public or private sector; improve productivity through investments in education, vocational training, and skills development; increase the availability of credit, technical assistance, and other factors of production for small/medium producers; and expand access to health care and other basic social infrastructure (Pinstrup-Andersen 1988: 44–5).

Other critics have called not only for compensatory policies, but also for fundamental changes in the direction of SAPs (e.g., Helleiner, Cornia, and Jolly 1991; El-Naggar 1987; Eshag 1989; Jolly and van der Hoeven 1991; Stewart 1987, 1991; Streeten 1987). They contend that the inclusion of poverty and other social concerns in development programs necessitates an integral approach pertaining to all adjustment measures rather than the mere addition of supplementary policies. Because SAPs remain narrowly focused on macroeconomic preoccupations to the exclusion of structural aspects of development, the IMF and World Bank can only make palliative recommendations concerning the poor as target groups rather than attacking the forces that make them poor in the first place. Needed structural changes to reduce inequalities and poverty have been neglected, while social expenditures remain inadequate to improve human capabilities and standards of living. Moreover, the design of the macropolicies of SAPs themselves has often exacerbated inequalities and other structural constraints that block more balanced and sustainable forms of development. Adjustment that protects the human dimension and supports structural change in developing countries needs to be incorporated into the design of both macro- and mesopolicies. Add-on programs are virtually certain to be inadequate.

A critical weakness of SAPs has been their failure to coordinate the long-term needs of structural transformation in developing countries with shorter-term macroeconomic considerations. As we saw in the previous chapter, a key factor propelling recent development in the Asian NICs was the successful coordination of various policy measures to promote broad goals of structural transformation, balanced and participatory development, and national unity. The state pursued a long-term vision of economic growth and development that was formalized in comprehensive development plans extending over five years or more. While the NICs implemented policies designed to foster macroeconomic stability and to create a hospitable environment for private-sector investment, they also invested heavily in the basic infrastructure and human-resource development needed to facilitate structural change. Sectoral and meso-level policies were coordinated with macroeconomic measures to improve internal linkages (e.g., rural–urban, agricultural–industrial, ISI–EOI) important for balanced development, economic diversification, and the participation of various sectors and social groups in economic growth. While policies paid attention to capital's requirements for investment and accumulation, the needs of capital were subsumed within the broader objectives of a long-term comprehensive development strategy.

The balanced, highly coordinated, long-term development planning of the Asian NICs may be contrasted with the obsession of SAPs with short-term macroeconomic indicators. The experience of the Asian NICs shows that development strategies for the late industrializing countries of the South need to incorporate a broad range of objectives that go well beyond immediate goals of deficit reduction and GDP growth. These broad development objectives (e.g., poverty elimination, employment generation, balanced growth, improved income distribution, structural change) cannot be achieved simply by improving fiscal balances or increasing exports and growth rates. Nor can they be achieved via add-on programs that seek to compensate for the negative consequences caused by the central components of a development strategy. Instead, policies must be designed to include these objectives as integral parts of a comprehensive development project.

An important initial step in this process would be to reassess the macroeconomic measures of SAPs in terms of both their growth potential and their effect on income distribution, poverty, and basic needs provisions. Rather than simply using blunt macroeconomic measures to reduce aggregate demand, more selective policies could differentiate between basic necessities and inessential or luxury goods and services. Given the extreme inequalities in most Third World societies, fiscal measures ought to be chiefly aimed at decreasing inessential public expenditures and private luxury consumption so as to minimize the effect of economic contraction on the basic needs of the poor. For example, raising indirect tax rates on luxuries or increasing direct tax revenues collected from the wealthy could curb private luxury consumption (Eshag 1989).

At the same time, complementary fiscal measures could be designed to redistribute income and raise demand for domestically produced wage-goods, particularly those produced by small/medium-scale, labor-intensive sectors. This might help to close the gap between increases in production and improved income distribution that many analysts contend is a basic weakness of economic restructuring under SAPs. Income redistribution in favor of the poor and wage-earners could modify patterns of internal demand to create 'virtuous circles' of economic growth and diminishing inequalities (García et al. 1989: 482). Progress could be made toward raising levels of 'social articulation' within Third World economies (e.g., de Janvry 1981; Dutt 1990), whereby more egalitarian income distribution stimulates demand for mass-consumption goods, which, in turn, creates new sources of employment and income for workers.

Socially articulated economies require policies that not only support progressive income distribution, but also direct public and private investment toward sectors producing goods and services for popular consumption. Rising production and productivity resulting from these investments increase employment and incomes of workers who, in turn, form new sources of demand for additional economic expansion in a mutually reinforcing manner. Social articulation may be furthered by cutting wasteful or inessential public expenditures in favor of public investments targeted to accelerate structural change, increase economic participation, and generate broadly-based patterns of growth. Similarly, fiscal and other incentives should be provided for private investments in sectors that further these overall goals. At the same time, policies should discourage wasteful and purely speculative activities that make no contribution to development, as well as investments in highly exploitative sectors that increase societal polarization.

As the recent history of the Asian NICs demonstrates, development strategies designed to foster equitable growth and social articulation need not necessarily preclude export promotion, especially for sectors that are compatible with national objectives such as employment creation or accelerating structural change. Nor do such strategies mean subsidizing unviable economic sectors which have no potential to contribute to future growth and development. However, they do require selective and well-coordinated policies to direct investment into sectors which offer good prospects for stimulating rapid growth, accelerating processes of structural change, and creating internal linkages important to social articulation. Moreover, investment policies must be carefully coordinated with public expenditures on physical and social infrastructure to create conditions for increasing productivity and profitability.

At the same time, infrastructure provisions and other government spending programs ought to be aimed at spreading both the benefits and costs of development more evenly among classes and social groups, economic sectors, and spatial areas. The creation of a fairer pattern of development should pay dividends not only in providing conditions for more dynamic, internally coherent, self-sustaining growth, but also in generating broadly based consensual support for a national development strategy. Unfortunately, SAPs seem to have generated exactly the opposite reaction in many Third World countries. Increasing societal polarization has generated a widespread perception that an elite minority has monopolized the benefits of development under SAPs, while the popular majority has been forced to endure a disproportionate share of the costs.

5

The South (2): The Neglect of Politics and People

This chapter continues analysis of the neoliberal development experience in the South. Many of the specific shortcomings of neoliberal policies uncovered in the previous chapter are linked to the neglect of sociopolitical considerations. In particular, insufficient attention has been paid to factors which may affect the political feasibility of neoliberal measures. As a result, inappropriate policies have often undermined state legitimacy and fueled instability. Elements of an alternative approach to structural change include an emphasis on democratic participation and a more equitable sharing of development costs and benefits. This requires a move away from ready-made strategies and top-down planning methods. Instead, closer attention should be paid to the specific development conditions and special needs of various countries and peoples. Such concerns have an especially profound impact on the social and environmental sustainability of development initiatives.

The Neglect of the State and Political Considerations

Many analysts emphasize that political factors matter enormously to the outcome of SAPs in individual countries, but have been largely ignored by neoliberal policy-makers (see, e.g., Bernstein 1990; Biersteker 1990; Colclough and Green 1988; Greenaway and Morrissey 1993; Herbst 1990; Killick and Stevens 1991; Nelson 1989; Onis 1991; Stein 1992). Political considerations particularly affect outcomes with regard to: (1) who participates in the bargaining process over SAPs, (2) how the implementation of SAPs proceeds, and (3) what the objective and subjective impact of SAPs on various groups will be. The character of the state and of state–society relations varies substantially across the South. The existence of powerful groups, both within and outside the state, that use political action to defend their interests can render many policies unfeasible, ineffective, or undesirable. In many cases, the prospects of successfully carrying out SAPs depend on the kinds of coalitions that form within the state and between the state and non-state actors. It should not be forgotten that the economic variables upon which SAPs are usually focused (e.g., real wages, real exchange rates) also represent underlying socioeconomic interests and institutional arrangements. These cannot be determined by policy alone, but are subject to many other historically constituted intervening factors.

Political and Institutional Considerations

Political and institutional factors act as essential filters through which the concepts and policies of a development strategy impact on and are interpreted by various classes and social groups. This underscores the need to pay attention to such factors, both in the analysis of development problems and in the framing of policies and procedures to address these problems. The selection of policy choices within any development strategy should take into consideration prevailing conditions within a particular country – in terms not only of more conventional indices (e.g., factor endowments, size of the country), but also of the nature of the political and institutional heritage. Even the most well-conceived, internally coherent policies will normally be counterproductive, or at least ineffective, in the absence of an associated set of compatible institutional and political structures. Moreover, if organizational or institutional changes are needed to implement certain policies or programs, these changes can best be made once policymakers have a clear idea of the various structures and interests involved, both within the state itself and in society at large.

The previous chapter stressed the importance of particular institutional arrangements to the successful development performance of the Asian NICs. Important policy changes were quickly and efficiently carried out within a coherent institutional framework that strengthened administrative capacities and created opportunities for cooperation in national development planning. However, underlying this institutional framework were two vital conditions: state relative autonomy and close public–private cooperation. Although states may take actions to increase their relative autonomy or improve public–private cooperation, these conditions are also historically determined by many factors outside of immediate state control. In the case of the NICs, state relative autonomy was strengthened internally by the historical weakness of the capitalist and working classes and externally by the international (especially geopolitical) context of postwar East Asia. Consequently, policies that proved successful in the NICs might be quite unsuitable for states in other Third World countries operating under different historically determined conditions. Throughout Latin America, for example, there are many weak, dependent states that lack autonomy from either international capitals or powerful fractions of the domestic elite. It is not uncommon for powerful interest groups to 'capture' parts of the state apparatus. Typically, the interests of transnational agribusinesses and the landholding elite are expressed through the Ministry of Agriculture, those of monopolistic industrialists through the Ministry of Industry, and those of the the large private banks through the Central Bank and the Ministry of Finance (Jenkins 1991). Under such conditions, policies and institutional arrangements which successfully guided growth in the NICs according to broadly based, consensual development objectives could well be manipulated to serve the narrow interests of dominant classes and social groups.

The overly technical, economistic focus of SAPs has all but ignored these types of political and institutional concerns. Perhaps this should come as no surprise, given the dominant role that the IMF and World Bank has played in imposing SAPs on Third World countries. Both of these organizations present themselves as neutral, technical agencies that do not take stances with respect to the internal political configurations of the countries they advise. The IMF and World Bank cannot officially be seen as politically involved, even though it is common knowledge that they pay close attention to political factors and that their policy prescriptions favor certain kinds of regimes and disfavor others (Brett 1987). SAPs are therefore worded in a purely technical language and inevitably take a wholly economistic direction. Policy alternatives are assessed on abstract, technical grounds as if development takes place in a political vacuum.

If the Third World state is considered at all, it is normally depicted as a major obstacle to more rational, market-led development. Following the precepts of public choice theory, the political arena is portrayed as full of rent-seeking politicians, bureaucrats, and lobbyists whose self-interested behavior is the antithesis of a more rational, objective approach to development. The state is seen to be all pervasive, yet powerless to direct development in a more rational manner. Widespread interventionism has caused state structures to become too large and unwieldy. Government spending to support such interventionism has reached unsustainable levels. All interventionist policies are regarded as similarly distortive, despite any differences which might exist in political and institutional arrangements.

At the same time, Third World markets are idealistically depicted as purely competitive and necessarily benign to overall development interests. Widely acknowledged causes of market failure (e.g., barriers to entry, tendencies toward monopolization) are brushed aside as insignificant, while 'government failure' is made the centerpiece of analyses of development problems. In the current ideological climate, the contention that market failures are trivial, but government failures are enormous, becomes a powerful slogan. But as a focus for serious economic and political analysis, it is wholly inadequate to understanding the many interrelationships between market and government failures that underlie most Third World development problems. Moreover, it exonerates other major actors (e.g., transnational corporations, oligarchic Third World elites, large private banks, the IMF and World Bank themselves) from any responsibility for the development failures of the South. While this position may serve certain ideological interests, it offers only a simplistic, naive conceptual foundation for setting policies designed to address many quite intractable real-world development dilemmas. As Toye (1987: 67) notes: 'Over-simplified "solutions," resting on little more than the political preconceptions of a distant ideologue, are incapable of resolving the real dilemmas of development satisfactorily.'

Inattention to the Political Feasibility of SAPs

The simplistic 'state versus market' dichotomy of the neoliberals fails to address many critical issues and questions concerning the political feasibility of SAPs.

For purposes of analysis, the political feasibility of SAPs may be divided into two parts: (1) the compatibility of policies with the interests of important classes and social groups; and (2) the compatibility of policies with the institutional and organizational framework of the state and state–society relations. The former is crucial to the maintenance of political stability necessary to sustain policies over the long term, while the latter is vital to the efficiency with which policies can be implemented.

SAPs, or any other development program, require a sound political basis. They must be carefully crafted to fit the circumstances of a country, taking into account both the political and economic environments. According to Bourguignon et al. (1991: 1485), it is particularly important that 'adjustment programs . . . recognize the interdependence of the three criteria of efficiency, welfare, and political feasibility.' It does little good to design the 'right' development strategy, if it proves impossible to implement or sustain. One of the important functions of a development strategy is to bring rationality and consistency to economic policies. But another is to cultivate the political support necessary to carry out such policies. As Fishlow (1984: 982) notes: 'Potentially superior economic outcomes are relevant, but by no means the whole story. If they were, developing countries would face much easier choices than they actually do.'

Complex interrelationships among many development problems mean that clear distinctions can seldom if ever be drawn between 'economic' and 'political' considerations in the framing of development policies. Policy-makers must recognize the legitimate role that politics should play in the choice and implementation of economic policies. Effective policies can only be designed by working within the parameters of political feasibility. The failure of SAPs to address the political consequences of economic reforms is particularly surprising, given that the major instruments of structural adjustment (e.g., privatization and public-sector reform, currency devaluation, price liberalization and the elimination of state marketing boards, the removal of labor regulations) have a profound effect on state–society relations and the constituencies upon which governments depend for political support. In most cases, the economic reforms entailed in SAPs involve not only changing constituencies, but also altering the mechanisms by which governments relate to their clients and supporters (Herbst 1990). Structural adjustment almost always makes the political climate much riskier for governing parties and leaders, through weakening state structures and changing the state–society relations upon which governments have traditionally relied to stay in power.

Nelson (1989) contends that, given the momentous changes brought about by SAPs, the politics of adjustment must necessarily be seen as the 'politics of the long haul.' Governments must, therefore, search for and hold together reform-oriented coalitions under difficult circumstances in which they may have little to offer their supporters in terms of immediate benefits. Consequently, governments need to be acutely aware of the political as well as economic changes that reforms will bring to various classes and social groups. Following a balanced assessment of the likely consequences of reforms for different groups, governments must be willing to modify policies to maximize the benefits and minimize the economic and political damage caused by structural adjustment (Hawkins 1991). Moreover, an informed assessment of various alternatives for reform may provide a useful framework for dialogue and discussion between representatives of the state and different social sectors (White 1990). If it is designed to foster genuine, broadly based participation in decision-making, such discussion can present opportunities for creating politically important compromises and modifications to policies, as well as divising appropriate compensations to certain groups that may be particularly disadvantaged by the thrust of reforms.

The Need for Political Stability and Policy Continuity

In most countries undergoing SAPs, it appears that economic reforms have caused significant hardship to a broad range of classes and social groups which collectively have the ability to undermine political stability. As we saw earlier, the lower classes (e.g., working class, peasantry, informal sector) and especially the traditionally disadvantaged groups (e.g., poor women and children, the elderly, ethnic minorities) have borne the brunt of the social costs of SAPs. In addition, SAPs have also harmed much of the middle class (e.g., public-sector employees, artisans and other small/medium producers) as well as some upper-class elements (e.g., bourgeois producers oriented toward the domestic market). Following decades of steadily increasing state intervention in most countries, many of these groups had become dependent on various state policies and programs for their advancement and survival. In the process, a complex structure of ideological mechanisms (e.g., nationalism, statism) had been created to legitimize the continuation of particular forms of state intervention.

Under SAPs, much of this postwar continuity in state–society relations has been dramatically broken. Many classes and social groups feel suddenly alienated and under attack, as their hard-fought social gains achieved through previous struggles have been stripped away with little or no consultation. Moreover, many of these social sectors still retain relatively high levels of mobilization and political influence, accumulated through previous struggles. As a consequence, the political costs of SAPs have been unusually high for many governments. Spontaneous rioting and demonstrations have frequently broken out, highly mobilized groups have used their influence to spread political instability and sabotage the reform effort, and an increasing number of governments have been removed via coups or elections. It appears that this will be the principal legacy of SAPs in the South if the economic dictates of outside organizations such as the IMF and World Bank keep governments locked into politically unfeasible positions. In particular, governments must be permitted to find locally appropriate methods to allow the diverse organizations representing the popular majority to cooperate and participate in the framing and implementation of policies.

Much of the neoliberal agenda contained within SAPs has proved incompatible, not only with major political interests but also with the institutional and organizational framework of Third World polities. Many authors contend that inadequate consideration has been given to the lack of administrative capacity to implement and manage the reforms that outside agencies are prescribing (see, e.g., Greenaway and Morrissey 1993; Helleiner 1992; Nelson 1989; Rondinelli and Montgomery 1990; Schoenholtz 1987; White 1990). In many cases, it appears that the simultaneous imposition of a broad range of reform measures overwhelmed the capacity of the state to carry them out coherently and efficiently. Poorly coordinated and haphazardly implemented policies often seemed to be working at cross-purposes. Contradictions, delays, and policy reversals destroyed confidence in the predictability and sustainability of the reforms. As a result, the credibility of the overall adjustment process was often undermined, leading to rising political instability and a withdrawal of investment capital needed to generate future growth.

It is now widely acknowledged that inadequate policy coordination and institutional failure have had a major, perhaps decisive, impact on the poor performance of SAPs in many countries. Consequently, one of the most important recent thrusts in development and adjustment thinking is the increasing emphasis placed on policy coordination, institutional coherence, political stability, maintenance of credibility, and sustaining government initiative over the long term. Helleiner (1992: 785) asserts:

> More important than achieving policy 'perfection' at each point in time, whatever that might mean, is the creation and maintenance of a stable overall policy environment, and the creation and preservation of credibility for and confidence in an announced adjustment and development program.

SAPs require states to have an especially efficient bureaucratic and technical apparatus, as well as the political capacity to design and carry out effective policies. Therefore, neoliberals who anticipate a 'withering away of the state' via economic reforms are mistaken; in fact, reform initiatives need to promote stronger, more capable states that can understand and react effectively to changing conditions. It often takes greater discipline and self-confidence for states to liberalize previously controlled markets than to extend interventionist policies that have favored politically important groups (Lewis 1989; Nelson 1989). Given the enormous difficulties of sustaining structural adjustments, the modern, reformist state has to be more stable, efficient, and effective at communicating and governing. It is not simply the minimalist state envisioned by the neoliberals.

Thus, the state must continue to play a key role in SAPs, or any other development program, whatever the currently dominant ideology proclaims. There should be no question as to whether the state has a legitimate and central role to play in development. Instead, questions should address the nature, extent, and frequency of state interventions needed to accelerate development under different conditions in individual countries. In much of the South, there is a problem not so much with the size of the state (in fact, it is relatively small in most countries), but with the inefficiency and unproductiveness of state interventions. This renders the state incapable of fulfilling many diverse functions necessary for sustained, broadly based development. The more successful cases of structural adjustment are due neither to laissez faire, nor to centralized bureaucratic control, but to governments that understood which areas to intervene in and which to leave alone, and how to conduct interventions efficiently (Streeten 1987: 1478). At the same time, the frequent failures of SAPs illustrate not only excessive state intervention, but also unwise and inefficient intervention in some areas, and inadequate intervention in others (ibid.).

Not only these, but many other recent development experiences, particularly in the Asian NICs, show that the efficient use of market forces does not necessarily preclude state development planning, especially if it is indicative, decentralized, and focused on limited problem areas (Dietz and James 1990). The potential for beneficial externalities resulting from state intervention has been ignored in SAPs and other neoliberal programs. Despite compelling evidence from East Asia and other areas, the fact that state intervention may accelerate structural change, create dynamic comparative advantages, and broaden patterns of growth seems not to have found a place in neoliberal theory.

In fact, there are many ways in which selective and carefully coordinated state intervention can alter Third World markets so that they function more efficiently to serve broad development interests (see Streeten 1993: 1283–4). The state can provide a legal framework and maintain law and order, including the enforcement of contracts and property rights. It can pursue correct macroeconomic policies (e.g., with respect to exchange rates, interest rates, wage rates, trade policy) to promote high levels of employment and growth without inflation. It can safeguard competition (e.g., through anti-monopoly and anti-restrictive practices legislation) and intervene in processes of price formation, production, and finance to improve both the efficiency and distributive aspects of markets. It can tax activities that it wishes to discourage (e.g., short-term speculation in real estate, consumption of tobacco or gasoline, highly polluting industries) and subsidize activities it wishes to encourage (e.g., use of public transport, education, health care services). It can invest in physical infrastructure and human-resource development to improve profitability rates and 'crowd in' private investment to activities that further national development objectives (e.g., structural change, poverty alleviation). It can contribute to the effectiveness of price incentives (e.g., devaluations, market liberalization) by assisting in the design and strengthening of complementary institutions (such as for land reform, information, credit, marketing). It can implement urban/regional planning programs and other measures designed to promote selective growth and counteract tendencies toward socioeconomic and spatial inequalities.

The Need to Transcend the State-Versus-Market Dichotomy

However, to enable the Third World state to carry out these important functions, development strategies must transcend the sterile state-versus-market dichotomy of neoliberalism. The problem is finding the correct mixture of market orientation and state intervention, given divergent development conditions in individual countries, and then devising a set of institutional and organizational arrangements that are compatible with this particular mixture. The choice between free market and state intervention largely depends on timing and circumstances. Helleiner (1990: 145) states: 'The political and economic efficacy of markets and governments varies across

countries and in individual countries over time.' Abstract, universalistic economic models cannot, therefore, provide an invariable set of development
policies which will be appropriate to the varying conditions and needs of
individual countries at particular times. Instead, development strategies need
to pay close attention to the historical context within which development
is unfolding, including elements such as state structures, state–society relations, markets and ownership patterns, class and other social relations, and
ideological concerns. Neither the state nor markets are neutral institutions;
both can work for good or ill. The question for development strategies
should be under what conditions states and markets can work to serve broad
development objectives and how to bring about these conditions. Solutions
will necessarily be particular to individual countries and will involve more
than just economic considerations. As Toye (1987: 57) notes: 'The plain
fact of the matter is that no one has yet succeeded in devising a division of
functions between the public and private sectors which is both universally
applicable and defensible on economic, rather than political grounds.'

It has become apparent that wholesale liberalization is neither economically desirable nor politically feasible for many countries. There are
many reasons why interventionist policies may have been pursued by
Third World states, including: equity objectives (e.g., income redistribution,
job creation, regional development); infrastructure development and other
'lumpy' investments (e.g., steel, petrochemicals); collection of monopoly
rents (e.g., on minerals); filling in for a deficient or absent private sector;
countering capitalist monopolies; and strengthening economic sovereignty,
especially *vis-à-vis* transnational corporations (Bienen and Waterbury 1989:
618). These are real concerns in most developing countries, which governments cannot neglect without paying a high price in terms of economic
polarization, social unrest, political instability, and loss of national unity.
As a consequence, liberalization measures designed to reduce allocative
inefficiencies must always be shaped to fit the historically constituted
conditions and special needs of individual countries.

Particularly in severely polarized and underdeveloped countries, large
parts of the private sector are typically undynamic or have been incapacitated by decades of infrastructural neglect. This is especially the case
for many internally oriented sectors, such as domestic food production.
Under these circumstances, 'liberalization may result in a precarious vacuum
inviting anticompetitive behavior by the few who have the means to step in'
(Reusse 1987: 299). In most highly dependent economies, this has caused
widespread displacements of small/medium producers by transnational corporations and other large-scale capitals with extensive foreign connections.
As a consequence, liberalization measures have worsened already severe
problems of polarization and impoverishment, especially in many rural
areas, despite generally offering producers higher prices. However, as

Reusse (1987: 316–17) points out, this pattern may be avoided if the state complements liberalization measures with 'bridging interventions' targeted to assist traditionally disadvantaged producers, such as small/medium peasants. Such interventions might provide assistance in areas such as credit and financing, technological and structural improvement, access to production inputs and basic consumption goods, and transportation and marketing.

Rising concerns over anti-competitive behavior and tendencies toward monopolization in many sectors stress the need to take a more balanced approach to questions of liberalization and relations between the state and markets. The main questions for development strategies seem no longer to concern the extent of state intervention and/or the size of the public sector. Instead, the questions now being raised concern the comparative advantages of the public and private sectors, how these sectors may complement each other, and how their performance may best be improved (van Ginneken 1990: 443). The state should be asked only to do what it can do best and should stay out of other areas. Nevertheless, it can take many important measures to promote development of both the private sector and society at large according to broadly based objectives. While there are often costs involved in state interventions, unfettered markets normally exact even higher costs, especially among the most vulernable and disadvantaged groups of Third World societies.

The Subversion of Sovereignty and State Legitimacy

One of the central paradoxes of SAPs is that they require a strong state and political stability in order to be successfully sustained, but they systematically weaken the governments that must carry them out through the imposition of inappropriate policies. In a study of IMF/World Bank programs in Africa, Havnevik (1987: 423–4) arrives at three conclusions concerning SAPs: they subvert national sovereignty; they portray no understanding that policies must be based on political consensus; and they are imposed by international agencies that have neither global nor local legitimacy. SAPs focus on a ready-made set of policies designed to meet abstract 'textbook' criteria of economic allocative efficiency. Consequently, they restrict the ability of national governments to determine their own policies and shape their societies according to local conditions and needs. According to Ghai and Hewitt de Alcántara (1990: 422), this generates a 'conundrum of governability,' which begs the following questions:

> How are the wide range of conflicting interests associated with crisis and adjustment to be channelled and expressed within a stable political environment at a time when the legitimacy and efficacy of many states are being so

thoroughly undermined? How can a sufficient sense of cooperation and purpose be developed to permit an adequate collective response to the crisis?

As the focus of development has shifted toward promoting macroeconomic growth away from other, more traditional concerns (e.g., income distribution, basic needs provisions, poverty alleviation), development strategies have also adopted a more 'top-down' rather than 'bottom-up' approach (Rondinelli and Montgomery 1990: 74). SAPs have been imposed on the people of the South in a top-down manner in terms of both relations between the international financial institutions and Third World governments and relations between these governments and their people. While the rhetoric of the IMF and the World Bank clearly recognizes the need for local participation in programs and policies, structural adjustment packages, in practice, are presented to governments on a take-it-or-leave-it basis. Negotiations between the large, highly sophisticated international institutions and governments from small, underdeveloped countries are unequal and often confrontational. For the IMF and World Bank, little is lost if a small country rejects an agreement, while the costs to the country can be enormous.

Both the Fund and Bank have a wealth of skilled and well-educated personnel to argue their case. Government officials may intuitively reject some aspects, but usually can neither present their case with sufficient rigor nor provide the intellectual rationale and political bargaining power necessary to win the argument (Stewart 1987: 42). They lack the resources to present their case properly, and the macroeconomic models of the international institutions leave little room for policy alternatives that recognize the varying conditions and needs of Third World countries. The institutional structure and macroeconomic focus of the international institutions generate considerable 'inbreeding' and 'herd behavior' in the policy community that oversees SAPs from Washington, thereby limiting the variety of sources from which consensus views are developed. While it is widely acknowledged that Fund/Bank personnel are highly skilled, professional training cannot completely substitute for local knowledge in the formulation of appropriate policies for Third World countries. Even if it could, SAPs and other policies can seldom be successful if those who implement them do not believe in them or do not regard them as their own (Helleiner 1992: 787).

This conclusion applies to the top-down manner in which SAPs have been imposed not only at the global level, between international institutions and Third World governments, but also at the local level, between these governments and their people. Key macroeconomic and other policy decisions are usually made in great secrecy by a handful of actors – normally the president, finance minister, head of the central bank, and their top advisors, along with representatives of the international institutions (Ghosh 1991; Kaufman 1989; Kraus 1991; Stewart 1987). These decisions are then normally pushed through at breath-taking speed with minimal or no consultation from other members of government, opposition parties, associational groups, or popular organizations. Likewise, resultant policies and programs are implemented from the top down with little meaningful participation by various organizations whose members are often dramatically affected by the reforms. Little room is afforded for negotiations between different classes and social groups designed to create a consensus or 'social contract' behind the reform effort. Dialogue is acceptable as a means to explain policy, but ultimately major development factors (e.g., prices, incomes, the distribution of public benefits) should be set by the marketplace or by technocrats, not through negotiations between different interest groups. In fact, within this top-down approach to development, popular participation and organization are commonly perceived as a hindrance to rational development, rather than as a precondition for its success.

The top-down manner by which SAPs have been imposed has significantly undermined the legitimacy of many Third World governments in the eyes of their people. Neither the IMF nor the World Bank are in any way accountable to the people of Third World countries, whose lives are often being turned upside down by the effects of SAPs (Bernstein 1990). It is widely perceived that Fund/Bank policies place the interests of the big banks and rich Northern countries before the needs of the popular majority in the poor Southern countries. Feelings of animosity toward these international institutions inevitably spill over onto governments which are given the task of carrying out the austerity measures mandated by SAPs. The result has been rising social unrest and political instability in many countries. Massive protest demonstrations, spontaneous strikes, food riots, and other sharp outbreaks of violence have generated high political and economic costs for many countries undergoing SAPs, including Argentina, Bolivia, Brazil, Chile, the Dominican Republic, Egypt, Ghana, Mexico, Morocco, Nicaragua, Nigeria, Peru, Senegal, Sudan, Tunisia, Venezuela, and Zambia (Bienen and Gersovitz 1985; Kreye and Schubert 1988; Maralidharan 1991; Mengisteab and Logan 1990; Pastor 1987; Streeten 1987; Weissman 1990). Paradoxically, then, SAPs seem to have heightened social unrest and political instability, thereby undermining capital's confidence that the reforms can produce a stable, predictable environment for investment and accumulation. Without such confidence, investment drys up, capital flees, growth stagnates, and 'trickle-down' effects do not occur – thus negating the central neoliberal elements of SAPs.

Tendencies toward Authoritarianism and Repression

One of the most important tasks of governance is to create stability and national unity by accommodating and reconciling the divergent interests of various classes and social groups within a society. Indeed, economic growth itself should be seen as a means to create and maintain a harmonious society rather than as simply an end in itself. In the previous chapter, we saw that the development strategies of the Asian NICs recognized this. However, within SAPs and other neoliberal development programs, it has remained largely unrecognized. Rising social unrest generated by widening polarization, coupled with the lack of any consensus over the basic elements of the 'social contract,' has made the task of democratic governments virtually impossible (Ocampo 1990). Faced with urgent and conflicting demands, and with a fundamentally weakened capacity to meet these demands, governments have opted for an authoritarian and repressive solution to avoid the total collapse of public authority (Black 1991; Ghai and Hewitt de Alcántara 1990; Herbst 1990; Killick and Stevens 1991; Rausser and Thomas 1990). This has commonly deepened and extended the use of coercive measures, sometimes within a democratic façade, beyond those that took place prior to the adjustment process. In particular, it appears that the working class, peasantry, and other major elements of the popular sectors have often been forcefully expelled from the political arena through the use of systematic repression and other forms of pressure against community groups, unions, peasant organizations, parties, associational groups, and other popular organizations.

Given the close association of authoritarian regimes with SAPs and other neoliberal programs, many analysts have suggested that authoritarianism may well be necessary to the sustainability of market-led development strategies in general (see Handelman and Baer 1989; Killick and Stevens 1991; Kohli 1989). Indeed, a senior official of the World Bank (Lal 1983) has openly stated that SAPs may have to be implemented by 'courageous, ruthless, and possibly undemocratic governments' (in Bienefeld 1989: 37). The logic of this proposition rests both on the need to contain political pressures generated by the rising social costs of SAPs and on the need to provide a stable, predictable environment to attract investment. It has been noted that the success of SAPs, in the politically volatile countries of the South, has been 'inversely related to working class resistance' and has 'depend[ed], in the last instance, on the capacity to control the class struggle' (Pastor 1989: 104). Democratic regimes find it difficult, if not impossible, to sustain structural adjustment because, under pluralist forms of government, the working class and other popular sectors have a greater capacity to disrupt and defeat neoliberal policies. By contrast, authoritarianism facilitates effective management of SAPs because of the ability it gives the state to repress popular dissent and provide technocrats with the autonomy they need to carry out unpopular policies. Instead of reducing its role in development, as envisioned by neoliberal theorists, the authoritarian state under SAPs seems to have turned into a modern Leviathan, extending its domination into every sphere of society in order to stamp out dissent and increase economic efficiency.

Elements of an Alternative Approach to Development

Given tendencies toward polarization, social unrest, and authoritarianism under SAPs, it is clear that an alternative approach to development must be found if equity, social stability, and democratization are to remain serious objectives of development strategies. The first major task will be to create conditions in which strong social partners can participate in decision-making at the local, regional, and national levels to enable a consensus or 'social contract' to be constructed over how development should proceed. This means strengthening community groups, popular organizations, and other associational groups to enable them to take an active and responsible part in the decision-making process. A widely acknowledged and respected social contract cannot be achieved, in most highly politicized societies of the South, if important social groups are unable to exercise a decisive influence on governments to ensure that their concerns are taken into account by the political system. Since SAPs, or any other development program, necessarily involve difficult choices over how the costs and benefits of development are to be distributed, any meaningful development strategy must obviously be based on a fair degree of social consensus if it is to be successfully sustained without resort to authoritarianism. As Levitt (1990: 1594) remarks, 'development cannot be imposed from without' in a top-down manner, and 'is not [simply] about financial flows' and other macroeconomic considerations, but fundamentally 'concerns the capacity of a society to tap the root of popular creativity, to free up and empower people to exercise their intelligence and collective wisdom.'

The economistic focus of SAPs has largely brushed political considerations aside. However, these must be dealt with seriously if difficult, but necessary, structural transformations are to be sustained under demo-

---

1   Concertation, or *concertación*, is a concept that has been largely developed in some Latin American countries. It focuses on processes of communication, dialogue, and accommodation between the state and various popular organizations and other associational groups from the broader society. It offers opportunities for different social groups to actively participate in the framing and implementation of policies, thereby creating a more cooperative, rather than confrontational, atmosphere to provide political stability and consensual support for development strategies. The idea is that processes of concertation will, on the one hand, generate a better understanding of policies among important social groups and, on the other hand, produce policies which are more appropriate to the different needs and interests of such groups. While concertation cannot, of course, completely eliminate divergent interests, the alternative is thought to be increasing social unrest, political instability, and a drift toward authoritarianism.

cratic rather than authoritarian conditions. The political requirements for sustaining needed reforms are satisfied better in democracies where there is 'concertation' and broad participation (Bitar 1988).[1] At the same time, however, these same requirements dictate that economic policies be adopted that are acceptable to the popular majority. Within highly polarized Third World societies, the effect that such policies have on equity is of particular importance if social stability and democracy are to remain key objectives of development.

Given the varying conditions and needs of Third World countries, there can be no single model for carrying out needed structural change. It is therefore necessary to learn from experience, reject the universalistic model of SAPs, and act pragmatically to meet majority interests. While the specific measures adopted will vary among countries, the active participation of diverse social groups representing the popular majority is essential to ensure that structural transformations can proceed without sacrificing equity, stability, and democracy.

On the one hand, this requires conditions which allow people to identify the causes and find solutions to their own problems. Broad education and 'conscientization' are especially important to this process, both for mobilizing human resources and increasing participation in decision-making.[2] As Collier (1991: 117) notes: 'No international agency, however well-informed and well-intentioned, can substitute for a well-informed society: the time for secular gods is gone.' On the other hand, increased popular participation also requires institutional reform to provide opportunities for various groups to organize, represent themselves, and exert influence over decision-making at the local, regional, and national levels. Hierarchical institutional structures and elite-imposed development policies should be replaced by more democratic, two-way planning processes that empower people to design policies in their own interests and build on their own resources to overcome the problems that they will inevitably confront.

Inattention to the Particularities and Sustainability of Development

In the preface to his book on the neoliberal counterrevolution in development studies, Toye (1987: viii) notes that 'when economic thinking is connected up with political movements of the right or the left, it seems almost

---

[2]   Conscientization, or *conscientización*, is another concept that has been largely developed in Latin America, especially through the work of Paulo Friere and other activists devoted to improving methods of popular education. It involves helping people to become more conscious both of the root causes of their problems and of devising their own solutions to these problems by using their indigenous capabilities and resources.

impossible to avoid the ill consequences of over-simplification.' Within the rightist strategy of neoliberalism, the problem of over-simplification appears most prominently in the narrow focus of SAPs on liberalization measures and short-term macroeconomic indicators. Liberalization is offered as a panacea for the macroeconomic ills of Third World countries, regardless of their particular historical backgrounds or institutional arrangements, and regardless even of the costs involved in the cure (Banuri 1991: 1–2).

The broad agenda for policy debate on development has given way to a narrow, technical focus on the means and the speed with which liberalization measures ought to be implemented. Little attention has been given to issues such as the environmental and political sustainability of the reforms, the nature and quality of popular participation in the decision-making process, and the appropriateness of policies to the special needs and interests of individual countries. Attempts to treat the diverse and multifaceted development problems of Third World countries with a one-dimensional, universalist solution have proven costly and ineffective, resulting in what Iglesias (1985) has termed 'the crisis of ideologized macroeconomics' (in Tokman 1986). Rather than remaining obsessed with the short-term mechanics of liberalization, development strategists ought to broaden their focus to the particular historical features and long-term development needs of different societies. Fixed, ideological conceptions and ready-made 'single objective' approaches to development are incapable of understanding the complexities of Third World countries, the range of their development prospects, and the feasibility and desirability of various policies and institutional changes.

The Neglect of Long-term Development Needs

The narrow, economistic focus of SAPs has resulted in what many Latin Americans now refer to as *cortoplazismo*, meaning 'short-termism' or 'that pervasive mix of chronic anxiety and skepticism that leads to an inability to plan beyond the next week' (Rodrik 1990: 936). The macroeconomic calculations according to which SAPs are typically evaluated neglect many development issues (e.g., technological change, human-resource development, structural transformation of agriculture and industry, equity and social justice, democratization, sustainability) that are of longer-term importance to developing countries. Indeed, many analysts contend that the short-term macroeconomic considerations that drive SAPs often not only neglect, but are fundamentally incompatible with, the longer-term development needs of Third World countries (see, e.g., Cheru 1992; Ocampo 1990; Rodrik 1990; Seidman 1989). As Helleiner (1992: 779) notes: 'If a long time horizon is generally regarded as appropriate [for development strategies], it is not helpful to undertake repeated evaluations of the adjusting countries' progress according to short-term monetary/credit targets, balance-of-payments performance, conventional growth measures, or social indicators, still less by one agency after another.'

A consensus seems to be emerging among development theorists that processes of structural adjustment normally take much longer than was originally envisioned. Therefore, issues of a long-term nature ought to be addressed in the design of SAPs, or any other development program, instead of being ignored or treated peripherally through add-on measures. The attainment of long-term development goals will require a rather different set of priorities and objectives from those of merely 'getting the prices right.' Policies ought to address the structural realities of development in various countries that underlie and shape surface features such as prices.

Ultimately, there is a need to reconsider what development is all about. The approach that the international financial institutions have taken with the implementation of SAPs views development simply in terms of financial flows and of other macroeconomic indicators to be programmed and targeted. However, development is fundamentally about people and societies. The exclusive macroeconomic focus and top-down implementation of SAPs have prevented them from satisfactorily addressing many issues around which a more people-oriented and sustainable approach to development would be constructed. These include equitable income distribution, basic-needs provisions, societal cohesion and national unity, popular participation and democratization, human-resource development, structural transformation of industry and agriculture, socially and spatially balanced growth, and the cultural and environmental sustainability of development.

Inattention to Local Conditions and Social Relations

Given the tremendous diversity within and among Third World countries, these types of issues cannot be addressed through universalistic, *passe partout* development programs such as SAPs. Most important elements of development strategies are not directly transferable; their design must be country specific. Strategies should be tailored to take into account the prevailing sociocultural, political, economic, and environmental conditions found within different countries. As Kearney (1990: 200) reminds us, 'The cards dealt by history [to particular countries] cannot be turned in for a new hand; however poor the deal, it must be played as well as possible.' This means that attention should be paid to the positive foundations of development, as well as the shortcomings and lacunae that have been generated in individual countries by varying historical and geographical circumstances. Through careful analysis of local conditions, strategies may be constructed to preserve and build on the positive foundations of development in each country, while seeking to overcome certain shortcomings or contradictions without causing undue disruption and hardship.

Within IMF/World Bank SAPs, Sarkar (1991: 2309) reports that the policies recommended to various countries 'show a 70–80% overlap of identity.' Such policy homogeneity, irrespective of the individual circumstances of the countries to which SAPs are being applied, has several causes, according to Glover (1991: 179–80). One is the limited knowledge that IMF/World Bank economists typically have of these countries. Albert Hirschman has termed this problem the 'visiting-economist syndrome' or the 'habit of issuing peremptory advice and prescription by calling on universally valid economic principles and remedies – be they old or brand new – after a strictly minimal acquaintance with the "patient"' (in Meier and Seers 1984: 93). A second reason is the ideological bias of the programs toward the interests of the core capitalist countries, international financial capital, and global capitalism in general. A third is the desire to provide a simple recipe of economic reforms that, it is believed, might offer some immunity to political interference or administrative failings. In practice, this means devising a simplified market solution for development problems. A fourth factor is the need to impose a fixed set of policies on all countries to avoid the appearance of unequal treatment or favoritism. A fifth is the lack of coherent country-specific alternatives to the Fund/Bank programs. As Glover (1991: 180) notes: 'If countries cannot provide tailor-made programs, it is difficult to see how external agencies could do better.'

However, given the widely acknowledged failures of SAPs, Third World countries have little choice but to devise development strategies more appropriate to their own needs and interests. A basic problem of SAPs, which is characteristic of mainstream development strategies in general, arises from the imposition of Western models on societies whose cultural values and traditions, social and political arrangements, and methods for carrying out economic activities make the absorption of these models neither possible nor desired. The resulting displacement process inexorably creates uncertainty, anomie, tension, and conflict – thereby undermining efforts to bring about needed socioeconomic change. Indiscriminate modernization, whether via SAPs or some other equally alien model, often fundamentally weakens the social fabric of Third World countries. This greatly adds to the difficulty of designing and implementing a process of cumulative social change and economic transformation that most analysts agree must form the centerpiece of any long-term development strategy to improve Third World standards of living. Broadly based development requires that economic processes be compatible with prevailing social and cultural conditions; but, at the same time, development must generate change in these conditions as part of the overall process of structural transformation needed for developing countries to sustain economic growth in the modern world system. Therefore, Third World strategies ought neither to neglect the real development problems of their own societies nor to seek a remedy to these problems by adopting inappropriate outside models. Chidzero (1987: 140) emphasizes both of these points in an article on African development, which is equally applicable to the rest of the South:

> Africa must see itself without the fig-leaf and not copy external models blindly. African countries have no choice but to examine and analyze their respective concrete situations, fashion development plans and programs accordingly, and craft effective policies to that end.

Most recent analysis of SAPs has concentrated on economic criteria and has been largely policy focused and sectorally oriented. At best, such analysis has included some mention of state structures and institutional characteristics. However, it has typically paid little attention to the broader features and particularities of the societies within which SAPs are being carried out. As a result, we know relatively little about how SAPs are affecting the problems, needs, and aspirations of the popular majority in most Third World countries, whose material and spiritual well-being must be a central concern for any broadly based development strategy. Unfortunately, academic research on SAPs seems to be 'reproducing,' in new intellectual forms, the marginalization process to which the popular sectors have been subjected by the development programs themselves.

Much of this research has replicated the obsession of SAPs with macro-economic indicators and 'getting the prices right.' Characteristically, one sphere of economic activity – that of exchange – has been abstracted from the totality of relations of production and power, thereby inevitably generating simplistic, partial analyses and distorted results. Such problems have been further compounded by the substitution of an overarching ideological conception of the market for careful analytical and empirical investigations into different kinds of markets and the mechanisms by which they operate. In particular, much of this research has neglected historical patterns of commoditization in Third World countries, which have been shaped according to the particular evolution of sociocultural, political, and economic relations. It is only through analysis of this complex realm of societal relations that policies can be developed to address many of the structural causes of persistent Third World development problems, (e.g., underdeveloped forces of production, widespread polarization and impoverishment, economic stagnation). The failure to do so represents one of the most serious shortcomings of most current policy-oriented research in the South.

Low levels of technological diffusion, underdeveloped forces of production, and, as a consequence, inferior rates of productivity have traditionally hampered development within many Third World agricultural sectors. The root causes of such problems are often located in a series of structural constraints (e.g., the extreme concentration of land and other means of production; tendencies toward financial and commercial monopolization; the continuing presence of sharecropping, usury, and other precapitalist rent-extracting arrangements) that have distorted rural markets and provided disincentives to productivity-enhancing investments, especially by peasants and other small/medium farmers. In order to understand these structural constraints and develop viable policies to overcome them, studies need to analyze underlying patterns of social relations within several interlocking areas. These include relations of production and social reproduction, both inside and outside of households; relations of commercialization and circulation; and relations of power at various levels.

A study by Koopman (1993) of rural areas in Africa found that small/medium agricultural production takes place in the context of at least two distinct sets of social relations of production: first, simple commodity production constrained by the dominance of monopolistic state policies and/or capitalist market power; and, secondly, simple commodity production and subsistence production structured by patriarchal relations of production among household members. What many conventional neo-classical agricultural models fail to recognize is that access to resources, productive services, and markets varies significantly according to the position that rural producers occupy within these distinctive sets of social relations. It is important to point out that the neglect of social relations has not only hindered academic research, but has also had a powerful practical impact on the design and implementation of rural development projects. Development strategies which conflate, for example, small-scale women producers, who often dominate domestic food production,[3] with large export-oriented agribusinesses are conceptually incapable of devising policies appropriate to the divergent needs of these two sectors. The special needs of particular rural sectors cannot be subsumed within some generalized conception of the interests of the agricultural sector as a whole (e.g., liberalization of agricultural prices). Nor can these special needs be addressed by policies which resort to market mechanisms alone. Instead,

---

[3]   In African countries, Cheru (1992: 508) reports that women are responsible for 60–90 percent of the production, processing, and marketing of domestic foodstuffs. Yet women have the least access to improved technology, credit, extension services, and land. Thus, rural development programs designed to increase food security have little chance for success if efforts are not made to improve women's access to productive resources and to reorient agricultural training and other supportive services to meet the special needs of women farmers.

they call for broader measures based on careful analyses of historical patterns of social relations. Without such analysis, development strategies can only offer piecemeal and palliative solutions to many deep-seated Third World development problems; policies cannot be effectively designed to address the structural factors that created these problems in the first place.

Weak and highly polarized production structures, non-competitive markets, and skewed power relations have rendered the market-based thrust of SAPs and other neoliberal strategies largely irrelevant to many of the most pressing development problems of Third World countries. As Streeten (1993: 1295) notes, such problems may often be rooted not in market failure, but in market success. If the signals propagated by the market are based on an unequal distribution of income, land, and other assets, it may be market success in responding to these signals that causes problems. The analysis of famines by Sen (1981, 1989) demonstrates that total food supply has often been adequate, but that the purchasing power (or, more generally, the entitlements) of the poor and other disadvantaged groups has normally declined and generated widespread starvation. In these cases, the market successfully responded to its signals, incentives, and allocations, while large numbers of people starved. Development policies designed to address such problems obviously cannot simply rely on market forces, but must bring about needed structural changes in underlying social relations, involving a redistribution of both productive assets and access to power.

The Neglect of the Human Dimension of Development

Given the exigencies of growth and development within the contemporary global economy, strategies that disregard the need for macroeconomic balance and allocative efficiency are bound to fail. However, these economic imperatives cannot be allowed to override the broader, long-term requirements of development which, in most countries, are based in fundamental redistributive reforms, human-resource development, and structural change. The primary objectives of long-term development can be summarized in very general terms as sustainable economic growth combined with social justice (Stewart 1991: 415). In most Third World countries, both of these objectives necessitate creating conditions to improve economic opportunities, develop human capabilities, and enhance social cooperation via the structural transformation of the economy and society. SAPs and other liberalization measures pay little attention to issues of social justice, the economic rights of the popular majority, and extending resource access to the currently deprived elements of most Third World societies. In fact, SAPs tend to sacrifice such concerns to the accumulation demands of an elite minority. In so doing, they often neglect general welfare and social cohesion for short-term profits and unsustainable, imbalanced growth. Furthermore, the macroeconomic focus of SAPs delays the search for and implementation of more sweeping changes needed for sustainable, majority-based development.

In spite of the experience of the Asian NICs, SAPs and other neoliberal strategies seem not yet to have grasped the essence of modern development, which lies not in short-term financial flows and other economic indicators, but in creating conditions for a more people-oriented development founded on social cohesion and the continual advancement of human resources. SAPs may increase the short-run 'efficiency' of some resource allocations, particularly from the point of view of the big banks, TNCs, and other large-scale capitals. However, while they may advance the short-term profitability and accumulation requirements of a minority elite, SAPs also commonly subvert national sovereignty, diminish local control and experience, damage domestic stability, and undermine social and economic infrastructure. In so doing, SAPs may be sacrificing any possibility for achieving a more sustainable and broadly based development trajectory in the South which, recent experience has shown, must be based on social and political stability and the enhancement of human resources needed to create dynamic comparative advantages within globally interconnected markets.

Inattention to the Environment and Issues of Sustainability

Another critical element of sustainable development that has been largely ignored by SAPs is environmental soundness. Many analysts contend that the fixation of SAPs on short-term growth has generated unacceptably high levels of environmental destruction in many Third World countries (see, e.g., Barkin 1990; Cheru 1992; Green 1991; Helleiner 1989; Munasinghe 1993; Riddell 1992). Deregulation, liberalization measures, and outward-oriented policies have accelerated the destruction of non-renewable resources and have created 'pollution havens' in the South for TNCs with production processes too toxic to locate in the North. Moreover, environmental concerns have normally been excluded from the standard accounting techniques (e.g., measurements of GDP growth) used to assess the performance of SAPs. Little or no attention has been given, for example, to the effects of environmental depreciation, which may appear in the form of added costs for the long-term maintenance of both physical and human capital. Irreversible environmental damage may be generating a substantial redistribution of such costs, both over time (i.e., intergenerational transfers) and over space (i.e., transfers from the North to the South).

SAPs and other mainstream development strategies have, until recently, largely ignored connections between environmental processes and issues of equity and social justice. However, as the long-term consequences of environmental degradation have become better understood, it is now evident that any viable strategy of sustainable development must incorporate a systematic analysis of the environmental impact of social and economic change, particularly for the poor majority of the world's population in the South. Within our increasingly interconnected world, environmental interdependence transcends political borders and divisions between classes and social groups. Environmental sustainability, as a social and political process, is about human beings as well as ecology in the narrower sense. Just as sustained wealth and economic growth are, in the end, incompatible with widespread poverty, no ecological system can be sustainable if the majority of its inhabitants are forced to exist in misery and extreme need.

While some neoliberal policy-makers (e.g., in the World Bank) have begun to pay more attention to the problem of environmental deterioration, they have yet to acknowledge that, without active intervention by the state and popular organizations, market forces are incapable of generating sustainable development. Given imbalances of power within and among societies, the market-led thrust of SAPs has presented opportunities for dominant economic and political groups to put their individual, short-term accumulation interests ahead of the collective, long-term interests of the popular majority in a sustainable social and physical environment. Deregulation, liberalization, and the dismantlement of participatory means of decision-making have rendered the popular majority virtually defenseless against environmental damage by TNCs and other large corporations. In many cases, large-scale resource extraction, industrialization, and agribusiness operations have caused irreversible damage to the environments inhabited by the poor, ethnic minorities, and other disadvantaged groups (see, e.g., Amin 1993; Barkin 1990; Batie 1989; Broad and Cavanaugh 1989; Dietz and van der Straaten 1992). In other cases, the special development needs of such groups have been ignored, leading to widespread 'need-driven' environmental destruction, such as the overcollection of brush and overcutting of trees, overgrazing, and overly intensive crop cultivation (see, e.g., Barham et al. 1992; Cheru 1992; Green 1991; Shaw 1991). The human environment in many underdeveloped countries is characterized by overcrowding, poverty, and the desperate search for dwindling resources to meet basic needs. Problems of unequal access to resources, diminishing shared resources, and environmental degradation are often important root causes of the poverty complex and are increasingly linked to violent conflict. Development strategies that neglect such problems stand little chance of being sustainable, in either social or ecological terms.

The Need to Transcend Development Orthodoxies

Pervasive environmental degradation and other urgent new development problems require that academic analysts and policy-makers reject the failed explanations and prescriptions of past development frameworks in favor of new, more appropriate alternatives. However, the current rush to the market by neoliberals represents neither a careful reflection on past strategies nor a thoughtful search for new innovative solutions. Instead, as Brett (1987: 35) notes: 'Neoliberals are returning to the [neoclassical economic] recipes of the past not because they have been tried and succeeded, but because other nostrums [i.e., the state *dirigiste* model] have been tried and failed and no-one can think of any alternative.' Neoliberalism represents the latest example of what Emmerij (1987: 16) terms 'conservative modernization' – 'problems are identified, policies are introduced and they fail, but the fundamental changes needed for a solution are avoided in favor of trying the failed policies once again.'

It must be emphasized that neoclassical doctrines are not scientific truths, despite neoliberal attempts to portray them as such. In fact, ideological concerns have played a dominant role in the resurgence of neoclassical doctrines under the guise of neoliberalism. For Apter (1987: 295), neoliberalism offers an example of a modern ideological doctrine which consists of 'various mixtures of myth and theory, which, over time, have a tendency to be transformed into each other. Myth becomes theory; theory myth.[4] According to Hirschman (1987: 34), the ideologically inspired rise of neoliberalism has produced a 'strange switch' in development theory: 'North Americans, so proud not long ago of their pragmatism, have taken an ideological turn, while [people in the South] have become skeptical of their former sets of certainties and "solutions" and are naturally exasperated by the neophytes from the North who pretend to teach them yet another set.'

Moving beyond outmoded development models that have outlived their historical usefulness necessitates transcending a series of false dichotomies that have traditionally polarized postwar development studies. Examples of

---

4   In addition to neoliberalism, Apter also regards orthodox Marxism as an ideologically driven or 'mytho-logical' theoretical system. He notes: 'Theory provides a logic for the resolving of certain political problems and their projective transcendence. Myth does the same by means of "overcomings" that defy ordinary logic . . . Theory is embedded in the representation of the state, projective, logical and teleological. It defines a negative pole and provides a method and an instruction for transcending it. Liberalism and Marxism in various versions and mixtures are examples of such theoretical systems. Each offers a complete corpus, a language, and a method of closure, as well as an interpretative frame for the analysis of events. Both have the capacity to produce myths, expecially in the context of their surrogate states, just as such myths of the state help create the space for them as theories. This is why . . . , in the context of the state, mytho-logics constitutes both an interpretative field and a system of obligation' (pp. 295, 302).

these dichotomies include state planning versus the market, centralization and professionalism versus decentralization and mass participation, large-scale versus small-scale projects, the latest technologies versus intermediate or appropriate technologies, industry versus agriculture, investment and growth versus consumption and basic needs, import substitution versus export promotion, and protectionism and inward-oriented development versus free trade and outward-oriented development. In order to transcend these dichotomies, a new approach is needed which avoids framing development issues and questions as either/or choices according to preconceived theories and models. Instead, the case for or against a particular strategy should largely depend on the historical and geographical conditions, the sociocultural and political institutions, and the specific needs and interests of individual countries.

This new approach to development will necessarily stress pragmatism, flexibility, and the contextuality of development. It will involve freeing up our minds and searching for innovative solutions, because the stale, ideologically driven debates to which we have become accustomed have lost their relevance. No development orthodoxy, whether that of market-led neoliberalism or state-centered Keynesianism, can provide blanket solutions to the problems of all countries at all times. Rather, strategies must address the contextuality of development, which is the product of specific historical and geographical conditions. The failure to understand the special opportunities and constraints presented by such conditions renders neoliberalism and other universalistic strategies irrelevant to the real needs and problems of Third World countries.

Part Two

Alternative Theories and Practices

6

Refocusing on Needs

While neoliberalism and other mainstream models have had the greatest influence on postwar development, a range of alternative frameworks has also emerged in recent years. Given the shortcomings of the mainstream models, these alternative strategies are receiving increased attention. Analysis of the key elements of alternative approaches can provide new insights into the theory and practice of development which may help overcome the major failings of the mainstream frameworks.

At the same time, however, the alternative development tradition is not without its own contradictions, which must be resolved if it is to play a larger role in development processes. This chapter analyzes these issues by examining the refocusing of alternative approaches away from simple economic growth toward broader considerations of equitable development and meeting human needs.

Redistribution, Basic Needs, and the Origins of Alternative Development

Since the early postwar period, the central focus of mainstream development strategies has been economic growth and the top-down diffusion of development impulses. Growth is treated mainly as a function of investment – a simple formula links appropriate levels of investment with the incremental capital–output ratio and desired growth rates. The process of economic growth is characteristically thought to follow a series of 'stages' which would ultimately spread benefits to all, thereby alleviating poverty and inequality. The diffusion of technology and the other attributes of modernization would allow the benefits of development to trickle down to the neediest sectors of society. Development is viewed as a top-down process in which important decision-making is controlled by major international institutions in cooperation with local Third World elites. Typically, international and national 'experts' have conceived and designed development projects from the outside. The people to whom these projects are supposedly directed exist mainly in the abstract as socioeconomic indicators. Popular participation is normally restricted to some hastily organized meetings in which outside experts 'brief' local people about the objectives and activities of the projects.

Dissatisfaction with Mainstream Development Strategies

By the end of the 1960s, many analysts began to notice that economic growth was not necessarily correlated with other development objectives, such as rapid employment creation, the reduction of poverty and inequalities, and the provision of basic needs. Even in some countries (e.g., Brazil, Iran, Kenya, Mexico, Nicaragua, Pakistan, South Africa) in which rapid economic growth had been attained, severe 'maldevelopment' problems were appearing. Growth was not eradicating poverty or providing jobs at the speed anticipated and, in many cases, income inequalities were increasing. By 1970, some 944 million people, or 52 percent of the total population of the South, were still living in absolute poverty (United Nations 1989: table 24, p. 39),[1] despite the development efforts of the previous decades. Moreover, evidence was accumulating of growing labor underemployment, especially in agriculture (e.g., Turnham 1971), and rising inequalities in income distribution (Adelman and Morris 1973; Fishlow 1972; Griffin 1969; Griffin and Khan 1972). Indeed, by the early 1970s, it had become 'a commonplace to argue that throughout much of the Third World growth was accompanied by increased inequality' (Griffin 1989: 165).

The experience of the 1950s and 1960s suggested that, while growth was important, it was by no means a sufficient condition to induce broadly based development. In fact, growth could be impoverishing for a significant section of the population if it was paid for by a steady deterioration in the distribution of income and assets. In many countries, growth had been accompanied by declining standards of living and decreased access to productive resources for large numbers of people, including landless farmworkers, peasant cultivators, and many informal-sector workers. As

---

[1]   These figures are for the whole of the South, excluding China. The absolute poverty line is defined as the income level below which a nutritionally adequate diet and essential non-food items (i.e., clothing, shelter) are not affordable. By 1985, the United Nations reported that the poverty rate in the South had declined somewhat to 44 percent, while the absolute number of impoverished people had increased to 1,156 million. Between 1970 and 1985, the figures show a declining rate of poverty in the South's major regions, except Africa. However, in terms of absolute numbers, poverty increased substantially in all regions of the South (ibid.).

Chenery et al. (1974: xiii) noted, 'It is now clear that more than a decade of rapid growth in underdeveloped countries has been of little or no benefit to perhaps a third of their population.' In fact, ample evidence was available to contradict the notion that top-down development and trickle-down strategies, whether based on industrial or agricultural growth, would alleviate widespread impoverishment. Economic growth had simply failed to filter down – a fact that led many analysts to conclude that the nature, rather than the pace, of growth was the crucial factor for development. Mahbab ul Haq (1976: 24–5), a Pakistani economist and World Bank official, remarked:

> In country after country, economic growth is being accompanied by rising disparities . . . the masses are complaining that development has not touched their ordinary lives. Very often, economic growth has meant very little social justice. It has been accompanied by rising unemployment, worsening social services, and increasing absolute and relative poverty.

As the United Nations First Development Decade (1961–71) gave way to the second, a growing number of theorists and practitioners of development concluded that the focus of development on macroeconomic growth had been misplaced. Rather, they argued, the focus should be on the 'animate' instead of the 'inanimate' – on human resources, as measured by quality-of-life considerations, rather than on material resources, as measured by GNP figures (Black 1991: 20–1). Successful development should be measured not in abstract, aggregate growth indices, but according to other people-oriented criteria, such as the universal provision of basic needs, the promotion of social equity, the enhancement of human productive and creative capabilities, and the capacity of communities to set and meet their own development goals. New development approaches should be oriented toward the satisfaction of basic human needs and desires, particularly at the local community level; and development projects should 'build development around people rather than people around development' (ul Haq 1976: 27–8).

Redistributive and Basic-Needs Strategies

As dissatisfaction with the mainstream models became widespread within the development community during the early 1970s, many international and bilateral aid agencies began searching for alternative, more people-oriented approaches. Efforts were made to uncouple the direct, exclusive relationship between growth and development to make room in development programs for other considerations, such as distributional equity and poverty alleviation, basic-needs provisions, and the adoption of appropriate technologies. Programs promoting decentralized patterns of development were given prominence, and emphasis was shifted to projects which directly targeted the poor, especially in rural areas. Many organizations adopted a rather broad, eclectic, and loosely defined 'neopopulist' ideology (see Kitching 1982), in contrast to the well-structured, but narrow, theoretical base offered by neoclassical economics. The effect was to redefine the aims of development toward fostering fairer distributions of income and resources, encouraging local participation, and promoting small-scale projects employing socially and environmentally appropriate technologies. It was thought that by targeting the poor and adapting programs to suit local conditions and needs, growth and development would proceed in a dispersed manner 'from below' (Stöhr and Taylor 1981), rather than following the conventional top-down, concentrated pattern. Through encouraging 'self-help' and participatory decision-making, the latent energies and creativity of the poor could be directed toward rapid and more appropriate forms of development.

Some isolated, halting efforts had been made to initiate alternative development projects in a few Third World countries. Bernstein and Campbell (1985), for example, report that a 'populist movement,' which stressed local farming practices and indigenous forms of knowledge, gained some support within the British Colonial Office during the 1930s. Similarly, Moser (1989) traces the origins of the concept of 'community development' to the British, who used it to develop basic education and social welfare in some colonial areas. In the 1950s and 1960s, some small-farm development projects, food-for-work, and labor-intensive public works programs were begun in a few countries (Peek 1988). However, the impetus for alternative development projects really began in the early 1970s when many large international organizations (such as the World Bank, ILO, UNEP, UNICEF) and bilateral aid agencies (e.g., USAID, CIDA) became involved.

In his presidential address to the 1973 World Bank annual meeting, Robert McNamara expressed the view that the mainstream development strategies of the 1950s and 1960s had made an unacceptably small impact on Third World poverty and inequalities. Largely through the prodding of its Development Research Center under Hollis Chernery, the World Bank began to adopt a new development approach, termed 'redistribution with growth' (Chenery et al. 1974). Redistribution and growth were treated as complementary rather than contradictory elements of development; sustainable growth would require redistributive policies and targeted programs for the poor during the initial stages of development, instead of simply relying on trickle-down mechanisms to eventually spread the benefits of growth. Priority was given to employment creation and basic-needs provisions rather than economic growth per se. In order to maximize job creation, emphasis was placed on small/medium agriculture and the informal sector. Similarly, basic-needs provisions were targeted for poorer and severely underserviced areas, such as outlying regions dominated by peasant cultivators.

Although it contained these new emphases, redistribution with growth represented a modification rather than a clear break with previous mainstream development strategies. It retained much of the optimism of the earlier models in its promotion of the benefits of market-led growth. Redistribution of income toward the poor essentially remained tied to rapid economic growth; the traditional recipe of balanced growth was simply extended to cover social as well as economic development (Hettne 1990: 57). Although there was considerable evidence that the poor formed the majority in most countries, the Bank tended to group the poor into administratively convenient 'pockets of poverty' (Friedmann 1992: 58), which could be treated by targeted programs rather than more fundamental changes within macro-policies. Moreover, despite the rhetorical stress placed on community participation, a top-down, social-engineering approach continued to characterize the Bank's development projects. The political dimension of development was given some recognition, especially the stabilizing impact that increased consumption levels among the poor would have on long-term development. However, concrete measures to empower popular organizations to take a more active role in political and economic decision-making were largely avoided.

Parallel to the World Bank's strategy of redistribution with growth, the International Labor Organization (ILO) adopted a basic-needs approach during the 1970s. The idea of basic needs may have originated in a report by a group of Latin American theorists (Herrera et al. 1976), prepared for the Bariloche Foundation, Canada (see Preston 1986: 109). However, the basic-needs concept was formally placed on the international development agenda at a 1976 ILO World Employment Conference in which the participants adopted a 'Declaration of Principles and Program of Action for a Basic Needs Strategy of Development' (in ILO 1976: 189–214). As it was elaborated by the ILO, the basic-needs concept put equal emphasis on growth and redistribution. Although it focused on the needs of the poor, it was not opposed to rapid growth in the modern sector; rather, it sought to strike a more balanced approach to development (Emmerij 1987). Basic needs were defined to include the following elements: minimum requirements of private consumption (e.g., food, shelter, clothing); essential services of collective consumption (e.g., electricity, water, sanitation, health care, education, public transport); participation of people in decisions affecting their lives; satisfaction of basic needs within a broader framework of basic human rights; and employment as both a means and an end (see Ghai 1977).

The ILO's version of basic needs concentrated on harnessing local resources and providing the poor with the means to fulfill their development potential. It attempted to define basic needs in operational terms and it established performance criteria and targets for countries. It also acknowledged the need for structural (internal) change in the development patterns of Third

World societies to meet the basic needs of the poor (Streeten 1981). Even though the ILO regarded its version as a minimum definition of basic needs, it nevertheless created much controversy at the conference (Friedmann 1992: 60). Some employers' delegates and representatives from core capitalist countries believed that the ILO was over-emphasizing the need for structural change and redistributive measures; instead, they called for attention to be placed on rapid economic growth as the most important remedy for problems of unemployment and poverty. Other delegates regarded basic needs as a key unifying theme around which a new alternative approach to development could be constructed which would be radically different from previous models that stressed top-down growth and capitalist modernization. Such controversy has continued to swirl around the basic-needs approach.

In addition to the World Bank and ILO, a number of United Nations organizations figured prominently in the creation of an alternative development agenda in the 1970s, including the UN Environment Program (UNEP), UN Conference on Trade and Development (UNCTAD), UN Development Program (UNDP), UN Research Institute for Social Development (UNRISD), UN Children's Emergency Fund (UNICEF), World Health Organization (WHO), and Food and Agriculture Organization (FAO). During the 1970s, these organizations held a number of landmark meetings dedicated to reformulating the development agenda. Particularly important was a symposium on 'Patterns of Resource Use, Environment and Development Strategies' convened by the UNEP and UNCTAD in Cocoyoc, Mexico in 1974. The Cocoyoc meetings brought together a wide range of development experts from all parts of the world, who represented two major strands of the nascent alternative development movement: those who argued that priority attention should be given to satisfying people's basic needs rather than simple growth maximization, and those who were concerned with the 'outer limits' of the world's ecological capabilities to sustain growth (Friedmann 1992: 2). At the end of the meetings, the participants issued a manifesto, the 'Cocoyoc Declaration,' which stated:

> Thirty years have passed since the signing of the United Nations Charter launched the effort to establish a new international order. Today that order has reached a critical turning point. Its hopes of creating a better life for the whole human family have been largely frustrated. It has proved impossible to meet the 'inner limits' of satisfying fundamental human needs. On the contrary, more people are hungry, sick, shelterless and illiterate today than when the United Nations was first set up.
>
> At the same time, new and unforeseen concerns have begun to darken the international prospects. Environmental degradation and the rising pressure on resources raise the question whether the 'outer limits' of the planet's physical integrity may not be at risk. (Cocoyoc Declaration 1974: 170)

The Cocoyoc Declaration went on to note that any process of growth that did not lead to the fulfillment of basic human needs was a distortion of development. Moreover, development should address not just basic human needs, but other considerations, such as freedom of expression and self-realization in work. 'Overconsumptive' types of development that violate the 'inner limits' of humans and the 'outer limits' of nature should be avoided. More people-centered and environmentally sustainable forms of development would require increased self-reliance. The South should work to establish a new international economic order (NIEO) in which mutual benefits of trade and cooperation would be derived from more symmetric global relations. However, until a more just international economic order could be created, Third World countries might want to pursue more self-reliance through 'a temporary detachment from the present system.' In the end, it would be futile 'to develop self-reliance through full participation in a system that perpetuates economic dependence' (Cocoyoc Declaration 1974: 174).

For Hettne (1990: 152), the Cocoyoc Declaration marked 'the birth of an "alternative" trend in development theory.' Alternative development approaches received additional attention in the mid-1970s, especially from two other sources. The first was the Swedish Dag Hammarskjöld Foundation, which published a document in 1975 entitled *What Now: Another Development*.[2] Mainstream development models were criticized for neglecting issues of mass poverty and sustainability. A broadly humanist approach to development was outlined, which advocated 'development geared to the satisfaction of needs, beginning with the basic needs of the poor who constitute the world's majority; at the same time, development to ensure the humanization of man by the satisfaction of his needs for expression, creativity, conviviality, and for deciding his own destiny' (Dag Hammarskjöld Foundation 1975: 7). The second source was the International Foundation for Development Alternatives (IFDA), which was established in 1976 in Nyon, Switzerland. The principal purpose of the IFDA was to promote an alternative, bottom-up approach to development, termed the 'Third System Project,' which was dedicated to exploring new methods of raising consciousness and increasing participation by grassroots movements in development decision-making. A distinction was drawn

[2]   This document was prepared by the Dag Hammarskjöld Foundation for the Seventh Special Session of the UN General Assembly and was further elaborated in the journal *Development Dialogue*. Another journal, *Alternatives*, published by the Institute for World Order (New York), also became a forum for alternative development approaches. While most of this work took place in the North, some research centers for alternative development also began to appear in the South, such as the Center for Developing Societies (New Delhi, India).

between the First System of political power (dominated by the state), the Second System of economic power (dominated by transnational capital), and the Third System of people's power, based on voluntary organization, consciousness raising, and local action. The Third System was composed of people acting individually and collectively through voluntary institutions and social movements. It was regarded as the principal source of new values and visions and thus held the most potential for engendering meaningful change. According to the IFDA (1980: 69–70):

> The 'third system' is that part of the people which is reaching a critical consciousness of their role. It is not a party or an organization; it constitutes a movement of those free associations, citizens and militants, who perceive that the essence of history is the endless struggle by which people try to master their own destiny – the process of humanization of man. The third system includes groupings actively serving people's aims and interests, as well as political and cultural militants who, while not belonging directly to the grassroots, endeavour to express people's views and to join their struggle. This movement tries to assert itself in all spaces of decision making by putting pressure on the state and economic power and by organizing to expand the autonomous power of people.

The Evolution of the Basic-Needs Concept

Perhaps the central focus of alternative development approaches in the 1970s was on basic needs. The basic-needs debate drew a fundamental distinction between economic growth and the satisfaction of basic needs. As a result, more direct, targeted methods were proposed for poverty alleviation, rather than the indirect approach of reliance on growth and trickle-down mechanisms to eventually benefit the poor. The satisfaction of basic needs figured prominently in the alternative approaches adopted by several international organizations, such as the World Bank, ILO, and UNEP, as well as many bilateral aid agencies, nongovernmental organizations, and independent development institutes. However, these approaches also differed substantially in their conceptualization of basic needs and the methods proposed for fulfilling basic-needs requirements. As the basic-needs concept developed, two schools of thought evolved which distinguished between a universal and objective interpretation of needs, on the one hand, and a more subjective and historically contingent interpretation based in the context of particular social systems, on the other (Lederer 1980).

The first approach was mainly associated with the World Bank, ILO, and some bilateral aid agencies of core capitalist countries. It concentrated on needs which in all societies are necessary for physical reproduction. It took a positive, quantifiable view of basic needs. According to the

ILO (1976: 32), for example, basic needs were defined as 'the minimum requirements of a family for private consumption,' especially food, clothing, and shelter, as well as 'essential services provided by and for the community at large, such as safe drinking water, sanitation, public transport, and health and education facilities.' The ILO and other international organizations struggled mightily to define hierarchies of basic needs so that development programs might be targeted to maximize benefits for the poor. Streeten and Burki (1978), for example, distinguished between 'core needs' (food, water, clothing, and shelter) and all other needs. Friedmann (1992: 63) notes that this definition of core needs was probably influenced by the focus of the international agencies on immediate, tangible needs (e.g., food, housing) and the desire to avoid more troublesome philosophical considerations (e.g., in which food requirements might have to be weighed in comparison with liberty or other concerns). For Griffin (1989: 172), this approach was rather modest in its targets. It did not hold out any immediate hope of providing people with fuller, more meaningful lives, but merely indicated what would be required to attain certain minimum basic-needs objectives over the next generation.

The second approach to basic needs was largely associated with the work of some UN agencies, nongovernmental organizations, and independent development institutes. It tended to concentrate on the more subjective concerns of what makes life worth living in different cultures and societies, and was much more normative and qualitative than the first approach. Some analysts may have thought of basic needs as consisting of a 'shopping list' of essential private-consumption goods and public services. But the second approach to basic needs clearly distanced itself from this interpretation. Priority was given not only to the minimum physical requirements for human subsistence, but also to a range of other less tangible needs, such as protection, affection, understanding, participation, leisure, creation, identity, and freedom (see Max-Neef 1986). Rather than being universal, finite, and quantifiable, these latter needs are infinite, qualitative, and subject to change across cultures and societies. Instead of concentrating merely on 'things,' they include the broader needs and desires which make life meaningful to people within particular historically constituted contexts.

From this perspective, basic needs are transformed into political claims for entitlements. The poor are no longer regarded simply as victims or passive recipients of outside aid, but as people who, despite enormous constraints, are actively engaged in the struggle to define their own lives and means of livelihood. For poverty programs to be effective, the poor must take an active role in the provisioning of their own needs, rather than simply relying on the state or outside organizations to solve their problems. However, in order to participate in this process, the poor must often receive some initial assistance to provide them with the means to help themselves. The basic-needs approach does not necessarily require more aid or higher levels of public spending; indeed, government expenditures already account for a relatively high proportion of GNP in most Third World countries (Griffin 1989: 173). Instead, the approach demands greater popular participation. On the one hand, this requires a change in the organization of institutions and agencies involved in development programs, and on the other, the attainment by the poor of the means to become more organized and self-reliant. At both of these levels, basic needs tend to become highly politicized, inextricably tied to issues of representation, participation, and empowerment, over which different classes and social groups contend within the political arena. As Conger Lind (1992: 144) notes in an article on poor women's movements and basic needs:

> The politicization of 'basic needs' demonstrates the way in which such 'needs' are actually much more than just the desire for bread and water. As poor women base their politics on their reproductive roles, they challenge the meaning of ascribed gender roles as well as the implications these roles have in the reproduction of society. They are not only struggling for access to resources, they are also challenging dominant representations of gender and incorporating this into their politics. State policy, then, cannot easily fulfill poor women's 'needs' simply by providing them with economic resources.

The Focus on Rural Development

Beginning in the early 1970s, many of the alternative development programs of various aid agencies and international organizations focused on rural development as the key to reducing levels of underemployment, increasing access to public goods and services, and lowering poverty and income inequalities in most developing countries. Rural development emerged as a major issue as it became clear that previous development approaches had generally failed to improve the well-being of the rural population. Aside from some highly urbanized parts of Asia and Latin America, the majority of the poor in most Third World countries continue to reside in the countryside. This is particularly true for many of the poorest countries, in which some 80 percent or more of the population may be rural (Grindle 1988). As a result, over vast areas of the South, particularly in sub-Saharan Africa, the most severe cases of absolute poverty remain largely a rural phenomenon (von Braun and Paulina 1990). Moreover, access to basic public goods and services (e.g., health care, and education) is usually most inadequate in outlying rural areas, thereby compounding problems commonly associated with poverty, such as high morbidity and child mortality and low life expectancy.

The focus on rural development inevitably raised many important policy issues concerning the relative neglect of the countryside as resources were concentrated in urban areas, the bias of development efforts toward industry at the expense of agriculture, and the manipulation of internal terms of trade against agriculture, particularly domestic food production. These issues were grouped together within the broad concept of 'urban bias,' which became an early theme of organizations such as the ILO and World Bank. It was asserted that an urban bias in the development programs of many countries had drawn away resource allocations from rural locations to meet urban and industrially based priorities, thereby adversely affecting both economic efficiency and distributional equity (e.g., Bates 1981; Lipton 1977). Accompanying their overall urban bias, development programs had particularly neglected the needs of small/medium agricultural producers, who continued to supply the bulk of domestic foodstuffs in most countries. Typically, development models had viewed rural development mainly from an urban perspective; in almost all countries, an urban-based network of public–private institutions had subsumed the dynamic of overall development to the rhythms and possibilities of growth in the urban sector. However, urban bias was regarded as particularly inappropriate and wasteful for many poorer countries in which the majority of people and economic activities continue to be concentrated in rural areas:

> Urban bias does not seem particularly helpful in the context of underdeveloped countries where upwards of 80 percent of the population live in rural areas, where agriculture constitutes the fundamental source of national wealth, where urban places have traditionally expropriated the rural surplus, and where the economy remains externally oriented. If development is to be for the people, there is a strong prima facie case that it should largely consist of development in and for rural areas: it is contended that urban biased strategies are incapable of achieving these aims. (Dattoo and Gray 1979: 261)

Not only had overall development efforts been marked by a pervasive urban bias, but agricultural programs had often reinforced structural inequalities in many rural sectors by promoting a 'bimodal' agricultural development strategy (Johnston and Kilby 1975; Thorbecke 1979). As a result, a bifurcated or dual pattern of rural development had been strengthened in which a few large-scale, capital-intensive farms, usually concentrated in the best agricultural zones and oriented toward agroexport production, were juxtaposed to a mass of small-scale peasants employing rudimentary techniques, often under precarious environmental conditions, to produce basic foodstuffs for their own subsistence and the local market. Although bimodal agricultural strategies became common in many parts of the South, they were particularly prevalent in Latin America, where they tended to reinforce and extend the traditional *latifundio–minifundio* pattern of rural development.[3] In most areas, bimodal strategies have been associated with policies favoring cash crop and agroexport production using relatively capital-intensive technologies, rather than basic foods production employing more labor-intensive techniques (Johnston and Clark 1982).

In addition to the agroexport model, the Green Revolution also contributed to the deepening of bimodal patterns of agricultural development (see, e.g., Conway and Barbier 1988; Johnston and Clark 1982; Shiva 1991). The Green Revolution focused on increasing food production, especially grains, rather than agroexports. It was envisioned that an increased supply of grains would lower the relative cost of food and, in turn, help to drive down per unit labor costs for both agriculture and industry. Lower unit costs would raise profitability levels, permitting higher rates of savings and investment and a faster rate of overall growth. Benefits would accrue not only to agriculture, but also to industries, especially those located in rural areas. Increased agricultural production would provide raw materials (e.g., for food-processing industries); stimulate the demand for agricultural inputs and intermediate goods (such as fertilizer, pesticides, irrigation machinery); and create an expanded rural market for simple consumption goods (e.g., bicycles, radios). Technological innovation was regarded as the key to accelerating agricultural growth, and it was from this focus that the Green Revolution strategy derived its name. Improved hybrid seed varieties, greater use of fertilizers and other chemical inputs, investment in irrigation systems, expanded agronomically based research and agricultural extension services, and increased rural credit were emphasized.

Thus, the Green Revolution adopted a very technocratic orientation. Relatively little emphasis was placed on the need for land redistribution and tenure reforms, institutional change, or the direct participation and mobilization of the rural population in development programs. While the Green Revolution enjoyed broad success in raising grain production, many analysts argue that this was accomplished at high social and environmental costs (e.g., Bartra and Otero 1987; Conway and Barbier 1988; Hazell and Ramasamy 1991; Shiva 1991; Zarkovic 1988). Many of the technological innovations proved to be labor-displacing and created a greater dependence on relatively costly external inputs, such as hybrid seeds, chemical fertilizers, pesticides, and irrigated water. Small/medium producers were often squeezed out of traditional markets by their inability to adopt the new

---

3   The origins of the *latifundio–minifundio* pattern of rural development in most Latin American countries can be traced back to the early colonial period. However, land concentrations accompanying the postwar expansion of agroexports significantly reinforced traditional dualistic rural structures. *Latifundios* are large-scale estates, while *minifundios* are small farms, often insufficient to meet the subsistence needs of a peasant family.

technologies (through lack of capital and rural credit) and by falling producer prices (resulting from the increased output of larger producers). Green
Revolution technologies have also often been linked to the penetration of
transnational corporations into rural areas (especially through increasing
dependence on hybrid seed types and chemical inputs) and to growing
ecological problems (e.g., soil erosion, chemical contamination of soils
and groundwater, depletion of water resources, greater crop vulnerability,
decreased genetic diversity). Moreover, because they have generally not been
accompanied by redistributive measures, these technologies have commonly
widened rural inequalities in both socioeconomic terms (favoring richer,
larger-scale farmers) and regional terms (favoring concentrations of large
producers in areas with irrigated water). Both of these tendencies have
accentuated the dominant bimodal pattern of rural development in most
countries.

In contrast to a bimodal agricultural strategy, proponents of redistributive
development approaches commonly advocate 'unimodal' strategies (e.g.,
Adelman 1975; Johnston and Kilby 1975; Johnston and Clark 1982).
A unimodal strategy seeks to modernize and raise production levels of
the entire agricultural sector, which is comprised of relatively uniform,
small/medium farm units. Perhaps the best examples of successful unimodal
strategies are found in South Korea and Taiwan, where extensive agrarian
reforms were followed by development programs designed to improve basic
social and economic infrastructure, raise levels of rural productivity and
standards of living, and integrate small/medium agricultural producers
into the dynamic of national economic growth. Unimodal strategies have
also shown success in a few other Third World areas – notably in the
Meseta Central of Costa Rica, where historically a relatively egalitarian pattern of rural development evolved in stark contrast to the polarized *latifundio–minifundio* pattern prevailing in most of the rest of Latin
America.[4]

Unimodal strategies not only stress an egalitarian structure of land tenure,
but also favor increased access to rural credit and other means of production, investments in basic social and economic infrastructure (education,
health care, transportation and communications networks, etc.), and agricultural extension and other programs designed to promote technological
change appropriate to smaller-scale, labor-intensive operations. Unimodal
strategies seek to create an internally articulated pattern of development
in which rising levels of production and consumption by small/medium
farmers are linked to other local economic sectors in a mutually reinforcing
manner through backward/forward linkages (e.g., supply of raw materials

[4] Although recent trends of increasing foreign ownership, land concentration, and peasant
displacements in the Meseta Central (the central plateau region of Costa Rica) have
begun to erode the traditional unimodal pattern of rural development in this area.

for agro-industries, demand for industrially produced agricultural inputs and consumption goods). However, despite theoretical arguments and empirical evidence supporting unimodal strategies, they continue to face considerable obstacles and have yet to be adopted in more than a handful of countries. Johnston and Clark (1982) list a number of these obstacles, virtually all of which have a common source – in political opposition rooted in the distribution of power relations in developing countries. In most cases, the adoption of a unimodal strategy would significantly weaken the economic and political power base of rural and urban elite groups, which usually exert a dominant influence on policy-making. This begs the question of how unimodal strategies can be adopted in most Third World countries simply through rural development initiatives and without complementary structural changes within overarching political and economic relations.

The origins of redistributive rural development initiatives in most areas of the South can be traced back to the agrarian reforms of the 1960s and 1970s. Agrarian reform has followed a number of different paths, but may be loosely defined as any state-sanctioned change in land tenure and associated institutions which ostensibly is designed to benefit small cultivators, landless laborers, or other groups of rural poor (e.g., Jones et al. 1982; Thiesenhusen 1989). Lipton (1993) identifies a number of different types of agrarian reforms, including those which focus on landownership ceilings and redistribution, the purchase of plots by tenants and share-croppers, restrictions on rental arrangements and other forms of tenancy, land-titling and freehold laws, the 'patrialization' of land from colonizers to the rural poor, and the creation or privatization of state and collective farms. Agrarian reforms have sometimes also been accompanied by resettlement strategies designed to relocate peasants into frontier areas (e.g., Indonesia's transmigration strategy, Amazonian colonization schemes in Brazil, Peru, and Ecuador) or to concentrate patterns of dispersed settlement by grouping peasants together into villages (e.g., the villagization movement in Tanzania and Mozambique). Perhaps the largest of these resettlement strategies took place in Tanzania where some 13 million peasants were transferred from their traditional plots into 7,000 'village communities' by the *ujamaa* movement during the 1970s (McCall and Skutsch 1983).

Despite much rhetoric to the contrary, agrarian reforms and resettlement schemes have generally been administered in a paternalistic, top-down manner which has systematically excluded local peasant organizations from meaningful participation in decision-making. Moreover, these rural development initiatives have attempted to affect a redistribution of rural income and resources, while leaving the thrust of national development strategies largely unchanged. Agrarian reform efforts have characteristically reflected a localized perspective in which development solutions are sought within the rural sector itself, but little recognition is afforded to the possible need for complementary systemwide structural change. In a broad analysis of Latin American agrarian reforms, de Janvry (1981: 203) concludes that the main economic thrust of the reforms was not actually aimed at the peasant sector at all, but was directed toward the capitalist transformation of the large-scale *latifundio* sector: by putting idle land into production, reorganizing precapitalist estates on a capitalist basis, and stimulating investments in capital-intensive technologies. Similarly, he finds that the principal political purpose of the reforms was to stabilize social relations under the peripheral capitalist model by removing state control from the landholding oligarchy, deepening rural proletarianization through the elimination of semifeudal labor exploitation, creating a stable class of middle peasants, and reinforcing an emergent class of capital-intensive bourgeois producers. The reforms were thus aimed not only at preempting a more radical solution to the 'agrarian question' in the wake of the Cuban Revolution, but, more fundamentally, at the consolidation of both capitalist agrarian relations and state control over the direction of rural development (ibid.).

By the early 1970s, interest in agrarian reform as a method for creating a more unimodal agricultural structure in areas such as Latin America was already declining. A two-pronged approach to rural development was adopted in many countries which institutionalized conventional 'dualist' notions of relatively autonomous modern and traditional sectors in agriculture (de Janvry 1981; Grindle 1986). On the one hand, integrated rural development (IRD) programs were established to assist small/medium farmers in producing a marketable surplus beyond their own subsistence needs. On the other hand, separate programs were created to provide incentives for larger producers to increase investments and expand output, especially of agroexports. This new rural development approach generally avoided the contentious issue of land redistribution; it also widened the scope for state intervention in various elements of rural production, circulation, and consumption (e.g., health care and education, agricultural extension services, rural credit, infrastructure provision, marketing). Politically, the extensive scope of IRD allowed the state to expand its influence among a broad range of rural client groups, while a de-emphasis on land redistribution avoided confrontations with powerful landowners and their allies.

Although the theoretical roots of IRD remained within traditional dualist notions of Third World agriculture, the approach also developed emphases which distinguished it from the conventional neoclassical agricultural models. Its programs were more specifically targeted toward problems of rural poverty and inequalities, and they expanded state intervention beyond the immediate realm of agricultural production per se (Harriss 1982; Lea and Chaudhri 1983). Generally, IRD may be defined as 'a process of combining multiple development services into a coherent delivery system with the aim of improving the well-being of rural populations' (Honadle et al. 1980: 4). More specifically, Cohen (1980) attributes the following characteristics to IRD projects: a particular geographic area focus; design and implementation by outside groups, typically a national development agency assisted by an international donor; concentration on coordinating provision of public goods and services; and use of a multisectoral orientation, while emphasizing agricultural production.

Generally, IRD viewed agricultural growth as a necessary but, by itself, insufficient condition for rural development. Emphasis was placed on both production and the spheres of circulation and consumption. The qualitative dimension of rural development, involving 'quality of life' considerations, local capacity-building, and improving access of the rural poor to basic goods and services was also stressed. IRD was envisioned as a multisectoral, multifunctional development initiative. It asserted that rural poverty stems from a host of interrelated problems requiring a package of coordinated responses – from increased agricultural extension services and rural credit, to more efficient distribution and marketing channels, to improvements in basic social infrastructure, such as health care and education. Because of their multifaceted nature, IRD projects often required a rather complex bureaucracy to administer them as special development enclaves. Typically, projects were identified and formulated by state agencies and international donors with little initiative from the local rural population. Moreover, most projects were separately funded and administered outside of regular state agencies, often with considerable assistance from foreign donors. This almost always placed IRD projects under tremendous pressure to show immediate tangible results, thereby negating possibilities for many longer-term development initiatives. As was common with redistributive strategies in general, the IRD approach was theoretically eclectic: it showed a respect for neoclassical economics, but was interdisciplinary and even borrowed some concepts from neo-Marxist thought (Thiesenhusen 1987). However, despite its somewhat eclectic nature, the IRD approach also re-created, in slightly altered form, conventional dualistic notions of modern and traditional rural sectors operating in virtual isolation from one another, for which separate programs might be successfully developed. The neglect of any serious analysis of the impact of overall development strategies or of interrelationships among rural classes and social groups prevented most IRD projects from addressing many of the root causes of rural poverty and inequalities. The best of them became showcases for international donors and the regimes in power, but most IRD projects failed to produce the desired results. As Friedmann (1992: 94) notes, 'In quantitative terms, they constitute[d] little more than a gesture.'

Major Elements of the Alternative Tradition

While no development strategy explicitly aims for inegalitarian development, mainstream strategies have implicitly assumed that inequalities, in both socioeconomic and regional terms, are a necessary price for growth. A trade-off is assumed between distribution and growth, especially during the early stages of development, so that redistributive measures may raise short-term consumption levels of the poor, but at the cost of reduced investment levels and diminished prospects for long-term growth. By contrast, the alternative development strategies that emerged from the 1970s were based on the assumption that there need be no conflict or trade-off between redistributive measures and other policies designed to accelerate growth. In fact, it was generally asserted that 'redistribution before growth,' and not vice versa, made more sense for development strategies. The alternative strategies, therefore, placed a higher priority on the need for immediate redistributive measures; they favored a direct, targeted approach to alleviating poverty and reducing inequalities rather than waiting for the 'trickle-down' effects of growth to occur. Because of their focus on direct redistributive measures, the alternative strategies have sometimes been called 'the Third World equivalent of Western social democracy' (Griffin 1989: 241).

As the alternative strategies developed during the 1970s, three distinct approaches evolved concerning redistributive measures (Emmerij 1987; Griffin 1989). Initially, there were strategists who stressed employment-intensive development measures for the working poor. Next, there were strategists who linked growth with redistribution by advocating measures that would transfer to the poor an increasing part of the increment to total income that would arise from accelerated growth. Such measures might take the form either of consumption transfers or of a redirection in investment toward the poor. Finally, there were strategists who focused on the elaboration of basic-needs approaches. Some concentrated on core basic needs (e.g., food, clothing, shelter) universally required for human subsistence, while others incorporated more culturally specific considerations (e.g., quality of life, identity, freedom) into the basic-needs approach. Especially for the latter, greater stress was placed on helping the poor to acquire more economic and political power. Within most polarized Third World societies, this would require a redistribution in the ownership of productive assets, especially land. Since most of the South's poor continue to be rural, it was generally believed that redistributive strategies ought to be based on a unimodal pattern of rural development dominated by small/medium farmers. Various rural development initiatives were promoted toward that end, including basic agrarian reforms (i.e., directed toward land redistribution) and more eclectic integrated rural development programs.

Complementary to their emphasis on redistributive measures, alternative strategies also tended to concentrate on relatively small-scale projects set at the local or community level. Within urban areas, informal sector activities and community-based groups commonly became focuses for attention; in rural areas, development efforts were usually directed at local organizations composed of farmers or agricultural workers. Typically, targeted programs for the poor were managed at the local level, particularly those stressing basic-needs provisions and investments in human capital (e.g., education, nutrition, health-care programs). Generally, alternative strategies regarded the local scale as especially important to a broad, human-centered approach to development in which non-material needs and quality-of-life considerations figured prominently alongside concerns for material welfare. Local organizations and primary communities were thought to be critical to people's 'creative unfolding,' through which a host of non-material needs (e.g., self-identity and expression, liberty, participation) might be more effectively pursued. It was at the local level (in neighborhoods, schools, parishes, sports clubs, women's organizations, and other community-based groups) that personal and societal developments were thought to interact most powerfully – thereby offering opportunities for direct, targeted development programs to achieve a maximum impact on changing at least some people's lives.

In one form or another, virtually all of the international organizations and bilateral aid agencies involved in alternative development projects also placed greater attention on questions of local participation – at least in theory, if not always in practice. Emphasis was particularly given to fostering local institutions to enhance people's participation in the selection, design, and management of development projects at the community level. Many analysts criticized tendencies to impose inappropriate, top-down projects without much public input from the affected communities themselves. Participation could be furthered both by decentralizing state agencies and other development institutions to make them more accountable to local groups and by helping the poor to represent themselves more effectively by strengthening community organizations and other local pressure groups. Participation was viewed as an important end in itself, but it was also linked to a number of instrumental values (Griffin 1989: 174). First, participation in community-based organizations could help to identify local priorities so that development projects might better reflect grassroots needs and wishes. Second, participation in popular organizations and groups (e.g., neighborhood groups, cooperatives, land reform committees, irrigation societies, women's organizations) might assist in mobilizing local support for development programs and projects. Third, increased local participation might reduce the costs of many public services and development projects by shifting more responsibility to grassroots organizations (e.g., by using the voluntary labor of the beneficiaries of projects, by employing local people rather than highly paid outsiders).

In conjunction with increased participation, many alternative strategies also called for greater self-reliance. More broadly based, participatory development strategies would, it was thought, lay the necessary foundations for a more autonomous and sustainable pattern of growth. Self-reliance implies making more effective use of a society's own strengths and resources, both human and natural. The movement toward self-reliance can also take place at a variety of scales, from the local to the regional or even the national. Within highly dependent, polarized societies, the concept of self-reliance, if followed to its logical end, implies structural changes to allow for genuine cooperation at various levels based on the principle of symmetry. The transformation from a dependent to a self-reliant society involves not only broad, structural changes, but also more localized, individual adjustments. For this reason, participation was stressed so that individuals and groups within local communities could understand and react appropriately to the changes taking place.

Although alternative strategies tended to focus on the community level, some attempts were also made to extend the concept of self-reliance into the international realm, especially through development of the complementary NIEO (New International Economic Order) perspective (see, e.g., Cocoyoc Declaration 1974; Dag Hammarskjöld Foundation 1975). While the NIEO perspective emphasized the need for more symmetric global relations and the mutual benefits that could be derived from increased North–South cooperation, the concept of self-reliance stressed the need to reduce external dependence by making more effective use of one's own resources.

Within at least some alternative approaches, these two perspectives converged in the 1970s to support more selective participation by Third World countries in the international economic system. Instead of continuing dependence on exogenous universal models, it was thought that different conditions among Third World countries necessitated a variety of strategies which, in turn, required exploring the possibilities for new forms of participation, cooperation, and more endogenous, self-reliant development.

To recap, then, alternative development strategies, although they were generally eclectic and did not depend on any well-defined theoretical base, were characterized by a number of common elements, including the following:

1   a move toward direct, redistributive measures targeting the poor, instead of continued reliance on the eventual indirect trickle-down effects of growth;

2   a focus on local, small-scale projects, often linked with either rural

development initiatives or urban, community-based development programs;

3   an emphasis on basic-needs and human-resource development, especially through the provision of public goods and services;

4   a refocusing away from a narrow, growth-first definition of development toward a more broadly based, human-centered conception;

5   a concern for local or community participation in the design and implementation of development projects; and

6   a stress on increased self-reliance, which might extend to a variety of scales, to reduce outside dependency and create the conditions for more cooperative, socially and environmentally sustainable development.

Criticisms of Alternative Strategies

As will be seen in later chapters, alternative development strategies have been substantially modified in recent years to include new emphases, concepts, and methods. Much of this modification resulted from a number of criticisms leveled at the alternative strategies of the 1970s. Such criticism essentially called into question the usefulness of such strategies and whether they really represented a viable alternative to mainstream development approaches. Further criticism was directed at the methods used to design and implement alternative development programs and projects. Many analysts questioned whether these methods were fundamentally different from those of the mainstream tradition. In many cases, it was concluded that a large gap existed between the theory or rhetoric of alternative development and the actual practice.

This gap between theory and practice was particularly noticeable in the area of participation. Many alternative programs and projects, especially in the sphere of rural development, were heavily criticized for paying only lip service to local participation (see, e.g., Brinkerhoff 1988; Daniel et al. 1985; Eckstein 1988; Esman and Uphoff 1984; Hyden 1980; Ngau 1987; Shao 1986), which supposedly was one of the key elements distinguishing alternative strategies from their mainstream counterparts. Indeed, many international organizations and state agencies involved in alternative development projects became increasingly self-critical. For example, a 1988 World Bank review of its Third World development projects candidly admitted, 'The principles guiding beneficiary participation in Bank-financed projects have been quite abstract and of limited operational impact. Beneficiaries were not assigned a role in the decision-making process, nor was their technological knowledge sought prior to designing project components' (World Bank 1988: 60). Similarly, a study by the Public Accounts Committee of the Indian Government into IRD projects found that 'the most important shortcoming in the programme is the absence of people's participation in it' (in Hirway 1988: A91). Alternative development programs and projects were commonly administered in a top-down, paternalistic manner that afforded little opportunity for local organizations to participate meaningfully in decision-making. Large bureaucracies staffed mainly by outside professionals often exerted fundamental tensions against the empowerment of local people. Typically, these tensions contributed to a distortion both of the nature of the projects (away from local concerns toward externally imposed ideas) and of the quality of local participation (a stifling of real participation in favor of paternalism and patronage). For Ngau (1987: 534), both of these outcomes ironically reflect 'departicipation or disempowerment at the grassroots level,' despite the frequent rhetoric of alternative strategies promoting genuine participation. The top-down, paternalistic manner in which they were administered, meant that many alternative projects systematically undermined indigenous forms of social organization and democratic political practice (Fox 1990). Local institutions often became dominated by 'imported' officials who failed to respond to local perceptions of development issues and problems (Daniel et al. 1985).

Frequently, external ideas and methods were imposed that paid scant attention to local particularities in terms of social relations, cultural traditions, spatial organization, and environmental conditions. In Tanzania, for example, the villagization program and associated rural development projects ignored 'existing patterns of settlement and land use that had been developed by the farmers after centuries of experience and of careful regard for the natural ecological conditions of their areas' (Shao 1986: 224). Commonly, alternative strategies tried to impose external development notions that were inappropriate to local conditions. Such was particularly the case for rural development initiatives that attempted to group peasant farmers into socialist-inspired collectives or production cooperatives in parts of Africa and elsewhere. As Texier (1974: 2) notes, these outside development models often produced consequences that were diametrically opposed to the professed goals of the programs:

> The actual consequences of trying to introduce into a traditional environment the cooperative system which [has been] successfully practised in Europe and in North America under very different circumstances and for the benefit of very different kinds of communities has in fact been to consolidate the traditional system and to encourage further social stratification by creating new opportunities for class exploitation by small privileged groups.

It has been pointed out that alternative strategies, despite their ostensibly progressive orientation, have normally generated rather conventional development projects that have replicated many of the problems of the mainstream approaches (Black 1991). Moreover, many of these problems have been glossed over by the tendency of many analysts to romanticize and simplify the alternative development tradition, especially in areas such as rural development (Burkey 1993). Ironically, it seems that one of the most consistent consequences of alternative development projects has not been increased local participation, but the extension of centralized state control through the establishment of patron–client relationships (see, e.g., Cheema 1985; de Janvry 1981; Grindle 1986; Mehta 1984). Through its involvement in various development programs and projects, the central state was often able to establish an extensive administrative and political apparatus in areas, especially outlying rural regions, in which its influence had previously been tenuous. This allowed the state, often in concert with outside aid agencies (e.g., USAID's Alliance for Progress program in Latin America), to manipulate and control, if not suppress, any development initiatives of the poor themselves (Black 1991). The steady extension of state control over various aspects of development transformed and coopted local identities and forms of political expression (Friedmann 1992).

Given this record and the limited scope of most alternative strategies, many analysts contend that preventing popular unrest and forestalling more revolutionary change, rather than reducing poverty and inequalities, actually represented the central goals of the state and external aid agencies in programs such as agrarian reform and integrated rural development (e.g., de Janvry 1981; Griffin 1976; Grindle 1986; Migdal 1974). Concerning the World Bank's poverty and rural development projects, Ayres (1983: 226) contends that the underlying rationale was 'political stability through defensive modernization. Political stability was seen primarily as an outcome of giving people a stake, however minimal, in the system. Defensive modernization aims at forestalling or preempting social and political pressures.' Paradoxically, then, it appears that alternative development programs were given impetus not by underdevelopment, but by the fear of development that might not be programmed from above (Black 1985: 529). For the state, there was often a 'fundamental contradiction between lip service to participation for reasons of political expediency, and real fear that grassroot organization will lead to the empowerment of local communities (Moser 1989: 118).

While alternative development programs generally failed to achieve their stated goals of more balanced and equitable development, they often provided the state with a wide array of mechanisms for directing economic growth, mediating social and political relations, and limiting the scope of social movements through processes of cooption, fragmentation, and control. However, as conditions continued to deteriorate for the poor in many countries, the state often became an increasingly visible target for social protest movements, as various elements of development became politicized. If widespread consciousness-raising and grassroots organization accompanied this process of politicization (as in the case, for example, of Nicaragua), a trajectory of rapid systemic destabilization and structural change could be created – ironically, the very scenario which the state's involvement in alternative development programs sought to forestall (Brohman 1989).

The top-down manner by which many alternative development programs were administered led to their frequent domination not only by the state, but also by local elite groups. In many Third World areas, local elites control land tenure systems, commercialization and economic terms of trade, allocations of credit, the electoral machinery and judicial system, and the principal means of coercion. As Cohen (1978: 42) notes, this has commonly resulted in the distortion of agrarian reforms and other development programs to serve the interests of local elites: 'Socio-political factors in many societies tend to condition a particular instrument [i.e., agrarian reform] in such a manner that the instrument, and the objectives to which the instrument was originally meant to contribute, are alienated from one another.' Typically, cooperatives and other local institutions, which alternative rural development initiatives created ostensibly to promote peasant participation and self-reliance, became instruments by which traditional patterns of domination and exploitation of peasants by rural elites were deepened and extended (see, e.g., Apthorpe 1972; Fals Borda 1971; Lele 1981).

Without a careful analysis of the social relations and structures within which development initiatives were being carried out, newly created local institutions, rather than representing instruments of progressive change, often became susceptible to cooption and manipulation by dominant classes and social groups. Local development projects would view communities in homogeneous terms, tending to deny possibilities for divergent needs and interests along class, gender, ethnic, or other factional lines (see, e.g., Fenster 1993; Moser 1989; Stone 1989). It was assumed that participation should take place through representatives of the whole community and that this process would empower everyone. However, in many fragmented and polarized communities, development projects quickly fell under the control of local elites, who used their influence to exclude the poorest sectors from meaningful participation (see, e.g., Black 1991; Burgess 1987; Ghose 1983; Peek 1988; Thiesenhusen 1989). Commonly, development projects were transformed to reflect the interests and perceptions of the local elites. Credit was extended only to those whose landholdings and other assets made them sufficiently creditworthy. Technological packages were employed that were appropriate only to medium/large-scale, capital-intensive operations. Infrastructure provisions (such as irrigation and transportation facilities) became monopolized by larger-scale producers. Projects designed to assist the poorest sectors (e.g., provisions of social amenities, labor-intensive programs) were systematically de-emphasized in favor of investments designed to boost the productivity of medium/large producers. As a result, elite groups were often able to consolidate their control over the direction of local development, thereby accelerating processes of social differentiation and deepening patterns of exploitation based on patron–client ties and other traditional social relations.

Support for alternative development strategies has diminished substantially in most countries since the early 1980s. On the one hand, the thrust of development efforts has turned away from state intervention and targeted programs toward macroeconomic considerations accompanying the rise of neoliberalism. On the other hand, the results of the alternative strategies generally failed to meet expectations. In many cases, alternative projects, especially in the rural sector, acquired a reputation for being relatively costly, difficult to monitor, too complex, and poorly designed (e.g., Brinkerhoff 1988; Lele and Adu-Nyako 1992). But, more fundamentally, it was gradually recognized that the limited scope of these projects prevented them from achieving their broad objectives. Typically, such projects created only an illusion of reform for a few people, rather than actual reform which could have addressed the structural roots of much wider problems of persistent poverty and growing inequalities. No matter how well intentioned, alternative projects proved incompatible with the overarching structural environment of many countries in which land and other productive resources were systematically skewed in favor of an elite minority. Projects were commonly administered on an ad hoc or temporary basis, with little complementarity or coordination, and could do little more than shift the location of poverty, while its underlying causes went untreated. In fact, many localized projects seemed to be working at cross-purposes with the central thrust of macroeconomic policies.

In the end, the bottom-up, localized approach of the alternative strategies amounted to swimming upstream against a strong current of economic and political power. Small gains in alleviating poverty or reducing inequalities were usually quickly overtaken by the force of larger processes. Although the proponents of alternative strategies mounted a rather effective critique of mainstream development models, no coherent, rigorous alternative analysis of the essential processes of Third World development was offered. Despite the fact that poverty alleviation was a central theme of most alternative strategies, no adequate conceptualization of poverty was forthcoming, leaving ad hoc projects to solve the problem of widespread impoverishment. Moreover, these projects characteristically reduced the poor to being passive recipients of often inappropriate forms of aid supplied in a top-down, centralized manner. In order to be more effective, it is generally acknowledged that alternative strategies should be altered in two basic ways: on the one hand, they need to 'scale-up' their approach beyond its focus on localized projects, and on the other, they must empower the popular sectors to find creative solutions to the sources of their problems. However, it should be noted that both of these requirements inevitably entail fundamental changes to the status quo and the distribution of economic and political power.

7

New Concepts of Planning

Complementing the thrust of alternative development strategies in general are a number of recently evolved derivative frameworks of regional and spatial planning. These alternative planning approaches stress the need to create a well-balanced, efficient, and locally suitable spatial organization to foster democratic participation and equitable growth. Decentralization measures are often advocated to promote more appropriate forms of development and assist in the mobilization of local human and material resources. Debates over decentralization and participation have shed light on the interwoven nature of social and spatial structures in development processes. In addition, the complex role that place and locality play in creating specific development contexts has been revealed. However, a number of theoretical and practical shortcomings have also marked alternative planning approaches, and these must be overcome if they are to be a vital component of development strategies.

Regional Decentralization and Alternative Spatial Strategies

The neoclassical paradigm has dominated regional and spatial planning in most Third World countries throughout the postwar period. A number of emphases of neoclassical planning theory (e.g., functional integration, hierarchical diffusion, polarization reversal through 'trickle-down' and 'spread' effects) have complemented the top-down thrust of the mainstream development frameworks – first with the modernization framework and more recently within neoliberal strategy. A 'functional' approach to regional development has been followed in which the development of various regions is essentially viewed as a function of their integration into the overall process of national economic development. Maximizing economic growth rates and establishing strong centralized control over investment are commonly emphasized. At the heart of this approach is the notion of polarized growth and the belief that a well-integrated system of urban-industrial growth poles, from which trickle-down and spread effects can emanate, is the best spatial solution to overcome initial problems of regional inequalities without sacrificing macroeconomic growth. It was thought that, following an initial period of regional divergence, market forces would inexorably produce growing convergence and restore regional equilibrium to patterns of development.

However, by the early 1970s, increasing criticism was directed toward this functional approach to regional development. Many analysts contended that just as mainstream development frameworks had increased socioeconomic inequalities in most countries, so too, accompanying neoclassical spatial strategies had widened regional inequalities, particularly between core urban areas and outlying rural regions (see, e.g., Brookfield 1975; Conroy 1973; Santos 1975; Slater 1975; Sunkel 1973). In many countries, uneven patterns of spatial development inherited from the colonial era were being reinforced, as major politico-administrative and commercial centers increased their dominance. Anticipated spread effects from urban-industrial growth poles had failed to materialize and given way to polarizing 'backwash' effects, accentuating both inter- and intra-regional inequalities. Rather than spreading development through positive urban–rural linkages, cities had become parasitic (see Hoselitz 1957), sucking up resources and surplus value from their surrounding regions. In particular, many outlying rural areas had been relegated to an ever more dependent and peripheral status – unable to generate their own internal development and increasingly subjugated to the dominant interests of the national elites in core urban areas. Characteristically, local services and infrastructure which were needed to overcome problems of economic stagnation in peripheral rural regions were virtually nonexistent.

Against this backdrop of widespread rural stagnation and growing regional inequalities, some alternative spatial strategies were developed in the 1970s. Complementary to alternative development approaches in general, these spatial strategies emphasized redistributive measures to promote more equitable growth; the provision of basic social and economic infrastructure; the creation of targeted programs to address the special needs of poorer, rural-oriented areas; and the use of decentralized planning methods. Rather than juxtaposing growth with equity, links were drawn between economic efficiency and equality in both the socioeconomic system and the spatial structure. An efficient, well-balanced spatial organization would maximize economic growth, which, in turn, would generate a more egalitarian pattern of income distribution. Two principal alternative spatial strategies emerged from the 1970s which have continued to receive considerable attention in subsequent years. The first has variously been termed integrated regional development planning (IRDP), urban functions in rural development (UFRD), or the 'concentrated decentralization' approach. The second has been called the territorial or 'agropolitan' development approach, and has frequently been associated with calls for the selective 'spatial closure' of peripheral regions from core areas.

Integrated Regional Development Planning

Integrated regional development planning (IRDP) emerged in the 1970s as an alternative spatial strategy which complemented the new focus of many international organizations and aid agencies on rural development initiatives in outlying Third World regions. Indeed, much of the work on the IRDP approach by Rondinelli, Ruddle, Evans, and others was carried out for the World Bank and USAID through projects in countries such as Bolivia, Burkina Faso, Ecuador, Malawi, and the Philippines. Whereas traditional neoclassical strategies had concentrated on the promotion of propulsive growth in a few large cities, a principal objective of IRDP was to curb continued metropolitan growth in favor of a more balanced spatial structure. The development of small and intermediate urban centers was emphasized, particularly as a crucial component of efforts to stimulate growth in peripheral, rural regions. It was argued that market towns, rural service centers, and intermediate cities could potentially play a key role in accelerating Third World rural development. However, in most countries, the settlement system was too poorly articulated and insufficiently integrated to generate growth impulses which might alleviate widespread stagnation and poverty in outlying, rural areas. The establishment of well-articulated regional hierarchies of spatially dispersed small/medium cities and market towns might stimulate more balanced and equitable growth, especially by offering improved rural–urban and inter-urban linkages. Among the major linkages noted by Rondinelli (1985: 143) are:

1   physical linkages: road networks, river and water transport networks, railroad networks, ecological interdependencies;
2   economic linkages: market patterns, raw materials and intermediate goods flows, capital flows, production linkages (backward, forward, and lateral), consumption and shopping patterns, income flows, sectoral and inter-regional commodity flows, 'cross linkages';
3   population movement linkages: migration (temporary and permanent), journeys to work;
4   technological linkages: technology interdependencies, irrigation systems, telecommunications systems;
5   social interaction linkages: visiting patterns; kinship patterns; rites, rituals, and religious activities; social group interaction;

6   service delivery linkages: energy flows and networks; credit and finan-
      cial networks; education, training, and agricultural extension linkages;
      health-service delivery systems; professional, commercial, and technical
      service patterns; transport service systems;
7   political, administrative, and organizational linkages: structural re-
      lationships, government budgetary flows, organizational interdep-
      endencies, authority-approval–supervision patterns, inter-jurisdictional
      transaction patterns, informal political decision chains.

According to IRDP theorists, the promotion of these types of linkages
through a well-integrated and regionally articulated system of urban centers
would serve a number of important development functions (e.g., Johnson
1970; Richardson 1977; Rondinelli 1983, 1985; Rondinelli and Evans
1983; Rondinelli and Ruddle 1978). First, it would relieve pressures on the
largest cities in housing, transport, pollution, job creation, and service provision. Second, it would reduce regional inequalities by spreading the benefits
of urbanization down the urban hierarchy to outlying areas. Third, it would
provide a more locally responsive and efficient politico-administrative system through regional decentralization measures. Fourth, it would help
to alleviate poverty, especially in peripheral regions where problems of
impoverishment and marginality are often most acute. Fifth, it would
stimulate rural economies by providing marketing, storage, and processing
facilities for agricultural products; by supplying credit, production inputs,
health care, education, and other goods and services to rural producers; and
by generating off-farm job opportunities to absorb the surplus labor that
might be created by rising agricultural productivity.
   The IRDP approach has continually expanded its focus to encompass
many aspects of Third World development. However, since its origins in
the work of Johnson (1970), IRDP has especially linked the development
prospects of rural peripheral regions with the increased commercialization
of agricultural production. Most elements of integrated regional development are, in the end, dependent on the ability of agricultural producers
to generate a commercial surplus to provide a critical initial stimulus to
peripheral economic growth – both supplying products for local agro-
industries and related sectors and creating demand for industrial goods and
services provided by local urban areas. The key constraints to agricultural
development, particularly among small/medium producers, are thought to
lie not within the realm of production per se, but within the sphere of
circulation. The absence of accessible market centers in many isolated rural
areas has left many producers prone to the monopolistic market practices
of local commercial intermediaries. This provides serious disincentives to
production increases because added surplus is merely extracted through
commercial exploitation. Therefore, the critical missing link within the

spatial structure of peripheral regions is most often thought to be a mid-level tier of well-distributed market towns and intermediate cities which would facilitate market access for the mass of rural producers and break their exploitation at the hands of rural merchants (see, e.g., Rondinelli and Ruddle 1978; Rondinelli and Evans 1983).

Another important component of the IRDP approach has been its emphasis on the decentralization of planning and other politico-administrative activities in order to facilitate local participation. If more equitable development is to be achieved, popular participation and the decentralization of authority ought to be encouraged (Rondinelli 1981). The alleviation of rural poverty, which has been the primary objective of most alternative development projects, is thought to have been hindered by the failure to decentralize decision-making and management structures (ibid.: 133). Moreover, increased decentralization could promote the rise of forceful and innovative local leaders who can identify with, and invest in, the regions in which they live (Rondinelli 1983). Among the ten factors that Rondinelli lists as basic requirements for more balanced and integrated regional development are the following: local leaders who identify their success with that of their city and region; public and private sectors which cooperate to promote economic activities that generate broad participation and distributive effects; and the willingness of local leaders to encourage social and behavioral changes responsive to new conditions and needs, and which are acceptable to local residents (ibid.).

While it has emphasized decentralization and local participation, the IRDP approach has also implicitly depended on a 'free-market,' capitalist framework to bring about progressive change in peripheral regions (see Slater 1989; Unwin 1989). This free-market orientation has recently been made more explicit in the work of Rondinelli et al. (1989), which links local participation to privatization measures advocated by public-choice theorists and other neoliberals. However, as Unwin (1989: 21) notes, 'In a free-market system, it is difficult to see how those with economic and political power are going to be persuaded to relinquish some of the advantages currently accruing to them for the benefit of the urban and rural poor.' Rather than stressing efforts to mobilize the poor through grassroots popular organizations, the IRDP approach focuses on local leaders to provide the impetus for progressive change. As we saw in the previous paragraph, many of Rondinelli's requirements for more balanced and equitable regional growth essentially depend on the altruism of local leaders in government and the private sector. A key problem with this strategy is that it pays scant attention to power structures in societies which are characteristically fragmented along class, gender, ethnic, and other factional lines. Given such fragmentation, is it realistic to depend on local leaders to be the key social agents to promote decentralization and popular participation? Moreover, how can the goal of economic growth be reconciled with the objectives of popular participation and decentralization within existing, highly polarized societal structures? For Slater (1989), a critical weakness of the IRDP framework is that it systematically bypasses these types of thorny political and economic questions in favor of a more technical approach to planning.

According to many analysts, this weakness of the IRDP approach stems from its failure to address the realm of underlying social relations in peripheral regions (e.g., Brohman 1989; Gore 1984, 1991; Painter 1987; Unwin 1989). In the absence of any serious social analysis, the nature of the settlement pattern becomes the critical factor which determines the outcome of development policies. Spatial organization is theorized per se, abstracted away from its historically contingent social construction. The IRDP approach is claimed to be appropriate regardless of the economic and political structures within which it is applied. Essentially, the approach rests on the twin propositions that an increased availability of goods and services (e.g., agricultural inputs, health-care facilities) will promote more rapid and balanced growth, and that the spatial distribution of goods and services provided by the settlement system constitutes the critical measure of supply. However, as Fass (1985: 377) remarks in a review of Rondinelli (1985), these propositions do not withstand serious scrutiny:

> The reader [should note] that the measure of supply is more conventionally understood as the price or (financial, time and opportunity) cost of obtaining a good or service, that location in space is one of several determinants of cost, that adding more goods and services to an existing distribution does not necessarily lower the cost, that lowering the cost does not necessarily yield net socioeconomic benefits, that implying a direct correspondence between supply and spatial distribution might be construed at best as an oversimplification and at worst as misplaced concreteness, and that although perceived inadequacies in number of places, distribution of services, and strength of linkages may be viewed as causal factors in rural income growth, the inadequacies may just as readily be viewed as the consequences of low income.

Persistent problems of economic stagnation and impoverishment in the peripheral regions of Third World countries are not just a function of market inaccessibility but, more fundamentally, are the result of underlying social relations and structures. There is no real evidence that merely modifying the settlement pattern to conform to some sort of classical rank–size distribution will stimulate more rapid development, nor is there evidence that such a pattern will necessarily benefit the poor and underprivileged (Unwin 1989: 20). In many countries, the profitability of the agricultural sector historically has been based on the extraction of surplus from the peasantry through a variety of mechanisms in the spheres of both production and circulation (e.g., land concentrations, the conversion of peasants into seasonal wage-laborers, monopolistic commercial relations and adverse terms of trade). Policies designed merely to alter the spatial distribution of marketing relations cannot adequately address the complex ensemble of rural social relations by which the peasantry has been systematically exploited. In effect, such policies attempt to extract spatial relations from their historically contingent setting within specific social formations. They therefore can offer only very partial and temporary solutions to the most critical development problems of peripheral regions. Until the spatial structure is thought of as an integral component of overall socioeconomic processes, spatial strategies are incapable of getting at the underlying determinants of development. Consequently, these strategies remain stuck at the level of simply trying to manipulate contingent spatial relationships without addressing the real causes of common development problems.

The Territorial Regional Planning Approach

The other alternative spatial strategy which has gained some prominence in the development literature over the past two decades is known as the territorial regional planning approach (e.g., Friedmann 1981, 1988, 1992; Friedmann and Douglass 1978; Friedmann and Weaver 1979; Stöhr 1981, Stöhr and Tödtling 1978; Stöhr and Taylor 1981). The territorial paradigm bears some resemblance to the IRDP approach in that it calls for decentralization measures to overcome problems of economic stagnation and underdevelopment in the rural peripheries of Third World countries. However, the territorial framework moves away from an emphasis on functional integration, which continues to characterize IRDP and other neoclassically driven spatial strategies, in favor of a radically different approach termed 'territorial integration.' It is contended that the major thrust of functional regional policies (i.e., promotion of efficient regional growth according to principles of comparative advantage) has typically caused the resources of peripheral regions to be exploited for outside interests. As a consequence, regional growth has failed to benefit the majority of the population and has led to little improvement in regional structures of production (Stöhr and Tödtling 1978). Therefore, rather than being subjected to outside domination through functional integration with core urban areas, peripheral rural regions should pursue a more endogenous form of development oriented toward their own territorial integration, defined as 'the use of an area's resources by its residents to meet their own needs' (Weaver 1981: 93).

Because it stresses decentralization and popular participation, self-reliance, basic-needs provisions, and more locally appropriate development, the territorial approach has often been associated with some of the alternative, 'bottom-up' development strategies that emerged in the late 1970s. Whereas the continuing focus of IRDP on functional integration fits more comfortably with the more conventional, within-the-system strategies developed by organizations such as the World Bank and USAID in the early 1970s, the emphasis of the territorial approach on endogenous development is closer to the more radical, neo-populist thrust of alternative strategies developed later in the 1970s, which stressed self-reliance and considered possibilities of selective withdrawal from the international economic system. Moreover, similar to neo-populist strategies in general, the territorial approach distances itself from neoclassical economics in favor of a more theoretically eclectic framework which often incorporates some Marxist elements associated with core–periphery concepts.

The territorial approach calls for a reorientation of thinking concerning the nature and process of development. Priority is given to promoting locally appropriate development by mobilizing the human, material, and institutional resources of regions to serve the needs of the popular majority. Policies should minimize inter-regional factor movements that do not directly meet the basic needs and interests of peripheral regions. Control should be exercised over resource use so that resources are valued according to the interests of the local community rather than by outside economic forces. Development should not be directed by abstract, universal concepts of economic organization, but should conform to the ethical and political values of the local population. More locally appropriate development can best be ensured by the regional decentralization of decision-making concerning resource allocations and the distribution of the benefits derived from regional growth. While there is no categorical recipe for territorial regional planning or 'development from below,' Stöhr (1981) lists eleven essential elements of the approach:

1   enabling broad access to land and other key forces of production (often involving land reform);
2   the creation or revival of territorially organized structures for equitable communal decision-making;
3   granting greater powers of self-determination for local communities and rural areas;
4   selecting regionally adequate and appropriate technologies;
5   giving priority to the satisfaction of basic needs;
6   introducing fairer pricing policies which offer more favorable terms of trade for agricultural goods and other peripheral products;
7   the use of external resources only when local control can be established and local resources are inadequate;
8   the production of exports only under conditions that contribute to a

broad improvement of the local quality of life;

9   restructuring urban and transport systems to improve and equalize access to all internal regions;

10   improving intra-regional and especially intra-rural transport and communications; and

11   encouraging egalitarian social structures and the rise of a collective consciousness.

The most concrete example of the territorial approach to regional development is offered by the model of the 'agropolitan' district (Friedmann and Douglass 1978; Friedmann 1981, 1988). The agropolitan district is a rural-oriented territorial entity of some 15,000 to 60,000 people, which is 'large enough to meet most of the basic needs of the population out of its own resources . . . [yet] small enough so that the entire population of the area might have reasonable physical access to the center for political decision making, planning, and administration' (Friedmann 1981: 248). Development impulses are assumed to emanate 'from below' and to 'filter up' from the local to the regional and, eventually, to the national level. Accordingly, development policies and planning should be decentralized and should reflect local agropolitan conditions. This should allow capital accumulation and reinvestment to conform to a more endogenous pattern which minimizes the leakage of surplus to external areas. Priority should be given to small-scale and labor-intensive activities which are oriented toward meeting specific human needs rather than the goal of economic growth per se. The 'active community' should direct development, emphasizing self-reliance, the satisfaction of basic needs, mass participation of the population, and harmony with nature. In order to achieve such ends, Friedmann and Douglass (1978) call for a reorientation of development strategies toward the following six elements:

1   replacing generalized wants by limited and specific human needs as the fundamental criterion of development;

2   focusing on agriculture as the propulsive sector of peripheral regional economies;

3   giving a high priority to the attainment of self-sufficiency in domestic food production;

4   reducing inequalities in income and living conditions among social classes and spatial areas;

5   encouraging the production of wage-goods for the domestic economy; and

6   adopting an industrial pattern of 'planned dualism,' in which small-scale producers for the domestic market are protected against competition from large-scale, capital-intensive enterprises.

Instead of depending on local leaders to carry out decentralization measures and other reforms, as in the IRDP approach, the territorial paradigm places much more emphasis on the participation of local communities and grassroots organizations in decision-making. Unless measures are taken to empower the popular majority in peripheral regions, decentralization initiatives may simply transfer power from one elite group to another, without bringing any real benefits to the poor. Development programs oriented toward alleviating poverty and reducing inequalities should be determined at the lowest feasible territorial scale; they should be motivated and directly controlled from the bottom (Stöhr 1981). Territorially based communities are regarded as the most important sites of popular participation, around which an alternative development strategy should revolve. According to Friedmann (1992: 73), the central participatory elements of the territorial approach entail 'the territorial character of an alternative development, greater autonomy over the life-spaces of the poor in the management of resources, collective self-empowerment, the importance of respecting cultural identities, and the democratic participation of the poor in all phases of development practice.' Alternative development strategies must stress territoriality for a number of reasons:

> Territory is coincident with life space, and most people seek to exercise a degree of autonomous control over these spaces. Territoriality exists at all scales, from the smallest to the largest, and we are simultaneously citizens of several territorial communities at different scales: our loyalties are always divided. Territoriality is one of the important sources of human bonding: it creates a commonweal, linking the present to the past as a fund of common memories (history) and to the future as common destiny. Territoriality nurtures an ethics of care and concern for our fellow citizens and for the environment we share with them. (Friedmann 1992: 133)

Connected with the reassertion of local control over development by territorially based communities is the concept of selective spatial closure. This is aimed at reducing external leakages (i.e., the outward transfer of resources and investible surplus) to create relatively self-sufficient, self-reliant territorial units. Backwash effects that drain surplus toward core areas should be minimized, while spread effects that generate more endogenous and regionally balanced development from either internal or external sources should be encouraged (Stöhr 1981). Selective spatial closure might be accomplished through some combination of supply- and demand-side policies, as well as political and administrative decentralization. It envisions the devolution of decision-making power over commodity or factor transfers (e.g., capital, technology), which is currently 'vested in functionally organized (vertical) units, back to territorially organized (horizontal) units at different spatial scales' (Stöhr and Tödtling 1978: 35). This should reduce external leakages and dependencies, and allow for the development of material and human resources according to local needs and interests.

While the concepts of territorially based communities and spatial closure may seem appealing to many proponents of alternative development strategies, they have also been the subject of much criticism. These concepts tend to emphasize the role of 'the political' in development processes much more than the neoclassical models of functional integration. However, the territorial approach has characteristically paid little attention to class, gender, ethnic, and other social relations which may be interrelated with the political sphere in various ways. The lack of a rigorous analysis of social relations within specific Third World contexts has left the territorial approach prone to several interconnected criticisms: that it is a rather idealistic neopopulist ideology in search of a workable methodology, that it invests territorially based communities with utopian qualities that often do not fit reality, and that it adheres to the spatial separatist theme in a new guise. Many of the objectives of territorial planning may prove unworkable in peripheral regions that are fragmented along class or other factional lines – especially those in which diverse forms of peasant exploitation are commonly based on traditional relations with local elite groups. Similarly, traditional social relations may form an important part of patterns of exploitation based in gender or ethnicity. Such relations are ignored by analyses which treat territorial units as 'organic wholes' and invest some sort of causal efficacy in them.

For Gore (1984: 230–1), this concept of territoriality takes on a sociobiological interpretation, which is inadequate for understanding and planning development processes for three reasons. First, territorial units at all scales (national, regional, district) are treated as organisms. However, to sustain the analogy, the country would have to be the organism and its various components would form organs. This leaves no room for autonomous organisms at the lower levels. Second, territorial communities are not objective organisms with a life of their own, but are subjectively and differentially defined according to social relations among individuals and groups. Third, the long-discredited biological notion of vitalism is at the core of the belief that restoring territorial integrity will release a force of 'willful action' that will enable local people to develop themselves. As Simon (1990: 15) notes, 'How important the organismic analogy in itself is to territorial regional planning, and hence how far we ought to be concerned with this aspect of Gore's critique, is debatable.'

However, as an element of a wider critique of spatial separatism, it underscores the neglect of social relations by the territorial approach, which typically treats regions and communities as undifferentiated wholes and therefore tends to assume away possibilities for internal fragmentation along class, gender, ethnic, and other lines. If this assumption cannot be sustained, the territorial approach to regional planning becomes much more complex and problematic.

Centralization versus Decentralization

The centralization–decentralization debate frames many of the key issues explored by alternative strategies of regional/spatial planning. Throughout the postwar era, there has been a tendency to counterpose centralization to decentralization, instead of treating them as integrative or, under some conditions, possibly complementary strategies (de Valk and Wekwete 1990). Most Third World countries emerged from the Second World War with highly centralized political and administrative structures that conformed to long established colonial patterns which fit well with the top-down thrust of Keynesian planning efforts. Conyers (1986) reports that some decentralization initiatives were introduced by British colonial administrations in the 1950s in an attempt to reproduce some elements of the British system of local governance in the colonies prior to their independence. However, in most countries, decentralization efforts began during the late 1960s and 1970s, as dissatisfaction with centralized Keynesian development planning became widespread. By the turn of the 1980s, when many of the newly implemented decentralization initiatives ran into political and administrative problems, and following the 'oil shocks' of the 1970s, there was a shift back toward more centralized control and management of resources. In recent years though, ideas of decentralization and local participation are again in favor, as notions of centralized government and planning are increasingly challenged, both in theory and in practice.

The centralization–decentralization debate has been made increasingly complex by the multiple meanings and emotive overtones of the terminology employed. In much of the development literature, the term 'centralization' has been given a negative association, while 'decentralization' has taken on positive connotations connected with objectives such as popular participation, local democracy, appropriate development, coordination and integration, and debureaucratization – all of which are equally complex and emotive terms (Conyers 1986: 595). However, decentralization initiatives have not always produced a desirable effect; their impact has depended on the type of program and the broader context within which it was implemented. Moreover, decentralization is not necessarily linked with popular participation and local democracy. As Slater (1989) argues, decentralization can be articulated into a neoliberal discourse, as in the case of many recent IMF or World Bank publications advocating privatization and market-led development, or it can be linked to an alternative discourse which combines ideas of popular participation, collective empowerment, and democratic socialism.

An examination of the recent development literature reveals many divergent meanings and objectives for decentralization. Decentralization has been variously defined as deconcentration (spatial relocation of decision-making), delegation (assignment of specific decision-making authority), or devolution (transfer of responsibility for governing understood more broadly) (Samoff 1990). Each definition reflects different objectives and distinct visions of the political arena – in terms of both what is possible and how governments are related to the people. Some analysts, particularly those associated with the IRDP approach, have been primarily concerned with administrative or institutional decentralization, while other more radical analysts have focused on political decentralization as a means to empower previously under-represented or marginal groups (Samoff 1990: 516–17).

Administrative decentralization is typically aimed at the wider distribution of infrastructure and services. It concentrates on barriers to change, ideal conditions for effective implementation, organizational capacities, and shaping positive attitudes. It constructs formal, positivistic models of spatial organization, and of the development process in general, that supposedly permit desired outcomes to be achieved through the correct manipulation of some known variables (though they may not be entirely predictable in the short term). The impetus for reform comes from external aid organizations and state agencies, often in cooperation with local leaders. The primary purpose of decentralization is not political; success is measured by improved access to various development factors, not by grassroots political mobilization. Increased local participation may be a consequence of the decentralization process, but decentralized administration is regarded as desirable on its own terms, with or without popular empowerment (Samoff 1990: 516).

By contrast, the central concern of political decentralization is not over the extent of administrative rearrangements, but over the actual transfer of decision-making authority to local representatives of the popular majority. The focus is on issues of empowerment. Decentralization efforts which merely concentrate on institutional reform are believed to be misdirected. It is contended that no real decentralization can be produced by institutional reform which does not empower under-represented and disadvantaged groups. Real decentralization does not automatically flow from organizational rearrangements. In fact, administrative decentralization (institutional reform that does not empower the disadvantaged) has often been used as a façade by the state to maintain or expand its central authority (Samoff 1990: 517). Therefore, from this perspective, it does not represent real decentralization at all.

Much of the recent development literature ascribes a variety of negative meanings to centralization. Yet, as Smith (1985) suggests, centralization might sometimes be considered a preferable strategy, especially if it is able to generate a more socially and spatially equitable pattern of development. In some instances, centralization may supply countervailing pressures at the national level, in cooperation with urban industrial and middle-class groups, to reduce the economic and political power of reactionary rural elites in peripheral regions (Islam 1992). In many Latin American countries (e.g., Argentina, Brazil, Mexico), centralized development planning oriented toward import-substitution industrialization was used to break the monopoly of power by the landholding oligarchy. Likewise, in South Korea and Taiwan, the central state implemented agrarian reforms and other redistributive measures which transformed a dualistic rural structure dominated by a feudalistic oligarchy into a more egalitarian, unimodal pattern of rural development. In other countries, particularly in Africa, centralized political systems were constructed to try to overcome traditional racial, ethnic, religious, or regional cleavages that were believed to be hindering national development efforts. Centralization was often promoted as a strategy for rationalizing the use of scarce resources needed to stimulate development. The central state or agents of the central state operating at the local/regional level might more efficiently provide some types of public goods and services. These might be 'network-based services that require large investments in capital equipment, and that must be linked together in order to operate effectively, those that have high political saliency or sensitivity, those from which a politically important group such as the poor or a minority would be excluded if they were provided privately, or those with strong implications for public health, safety, or welfare' (Rondinelli et al. 1989: 75).

In practice, however, centralization has been associated with a number of serious development problems. Prescriptions for central planning have often been used by Third World states to carry out sweeping nationalization measures and other programs which have placed as many economic activities as possible under the control of the central state (Rondinelli 1990). Commonly, the result has been inefficient and inappropriate forms of development that have neglected the needs of the popular majority in favor of the interests of national elite groups. Especially in many African countries, it has long been suspected that the centralized nature of the bureaucracy has contributed to the general underdevelopment of the rural sector and the stagnation of smallholder food production in particular. In many countries, locally responsive institutions were nationalized or eliminated, while local centers of power were subsumed within newly created political and administrative structures controlled by a distant central state. Even in countries where some efforts were made to decentralize the provision of public goods and services, access to decision-making was typically being steadily centralized.

Against this background, an increasing number of development analysts, representing a variety of perspectives from the Left to the Right, have begun stressing decentralization in recent years. Commonly cited reasons for decentralization include the following needs and desires: lowering the rate and scale of urbanization in the principal cities; diminishing regional and rural–urban inequalities; securing an adequate food supply and/or increased levels of agricultural exports; reducing inefficiencies, waste, and corruption within government and bureaucracy; promoting more appropriate development of human and natural resources; alleviating poverty through redistributive measures; and facilitating more effective policy implementation via improved local responsiveness and participation. In many cases, decentralization is viewed as a means to transform inherited (neo)colonial political and administrative systems to make them more efficient, more appropriate to local conditions, and more responsive to the changing needs of socioeconomic and spatial development (Conyers 1986: 598). Decentralization is regarded as vital to facilitate local participation in development projects, to increase the efficiency and flexibility of various development programs, and to create consensual support for development initiatives needed to promote national unity and political stability (Maro 1990: 673).

However, just as centralized development failed to meet the expectations of many development planners in the early postwar period, more recent decentralization initiatives have often failed to produce the desired results. In many cases, the actual practice of decentralization has not matched the rhetoric (e.g., Vengroff and Johnston 1989; Khan 1987; Samoff 1990; Simon 1990; Slater 1989). Whatever the rhetoric about people's power, local autonomy, and the need to transform unjust colonial structures, few genuine attempts have been made to implement democratic decentralization (Simon and Rakodi 1990). The political commitment for decentralization has been marginal, at best. There has been little effort to involve the popular majority in decision-making processes at the local/regional level. If some political decentralization has occurred, power has typically devolved to members of local elite groups and/or centrally recruited and centrally controlled civil servants. Popular participation has effectively been neutralized by elaborate mechanisms of centralized supervision and control. Moreover, most decentralized planning programs have been given inadequate human and material resources to fulfill their functions. Without the requisite resources, they remain empty shells. Experience has shown that the mere creation of decentralized planning structures does not guarantee better coordination and operational coherence, more effective local participation, or a more equitable distribution of resources in either socioeconomic or spatial terms.

In most instances, decentralization has primarily been viewed as a 'technical solution' to problems of bureaucratic inefficiencies and inflexibilities.

Typically, when development consultants analyze particular decentralization efforts, the measures of successful performance are those of administrative decentralization (Samoff 1990). Successful programs are those that improve service delivery, reduce waste and delays, and enhance cost recovery. The empowerment of disadvantaged groups is, at best, viewed as a secondary goal – not necessary to ensure the success of programs. Issues of empowerment are either dismissed peremptorily or are translated into a technical, functional language (i.e., local participation is desirable to the extent that it increases efficiency). This discourse sets a narrow, economistic agenda about what matters in decentralization. At the same time, it typically dismisses criticism based on other, more political concerns as unscientific and ideological.

However, as with all other aspects of development, decentralization is an inherently political process which takes on distinct institutional forms in different places at different times. For decentralization to be empowering, it must be linked to broader goals of economic and political democracy; participation cannot merely be functionalized as a technical planning exercise. Fundamentally, decentralization has to do with power: who rules and how that rule is carried out. Moreover, these questions are normally complex and do not lend themselves to universal or unidimensional solutions. Different groups may prevail in different issue arenas or in the same arena at different times. Coalitions and alliances are assembled at various scales, are dissolved, and become reconstructed in new forms. Within this context, a critical element of all decentralization initiatives is the specification of who is to rule in particular settings. Therefore, as Samoff (1990: 519) notes, 'To make sense of its forms and consequences in particular settings we need to understand decentralization as a political initiative, as a fundamentally political process, and consequently as a site for political struggle.'

Viewed from this perspective, neither decentralization nor centralization can be regarded as necessarily supportive of development goals such as increased popular participation, economic and political democracy, and social justice. There is little point in identifying particular institutional arrangements (such as decentralization) in advance as more democratic or egalitarian; the appropriate choice has to be made in light of local circumstances (Hindess 1991). The ability of some institutional arrangements to achieve certain goals cannot be guaranteed by the institutions themselves, as if they can be separated from the economic, political, and sociocultural setting in which they are constructed. The extent and form of (de)centralization that may be desirable can only be determined in concrete situations. Similarly, the measures of success for (de)centralization can only be specified in particular contexts.

Therefore, it is pointless to try to construct a general strategy for (de)centralization, or a universal, formal model of (de)centralized government, or a generic set of evaluative criteria that are applicable in all cases (Samoff 1990: 523). Just as centralization was generally viewed positively in the early postwar era, many analysts have recently attributed an absolute value to decentralization. However, there is no absolute value in either central direction or local autonomy (ibid.: 521). Both must coexist and both are more or less important in different places at different times. Neither is there one 'right' form of (de)centralization; different forms have advantages and disadvantages, and the choice over which to use must depend on the objectives and the particular setting (Conyers 1986: 598). What works reasonably well under some conditions may prove disastrous under others. Neither decentralization nor centralization in and of itself promotes balanced, democratic development. Given the inherently political nature of both strategies, the determination of preferences should be a function of (1) what priorities are to be assigned to the different interests involved (whose interests are to be given precedence), and (2) the characteristics of the particular situation (Samoff 1990: 522).

Shortcomings of Alternative Spatial Strategies

Despite their common focus on institutional reform, many alternative spatial strategies have failed to develop the institutional capacity at the local/regional level to plan, implement, and manage development activities properly. Rondinelli and Nellis (1986: 15) report that a 'kind of schizophrenia . . . about the desirability and feasibility of transferring powers and responsibilities away from the central government' has marked decentralized planning. In many cases, it appears that local/regional bodies have formally been given added authority over planning and other related development activities, but have not been provided with the resources needed to carry out their programs effectively. Typically, studies of decentralized planning initiatives have revealed inadequate human, material, and financial resources (e.g., Bienen et al. 1990; Jansen and van Hoof 1990; Rakodi 1990; Rondinelli and Nellis 1986; Sazanami and Newels 1990).

Moreover, decentralized planning has often been implemented in a piecemeal and disjointed manner, which has hindered efficiency, coordination, and continuity (Khan 1987; Simon and Rakodi 1990). Commonly, the control of the central government over planning and expenditures is weakened, systems of information flows between the center and peripheral areas are overburdened, various programs and projects sponsored by separate agencies and organizations are poorly coordinated, and administrative costs increase without additional sources of revenue. Such problems have often been particularly severe in programs and projects sponsored by international or bilateral aid agencies (Rakodi 1990; Rutten 1990). In many instances, this has led to the establishment of semi-autonomous 'donor republics,' each with their own sectoral preferences, organizational procedures, and short-term objectives. As Simon and Rakodi (1990: 255) remark, 'Planning has consequently often been reduced to the matching of donor preferences to projects needing funding, even if these projects are not high in national development priorities.' As a result, the long-term coherence and overall coordination of national planning efforts have frequently suffered.

Over and above such organizational problems, alternative spatial strategies have also been handicapped by some basic theoretical inadequacies. Many of these shortcomings stem from the tendency to conflate spatial problems with social problems. Ironically, it seems that alternative spatial strategies have become expressions of the same type of spatial separatism that has traditionally characterized more conventional neoclassical spatial frameworks (Forbes 1984; Gore 1984; Painter 1987; Simon 1990). An isolated conception of space has been constructed which divorces spatial organization from underlying social relations and other dimensions of development. Development projects using this approach have tended to focus on narrowly defined technical aspects of spatial organization, rather than examining interrelationships between spatial and social aspects of regional development. This has robbed alternative spatial strategies of much of their theoretical insight and practical utility, upon which their attractiveness was largely based in the first place.

Because elements of spatial organization are uncoupled from their social roots, alternative spatial strategies tend to invest abstract locational factors with an ahistorical, generalized causal efficacy. A direct correlation is established between spatial and social processes, so that problems of socioeconomic inequalities become translated into problems of spatial unevenness. For example, Rondinelli and Ruddle (1978: 175) observe: 'The failure of developing countries to achieve growth with equity can be attributed largely to their poorly articulated spatial systems.' In essence, the nature of the urban hierarchy and settlement pattern becomes the critical factor in determining the success of development programs. However, no real evidence has been provided that a balanced urban hierarchy will stimulate development, nor that growth in intermediate cities will be mutually beneficial to urban and rural areas, particularly for the poor in the latter (Unwin 1989: 21). Moreover, as Rakodi (1990: 149) notes, 'The conception of a national urban settlement policy in essentially physical terms, with insufficient understanding of the economic and political forces shaping settlement distributions, has led to over-optimistic expectations of the scope for change.'

Indeed, the conflation of spatial problems with social problems may inadvertently lead to the design of development programs that are harmful to the very people they are intended to help. For example, much evidence suggests that programs designed to alleviate rural poverty by promoting market integration and capitalist-oriented growth in peripheral regions have instead accentuated intra-regional inequalities, rural differentiation, and peasant marginalization (see, e.g., Brohman 1989; de Janvry 1981; Grindle 1986; Painter 1987). Such programs have commonly neglected the realm of rural social relations within which various mechanisms of peasant exploitation are based – not only in the sphere of production, but also in commercialization and circulation. As a result, the incorporation of peasant producers into rural markets by the spread of capitalist relations has often actually compounded problems of rural differentiation and impoverishment. Typically, new capitalist mechanisms have been added to more traditional means of exploitation by which large landowners and commercial intermediaries extract surplus from peasants.

Athough much of this criticism has been directed toward the IRDP approach, the territorial strategy of regional planning has also been marked by spatial separatism in a slightly different guise, which makes it susceptible to the same types of problems. The territorial approach typically treats regions as homogeneous, undifferentiated entities with a common collective interest (Gore 1984). In reality, though, regions are differentially and subjectively defined by people and social relations. They are also normally fragmented along class, gender, ethnic, and other lines which condition people's participation in development. By means of selective spatial closure, the territorial approach aims to reverse the underdevelopment of peripheral regions by severing chains of surplus extraction toward core areas. Because regions are conceptually aggregated into undifferentiated wholes, the principal processes of underdevelopment and exploitation can take place only at the inter-regional level; possibilities for intra-regional exploitation are systematically elided.

However, it is people rather than places that exploit and are exploited. It is relatively easy to identify places as being rural/urban or peripheral/core areas. But this is a much more difficult exercise when applied to people – and it is people, instead of places, that are responsible for different types of interaction over space (Unwin 1989: 17). What is needed is an analysis of how rural–urban, core–periphery, and other spatial interactions differentially affect various classes and social groups. This can only be accomplished by moving beyond a focus on space and place per se to address the great diversity of factors (e.g., economic, politico-ideological, sociocultural) that influence commodity flows and other forms of spatial interaction within and among regions. Factors such as people's perceptions, particular development ideologies, and different class and gender interests influence the direction of resource flows, and should be included in any analysis of spatial interaction. This makes generalizations about the role of space and place in development, based on whatever theoretical stance, hard to sustain.

The actual role of various components of spatial structure (e.g., land-use and tenure patterns, rural circulation systems, settlement hierarchies, core–periphery linkages) is difficult to gauge without an analysis of their concrete regional context, taking into account social relations of production and social reproduction as well as structures of political decision-making and control.

In recent years concepts of space, place, locality, and region have been afforded increasing attention in the social sciences and humanities, especially in the interdisciplinary field of urban and regional studies. Many analysts contend that the nature or character of regions must form an important part of how social processes are conceptualized within those regions. This, in turn, requires a social theory in which the regional setting is not treated simply as an abstraction or an a priori spatial given, but is viewed as the result of social processes that reflect and shape particular ideas about how the world is or should be (Murphy 1991: 23). From this perspective, regions become more than just backdrops for case studies; they are themselves part of the social dynamic of development. Communities and societies are largely territorially defined, and social identity is commonly tied to territorial affiliation. This is clearly seen in the case of many community or place-based social movements.

In an article on women's movements in the Southern Cone of Latin America, Scarpaci and Frazier (1993: 1) stress 'how the meanings of urban spaces are culturally and historically evoked' and 'the ways in which women protesters challenge and transform those meanings through action.' Similarly, a study by Routledge (1992: 600) of a peasant movement in Baliapal, India, emphasizes the importance of 'sense of place,' which was articulated through Hindu religious mythology and peasant folk culture and 'which refined and strengthened the economic motivation [for resistance] provided by the locale.'

Routledge concludes:

> The constituent elements of place, and their mediation of movement agency, provide important insights into the landscape of struggle and its place-specific character. A consideration of location provides us with the (regional) economic, political, military/strategic and geographical factors that have contributed to the emergence of the Baliapal movement. (ibid.: 602–3)

Since the early 1980s, interrelationships between space, place, and society have been explored in geography and related disciplines using a variety of theoretical approaches, including poststructuralist Marxism, feminism, structuration theory, and postmodernism.[1] Much of this research has focused on the interpenetration of social processes and spatial forms. In political geography, for example, the contributors to Wolch and Dear (1989) explored the complex roles that locality plays in both constraining and enabling social change. Instead of merely being portrayed as a passive container for social action, space was conceived as actively constituted through social processes along multiple, and potentially contradictory, axes. Similarly, Pratt and Hanson (1994: 250) contend: 'Places are constructed through social processes and, so too, social relations are constructed in and through place.'

Particularly in Britain, concerns for space and place have spawned a new research area termed 'locality studies,' which has focused on the ways in which the general dynamics of capitalist development have been conditioned by the context of localities. Closely related to this research, a new or 'reconstructed' regional geography has also been developed, which, according to Thrift (1983: 38), 'builds upon the strengths of traditional regional geography, for example, the feel for context, but that is bent towards theoretical and emancipatory aims.' For Agnew and Duncan (1989), recent theoretical advances in locality studies and the new regional geography have begun to reclaim the power of place-based experience.

In contrast to the rather economistic slant of most earlier spatial analyses, recent research in locality studies has focused on the important role of culture in creating regional and other spatial identities (e.g., Agnew 1987; Duncan 1989; Duncan and Savage 1991; Jackson 1989, 1991; Paasi 1991). Culture represents the way that the social relations of a group are structured and shaped; but it is also the way those 'shapes' are experienced, understood, and interpreted (Jackson 1989). Communities and regions are viewed as spatially constituted sociocultural structures, as well as centers of collective consciousness and socio-spatial identities. It is asserted that regional cultures and identities are formed by social practices and relations that are, at least partly, independent of the logic of capital

---

[1]   Within geography, the increasing attention given to issues of space, place, and locality by many analysts reflected a rejection of the universalizing claims of both neoclassical and structural Marxist spatial theory. It was contended that the economistic and class-reductionist biases of previous spatial frameworks had excluded much of the diversity of modern experience. For the post-structuralist Marxists, locality provided a basis from which to contest the universalizing claims of structuralism (e.g., Thrift 1983). Similarly, many feminists countered tendencies toward class reductionism by emphasizing the diversity of women's experiences based in place (e.g., Rose 1993; Pratt and Hanson 1994). The structurationist approach introduced the concept of 'locales,' which referred to 'the use of space to provide the settings of interaction [which are] essential to specifying its contextuality' (Giddens 1984: 118). In postmodern discourse, a focus on the 'spatiality of social life' was a means to approach the multiple and conflicting axes along which social power was exercised (Soja 1989).

(Duncan 1989). Moreover, struggles among various groups over resources and power are seldom restricted merely to economic and political issues, but normally also extend into cultural areas involving divergent lifestyles and meanings.

An important part of the way that localities become constituted takes place through everyday cultural practices and personal histories, which also help to shape processes of regional production, reproduction, and transformation. Concepts of place cannot be simply reduced to a specific site or scale. Instead, place is composed of situated episodes of life history which unavoidably have various geographical dimensions: real, imagined, or utopian (Paasi 1991: 248). As centers of meaning, places are constituted through personal histories; however, they also take on more permanence through collective action and institutional practices:

> The time-space-specific, socially contingent encounter of personal histories engenders collective forms of action and thought . . . [which over] generations bind people together with the episodes of history that affect a locality, and not merely local history, but also broader spatial and historical contexts. (Paasi 1991: 252)

This research stresses the historical creation of regional identities, which are constructed through a range of political, economic, and sociocultural forces. Once historically established through collective action and institutional practices, they often take on a more permanent and distinct presence. As a result, social relations take place not only at different scales, but also with different meanings, which are normally wrapped up in place-based identities and territoriality. From this point of view, territory is not just part of a functional or physical spatial system, but also takes on a sociocultural, experiential connotation (Vartiainen 1987).

This concept of territory stresses subjective as well as more objective elements of spatial and regional development. It moves away from the functionalist or sociobiological interpretations which, up to now, have characterized alternative spatial frameworks in development studies. It is also slowly being introduced to the development literature, principally by the work of some geographers. For example, Brown (1988: 264) calls for development research to take 'a less nomothetic, more idiographic perspective that takes account of places as entities in their own right and their experience, or perception of change – what might be seen as a behavioral geography of places.' Similarly, Peet and Watts (1993) advocate a basic rethinking of development theory to incorporate poststructural concerns for power, discourse, and cultural difference. At the regional level, they introduce the concept of 'regional discursive formations,' which are composed of 'certain modes of thought, logic, themes, and styles of expression which run through the discursive history of a region [and are] important both for the ideas [they] allow and disallow' (ibid.: 230).

Within alternative spatial strategies, the neglect of the realm of subjective meanings and identities, and of the general sociocultural sphere of regional development, has had a particularly serious effect on methods of local participation. Ideally, regional development planning should be a social learning process which makes room for the various ideas and perspectives of different individuals and groups. Therefore, the institutional design for the planning process ought to be open and capable of incorporating contributions from the widest possible variety of local sources. This implies abandoning the technocratic, top-down, exclusive nature of past development planning; designing procedures to permit modification according to divergent meanings, needs, and interests; and creating institutions to enable the largest possible number of individuals, social groups, and organizations to become actively involved in planning processes. Planning practices should facilitate a two-way learning process – on the one hand, disseminating outside information and techniques, but, on the other, permitting local ideas, perceptions, and methods to inform decision-making and implementation procedures. Emphasis should be placed on participatory methods designed to elicit the different perceptions, intersubjective meanings, and objectives of local actors concerning development – which may be quite distinct from those of outside development professionals and without which locally appropriate development programs cannot be created.

By contrast, alternative spatial strategies have typically been implemented in a top-down manner which excludes local perceptions and concerns over development, especially from the various social sectors comprising the popular majority. Characteristically, participation has assumed a rigorously formal character and has been limited to the well-organized population composed of representatives of clearly identifiable interests, organizations, or institutions (Allor 1984). This has often allowed elite groups to play a dominant role in the planning and management of local projects, thereby reinforcing their control over the direction of regional development (Bienen et al. 1990; de Valk and Wekwete 1990; Khan 1987; Slater 1989). In many cases, local and regional elites have formed alliances with outside state agencies to systematically bypass local input from the popular sectors. While the rhetoric of decentralization has proved attractive to international aid agencies and donors, the practice of decentralization has often been used to deepen the control of the state and elite groups over local/regional development processes (Conyers 1986; Gore 1984; Khan 1987; Rakodi 1990; Romein and Schuurman 1990; Slater 1989). Given the frequent domination by elite groups over development at various scales, Griffin (1981: 225) notes that no simple correlation can be drawn between decentralization and popular participation:

It is conceivable, even likely in many countries, that power at the local level is more concentrated, more elitist, and applied more ruthlessly against the poor than at the center . . . [Therefore,] greater decentralization does not necessarily imply greater democracy, let alone 'power to the people' – it depends on the circumstances under which decentralization occurs.

Such concerns underscore the fact that spatial/regional planning, like all aspects of development, is an inherently political process. Space is, in a very real sense, a politically contested domain. Planning programs cannot be effective if they attempt to isolate spatial issues and problems from the territorial expressions of power relations. Spatial problems are necessarily also political problems and, therefore, spatial strategies inevitably become political strategies. Consequently, discussions and debates over alternative spatial strategies should be informed by an analysis of the broader political context within which these strategies occur. This means transcending the conception of spatial planning as a purely technical, apolitical process. As Wood (1985: 348) puts it, we must learn to recognize 'the political in the apparently non-political.'

Although some attempts have been made to incorporate a political dimension into alternative spatial strategies, these have largely been framed by a rather simplistic 'core–periphery' perspective in which political relations are aggregated into undifferentiated regional wholes. Accordingly, major political conflicts are possible only at the inter-regional level. Even territorial approaches that stress democratic participation typically fail to analyze power relations adequately, so that possibilities of local institutions becoming coopted and dominated by elite groups are minimized. Little consideration is given to questions of who are going to be the actual social agents of decentralization and popular participation, and how these processes are going to be carried out in highly fragmented peripheral societies (Slater 1989: 517–18). However, as Samoff (1990: 524) notes, the failure to pose political questions is itself inescapably political: 'Not to ask who rules, or who benefits, is surely as political as posing those questions.'

The neglect of politics and power relations by alternative spatial strategies is part of a larger conceptual problem: the failure to situate space and place within the broader context of development processes conditioned by particular, historically constituted patterns of social relations. The internal dynamics of peripheral societies are considerably more complex than the alternative spatial frameworks would lead us to believe. Likewise, the nature of the peripheral state varies significantly among countries, as do international economic and geopolitical relationships. Spatial problems at the local/regional level normally reflect deeper socioeconomic difficulties, which cannot be addressed without an analysis of underlying social relations of production and social reproduction. Only this type of analysis can reveal the roots of many spatial/regional problems within patterns of class, gender, ethnic, and other social relations.

Moreover, alternative spatial frameworks need to grapple with larger questions of power and the national/international dimension of development. Issues such as the local/regional impacts of state policies, the growing penetration of transnational capital, and the overall development path adopted by particular countries need to be considered. All of these concerns are basic to an understanding of why the poor commonly remain economically disadvantaged and politically disenfranchised, despite the best efforts of alternative development programs. The continuing failure to address such concerns means that alternative spatial strategies have reached a theoretical impasse from which it is proving increasingly difficult to generate viable, practical solutions to the most serious development problems of the popular majority in the peripheral regions of the South.

8

Participation and Power

Since its origins in the early 1970s, the alternative development tradition has evolved considerably. Much recent progress has been made in filling in some conceptual gaps which marked earlier alternative frameworks. These efforts have been helped by new ideas and methods from a broad range of the social sciences and humanities. At the same time, new contradictions and problems have appeared which must be resolved. This chapter explores an especially critical area: popular participation and empowerment.

Local Participation, Popular Movements, and Empowerment

Given the shortcomings of top-down development efforts, participation has come to be recognized as an absolute imperative for development not only within the alternative tradition, but also in many mainstream strategies. Nevertheless, it has remained an elusive concept. In the last few decades, it has been given multiple meanings and been connected to multiple methods of implementation. At the same time, results of evaluations of participation have often been disappointing. Questions often remain over who participates (e.g., just an elite group or a broader range of people), what they participate in (e.g., a more limited or broader range of decision-making), how they participate (e.g., as benefit recipients or as project designers), and for what reasons they participate (e.g., as a means toward other objectives or as an end in itself). In the end, participation has become a complex issue because it is a multi-dimensional concept and because, as an inherently political act, it can never be neutral.

For most analysts, participation emphasizes the decision-making role of the community (Fleming 1991: 37). Such participation helps 'to improve the design of policies so that they correspond to the needs and conditions of the people to whom they are directed (Cornia et al. 1987: 163). While many development economists define community participation as the equitable sharing of the benefits of projects, social planners tend to define it as the community's contribution to decision-making (Fenster 1993: 190). According to Oakley and Marsden (1984), the more common interpretations of community participation may be represented as a continuum. At one end, participation may merely mean voluntary contributions to projects without any local influence over their shape. At the other, participation may be seen as an active process to increase local or community control.

For Paul (1986: 2), community participation is 'an active process by which beneficiaries influence the direction and execution of a development project with a view to enhancing their well-being in terms of income, personal growth, self-reliance or other values they cherish.' However, as Fenster (1993: 190) notes, this definition may refer to participation that is not spontaneous or bottom-up, but is induced, coerced, or top-down. Similarly, the United Nations (1981: 8) distinguishes between coerced participation, which it condemns, induced participation, which it regards as second best, and spontaneous or bottom-up participation, which it contends 'comes closest to the ideal mode of participation as it reflects voluntary and autonomous action on the part of the people to organize and deal with their problems unaided by governments or other external agencies.' Midgley (1986) makes a similar distinction between 'authentic' participation, in which local people democratically control project decision-making, and 'pseudo' participation, in which projects are carried out according to prior decisions made by outsiders.

A closely related distinction has been made between participation as a means to improve project results and participation as an end in itself (e.g., Conyers 1985; Moser 1989; UNCHS 1984). In the first definition, participation is regarded as a means of improving the quality and relevance of projects by facilitating their implementation (including the contribution of local resources) and acceptance (Conyers 1985: 8). For example, UNCHS (1984: 6) states, 'If people participate in the execution of projects by contributing their ingenuity, skills and other untapped resources, more people can benefit, implementation is facilitated, and the outcome responds better to the needs and priorities of the beneficiaries.' The second definition sees participation as an end in itself – as an essential component of a democratic society to ensure the well-being of individuals and communities (Conyers 1985: 8). As UNCHS (1984: 6) notes, 'People have the right and duty to participate in the execution (i.e., planning, implementation and management) of projects which profoundly affect their lives.' Accordingly, participation as an end is closely linked to questions of empowerment and control over decision-making. For Moser (1989: 85), 'Community participation at the outset, in decision making, is a precondition if the objective is empowerment. Where participation is a means to achieve a development object, it is usually included only at the implementation and maintenance level.'

The Role of Nongovernmental Organizations

Over the past two decades, nongovernmental organizations (NGOs) have played an increasingly important role in local and community-based development initiatives in many countries.[1] One study identified some 2,200 Northern NGOs operating in 1984, with total expenditures of about $4 billion for projects and institutions in the South. NGO expenditures had increased threefold over the previous decade and were estimated to equal 10–20 percent of the annual budgets of bilateral and multilateral aid agencies. The growth of Southern NGOs has been even more impressive; some 10,000 to 20,000 of them are presently working with Northern counterparts (Black 1991: 75). A major reason for the proliferation of NGOs has been the recent inclination of both international and bilateral aid agencies to contract out much of their fieldwork. In 1988, 62 percent of USAID employees were under contract – compared with about 25 percent in the early 1960s (ibid.). Similarly, the government aid agencies of Canada and many European countries have contracted out an increasing proportion of their development work to NGOs.

Much of the recent work of NGOs has been devoted to poverty alleviation or 'filling in the gaps' in social development left by structural adjustment programs (Fleming 1991). NGO projects tend to be small-scale and commonly stress the need for local participation, and the use of appropriate technologies, and of local knowledge and resources (Simon 1990). Many NGOs have made substantial contributions in areas of participatory development, innovative methods, institutional organization, and project implementation (Bebbington and Farrington 1993). Following a 1987 conference of NGOs convened in London by the World Bank and the Overseas Development Institute, a report was issued by a group of Third World NGOs. According to Drabek (1987), the principles it contained are as close as we have to a contemporary vision of alternative development from the South itself, as defined by development agents other than the state.

---

[1] The United Nations originally coined the term 'nongovernmental organization' to refer to organizations that had a consultative status and received some funding from that body. Subsequently, however, a nongovernmental organization has come to mean any private or community-based organization that may receive funding from governments or international organizations but is not a direct appendage of them. The terms 'nongovernmental organization' and 'private voluntary organization' are sometimes used interchangeably, especially in the US. Originally, the latter term was introduced into the literature by USAID to categorize private-sector nonprofit organizations receiving USAID contracts or grants (Black 1991: 74).

Friedmann (1992: 72) lists these principles, which seek to alleviate poverty and inequalities by:

1   permitting the poor to reacquire 'the power and control over their own lives and the natural and human resources that exist in their environment';
2   strengthening 'their inherent capability to define development goals, draw up strategies for self-reliance and be masters of their own destinies';
3   refusing 'to compromise on issues related to the social and cultural identity of [Third World] societies';
4   placing 'special emphasis on and attention to utilizing and developing the indigenous efforts, however small, that are promoting self-reliance';
5   uncoupling from development processes 'all aid which is intrinsically tied to the foreign policies of donor states';
6   recognizing 'that nongovernmental development organizations working with the poor and having an indigenous evolution are important vehicles for change in the development process and [that] support should be primarily provided to them'; and
7   acknowledging 'that all development efforts must have as equal partners women who have until now borne the burden of the anti-development processes.'

Support for NGOs has grown in recent years largely because international organizations and state agencies have recognized the cultural sensitivity, innovativeness, and dedication of such groups, as well as the advantages that they offer for decentralizing and debureaucratizing development initiatives. Contracting out development work has also been used to broaden the aid constituency and make official agencies less vulnerable to criticisms of bloated bureaucracy and inappropriate, top-down development efforts (Black 1991: 76–7). Bebbington and Farrington (1993: 207) link the innovativeness of many NGOs with characteristics, such as small size, flexibility, 'shallow hierarchies,' and short communication lines, which make effective collaboration, rapid decision-making, and quick responses easier. Likewise, Vivian (1994: 190) states, 'The greatest advantages that NGOs have over other development agencies . . . are their flexibility, speed of operation, and ability to respond quickly to changing circumstances.'

The small size and relatively autonomous organization of many NGOs has also encouraged direct links to local organizations and communities. A recent study of NGOs operating in Chile, for example, praised their efforts to cooperate with the popular organizations in poor communities to develop appropriate, bottom-up responses to local problems (Downs and Solimano 1989: 205). Without the traditional burdens of government responsibility, these NGOs were able to carry out extensive participatory fieldwork, which enabled them to provide some innovative solutions to local problems rather than the standardized responses of state institutions. The direct participation of many NGOs in the popular organizations of the communities in which they were based gave them a role not simply as providers of outside resources, but as partners in the struggle for survival (ibid.: 210). Instead of imposing outside ideas and solutions, the NGOs tried to understand local needs and interests as defined by members of the communities themselves. This allowed them to use cooperation and empowerment rather than top-down leadership.

Within this type of approach, development workers become catalysts for change and information brokers rather than decision-makers or information givers. Instead of increasing dependency, they seek to build confidence and self-reliance, raise consciousness, develop critical and analytical skills, and promote participatory dialogue and democratic practices. This is a difficult role which demands certain human qualities as much as special technical or organizational skills. For Ghai (1989: 232), these qualities include 'a deep understanding of the economy and society of the impoverished groups, compassion and sympathy with their plight, and an ability to inspire trust and confidence and to motivate and guide them, not in a paternalistic and authoritarian way, but in a manner to enhance their confidence and self-reliance.' Such qualities may help to reduce the distance between outside development workers and local community members. As Escobar (1992: 422) comments:

> [The outside intellectual or development 'expert'] is no longer seen as the rationalizer of the dominant model of power and knowledge, but rather as a person who is (or must be) committed to the generation of knowledge that serves the grassroots, precisely because s/he takes local knowledge seriously . . . This new shift in the character of intellectual work also restores a certain human dimension to knowledge and social action. Furthermore, the thick veneer that conventional knowledge has laid upon popular forms of knowledge starts to dissolve, bringing into view struggles, conflicts and forms of knowledge that were previously buried.

Because problems of impoverishment are particularly severe in the outlying rural areas of many poorer countries, a high proportion of NGO projects have focused on rural development.[2] Like the rural development efforts of organizations such as the World Bank and USAID, NGO projects commonly

2   A number of recent studies offer long lists of NGO-sponsored rural development initiatives in a range of Third World countries (e.g., Egger 1986, 1992; Ghai 1989; Uphoff 1993).

stress expanding sources of employment, providing alternative credit and financing channels, increasing linkages between primary agricultural production and other local economic activities, improving basic social and economic infrastructure, and offering more socially and environmentally appropriate methods of development.

Increasingly, NGOs have turned to providing alternative sources of credit as a key method for stimulating rural development. In most Third World countries, rural areas receive less than a quarter of total credit allocations and many farmers lack access to formal credit channels. Non-institutional or informal credit, which is often tied to usurious and highly exploitative relationships, accounts for more than half of Third World agricultural finance (Egger 1986: 448). As a result, many small/medium farmers lack the capital to make investments needed to improve and expand production, and diversify into promising new agricultural sectors and non-farm activities. Too poor to save and unable to secure credit, much of the peasantry has been excluded from the technological advances associated with the Green Revolution.

Given this situation, many NGOs have initiated innovative credit programs designed to provide small loans efficiently to the rural poor, of which perhaps the most well-known was begun in 1976 by the Grameen Bank of Bangladesh in cooperation with government and donor agencies such as the International Fund for Agricultural Development (IFAD), the Asian Development Bank, and the Ford Foundation (Ghai 1989: 219–20). By the early 1990s, the Grameen Bank had extended loans (averaging about $70 each) to over 1.2 million people. Repayment rates were close to 100 percent and the incomes of loan recipients commonly increased by 50 percent or more (Uphoff 1993: 617). Among the key principles of the bank have been the extension of credit to previously excluded groups, particularly poor women; replacing individual collateral requirements with group liability; and combining group supervision and 'peer pressure' with freedom for borrowers to choose their own uses for loans (Egger 1986: 450). According to Islam (1992: 118), important factors contributing to the success of the bank are 'the exercise of peer pressure on the members of the borrowers' group, intensive intragroup consultation for the identification and formulation of bankable projects, involvement of bank staff in the supervision and management of loans as well as subsidies from the government to cover the high costs of administration of small loans.'

Many NGO rural development initiatives, including the Grameen Bank, have been designed to enable the poor to increase their agricultural production, but also to diversify into other economic pursuits. Given land concentrations in many rural areas, an increasingly important source of income for rural households has become non-farm economic activities, such as artisanal sectors, agro-industries, trade and commerce, and services (Grabowski 1989; Lele and Adu-Nyako 1992; Stokke et al. 1991). Rather than merely representing a residual source of income, non-farm activities are often the principal means by which poor rural households support themselves. This is particularly true for many landless and land-poor households headed by women. Therefore, many NGOs have recently incorporated assistance for non-farm activities into rural projects in order to meet a number of interrelated development objectives. These include reducing rural poverty and inequalities; providing sources of income for traditionally disadvantaged groups; lowering rural–urban migration by offering rural employment opportunities; diversifying production structures by reviving labor-intensive artisanal and other activities; stimulating productivity in agriculture by improving backward linkages; raising rural consumption levels by providing affordable, locally available basic consumption goods; and generating more local value-added from agricultural products by improving processing facilities and other forward linkages.

A further focus for many NGO rural projects has been creating more socially and environmentally appropriate forms of development. Projects have typically stressed human-resource development via improvements in basic social infrastructure to allow the poor to participate in economic growth more fully. In addition, attention has been given to creating more balanced and environmentally sustainable forms of agricultural development by improving the technological basis of small farmers. Research initiated by NGOs, for example, has addressed peasant farming systems, agroecology, social forestry, and alternative technologies. In many cases, the adoption of an ecological and anthropological approach has substantially redefined the issues and methods by which rural development projects proceed. 'Learning-by-doing' methods and 'barefoot researchers' at the community level have been used to enhance popular participation, make projects more locally appropriate, and improve information and data gathering (Thomas-Slayter 1992). Many of these projects have combined goals of improving food security for the poor with solving environmental problems (Peet and Watts 1993). In this way, efforts to provide basic needs and promote more balanced social development have been combined with initiatives to create more environmentally sustainable forms of resource use. Increasingly, projects have been designed to address social and environmental problems as inextricably linked, so that solutions for the former demand solutions for the latter, and vice versa.

The Rise of Popular Movements

Accompanying the proliferation of NGOs, a growing number of popular movements oriented toward a broad range of issues have been established throughout the South in recent years.[3] According to Jelin (1986: 22), popular movements are 'forms of collective action with a high degree of popular participation, which use non-institutional channels, and which, at the same time that they formulate their demands, also find forms of action to advance those demands and to establish themselves as collective subjects, that is, as a group or a social category.' The term 'popular' is commonly used to represent a range of economically marginalized, politically disenfranchised, and culturally threatened groups, including many indigenous populations, informal sector members, landless and land-poor peasants, and the under- and unemployed (Stephen 1992). In the Latin American context, for example, Escobar (1989: 38) uses the term 'popular' to link all of those 'who populate the contradictory categories and identities created by the process of dependent capitalist development.'

The recent emergence of popular movements in many Third World countries has been correlated to mounting contradictions and crises in the mainstream development models, as well as the widely perceived failure of both ends of the ideological spectrum to adequately represent the interests of the popular majority (Escobar and Alvarez 1992; Petras and Morley 1990). Friedmann (1992: 1) notes, 'In a historical perspective, one can see in all of [the popular movements] the rise of civil society as a collective actor, working for political agendas outside the established framework of party politics.' Because political parties on both the Right and Left are commonly believed to have manipulated people's organizations and local groups for their own purposes, popular movements have been wary of party support. Moreover, most popular movements have not sought state power, but autonomy from the state (Fuentes and Frank 1989). Especially in the era of neoliberalism and SAPs, popular movements have often attempted to fill the void where the state has been unable or unwilling to act. For people who have lost faith in the ability of mainstream institutions to improve their well-being or defend their rights, popular movements seem to offer a viable bottom-up alternative.

---

3   In some countries, such as Chile, popular movements are regularly called 'popular economic organizations' (PEOs). The term 'new social movements' is probably more commonly used in the development literature. However, the term 'popular movement' is used here, on the one hand, to include organizations based in non-economic issues and, on the other, to exclude those which are not formed from the popular sectors themselves (e.g., the working class, peasantry, poor women and children, marginalized ethnic minorities) but may merely represent elite groups. In contrast to social movements in general, which may emerge from any sector of the population in relation to a particular social issue or perspective, popular movements are specifically based in, and take their name from, the popular sectors. Moreover, as Fuentes and Frank (1989: 179–80) note, many of these movements are not really new. 'Classical' working-class and union movements often date back to the nineteenth century, while other peasant, ethnic, religious, and women's groups have existed in various forms for centuries.

Throughout the South, popular movements display a tremendous diversity in their social constitution, organizational form, and issue orientation. Examples of popular movements include women's organizations of various kinds, ecology and 'green' movements, peasant and rural-based associations, working-class movements, groups of urban marginals and informal-sector workers, squatter movements, church-sponsored and religious organizations, community and civic groups of diverse nature and scope, student and youth movements, popular culture groups, and organizations of ethnic minorities and indigenous peoples.

Although it covers only a small fraction of the popular movements around the South, table 8.1 portrays the great number and diversity of such movements.

Table 8.1 Popular movements in selected Third World countries

Country	Description
Bangladesh	1,200 independent development organizations formed since 1971, which are particularly active in health care and income generation among the rural landless and land-poor.
Brazil	Rapid growth in community action since the early 1980s; 100,000 Christian Base Communities with 3 million members; 1,300 neighborhood associations in São Paulo; proliferation of organizations among the rural landless and land-poor; numerous indigenous and environmental groups in the Amazon Basin.
Burkina Faso	Naam peasant movement with 2,500 groups participating in dry-season self-help (similar movements in Mali, Mauretania, Niger, Senegal, and Togo).
Chile	1,446 Popular Economic Organizations (PEOs) in Santiago alone in 1986 with a membership of more than 25 percent of the households in the city's *poblaciones*; PEOs oriented toward a range of activities, including communal kitchens, wholesale food buying, organic gardens, health care, day care, and artisanal workshops.
Ecuador	126 native groups organized in the Confederation of Indigenous Nationalities in 1990 and primarily focused on cultural, environmental, and land-tenure issues.
India	Independent development organizations estimated at 12,000; local groups numbering the tens of thousands; many groups oriented by the Gandhian philosophy of self-help, social justice, and environmental sustainability; numerous indigenous and local environmental movements opposed to logging, mining, large-scale dams, etc.
Indonesia	600 independent groups working in various forms of environmental protection; proliferation of peasant irrigation societies.
Kenya	16,232 women's groups with 637,000 members registered in 1984, many of which began as savings clubs.

Mexico	At least 250 independent development organizations; massive grassroots movement among urban squatters; growth of consumer cooperatives and neighborhood organizations; rise of the *Zapatista* indigenous movement in Chiapas.
Nicaragua	Rapid growth of popular organizations since early 1980s; peasant organization is particularly strong, with a membership of more than 125,000 households.
Peru	300 independent development organizations; women's self-help movement with 1,500 communal kitchens in Lima's shantytowns.
Philippines	3,000–5,000 Christian Base Communities; numerous indigenous and local environmental movements opposed to logging and large-scale dams.
Sri Lanka	Over 8,000 villages representing one-third of the country's total, participating in the Sarvodaya Shramadana village awakening movement; 3 million people involved in a range of initiatives, including work parties, popular education, preventive health care, and cooperative crafts projects.
Thailand	Rapid growth of peasant movements opposed to logging, tree plantations, and other commercial development in forest reserves.
Zimbabwe	Small-farmer groups with an estimated membership of 400,000, 80 percent of whom are women; proliferation of urban community gardens.

Sources: Bebbington and Farrington (1993); Burger (1987); Colchester (1994); Durning (1989); Razeto (1991).

Fuentes and Frank (1989) point out that, while most of the 'new' social movements in the North have been primarily organized among middle-class sectors, the movements of the South typically have a more popular/working-class orientation. At the same time, however, the popular movements of the South have normally distanced themselves from mainstream labor organizations and traditional leftist political parties, both of which are commonly perceived to be unrepresentative of the popular majority and to offer no viable solutions to critical development problems. Moreover, many theorists argue that the participants in popular movements have primarily been motivated not by traditional class concerns, but by conjunctural and shifting subjectivities that seldom result in stable political positions (e.g., Escobar 1992; Laclau and Mouffe 1985; Stephen 1992; Touraine 1988). Not all of the diverse problems of marginalized and disenfranchised groups are centrally rooted in capitalist class relations. In many countries, for example, conflicts based in ethnicity, religion, gender, and the environment are much more pervasive and potentially explosive than the classical theories put forth by the Left have recognized (Basu 1987).

Given their tremendous diversity, recent thinking on popular movements has tended to move away from the restrictions of classical theories (Peet and Watts 1993). For many theorists, the rise of popular movements has encouraged the emergence of new subjectivities. As Escobar (1992: 423) notes, 'The categories of the "underdeveloped" and the "poor" are witnessing a hopeful fragmentation; women, indigenous peoples, the peasantry, various types of urban groups, the environment, questions of peace and security, and so forth, are achieving a new visibility which makes them important historical subjects in their own right.' Even if they are frequently motivated by economic reasons, the members of popular movements often do not conceive of their struggle in purely economic or class-based terms. Theoretically and practically, the result has been a rupture in the previously unified political space of the Left which was characterized by a single privileged subject – the working class. This unity has been shattered to make room for the plurality of collective actors that make up the popular sectors. New grassroots conceptions have been advanced which view society as an essentially plural and multi-sided entity which is produced, at least in part, through the construction of collective identities by diverse social actors according to multiple views and interests. These identities are necessarily unstable; they are historically formed through complex processes of articulation among various social groups. Furthermore, they have generated a whole new style of pluralistic, bottom-up, non-party political activity, which is slowly revitalizing the Left in many countries by transforming the very nature of what constitutes a progressive political practice (Basu 1987; Escobar 1992).

In many Third World countries, prolonged periods of authoritarianism have weakened traditional organizations such as labor unions and political parties. In addition, the market-oriented nature of neoliberalism has frequently undermined conventional political structures, rendering them unable to direct or intervene in the economy on behalf of the poor and disadvantaged. Both of these factors have contributed to the recent rise of decentralized, autonomous popular movements. Characteristically, these movements conform more to the principles of direct, participatory democracy than those of hierarchical, party-based, 'representative' political systems. For Escobar (1992), a new type of politics is emerging that is focused on 'autopoietic' (i.e., self-producing and self-organizing) movements which exercise power outside the state arena and which seek to create 'decentered autonomous spaces.' The aim of such movements is not power per se, and especially not state power, but the establishment of conditions (often involving non-formal or non-conventional forms of power) which would allow people to gain greater control over decisions affecting their lives. Sometimes movements may branch out horizontally or vertically, especially to form alliances or coalitions with other organizations over common concerns, but typically such 'scaling up' is done in a manner which stresses local participation and autonomy. According to Nerfin (1977), these movements collectively form a 'Third System politics' which represents a 'counter power' to the state and conventional political systems. To the extent that the Third System is growing, the traditional role of the state in development is diminishing.

In one of his books advocating an alternative, bottom-up approach to development, Friedmann (1988: 290–1) asserts, 'What is needed is a new set of rational beliefs that will express new sets of power relations. Such a counterhegemony . . . must necessarily arise from outside the existing structures of domination. It cannot merely be invented, but must arise from the everyday practices of ordinary people as they struggle to carve out a small niche for themselves in the incipient chaos.' This statement underscores the importance to alternative development of culture and cultural politics, which are primarily formed from the everyday practices of people at the local scale. Cultural politics may be defined as the domain in which meanings are constructed and negotiated – where relations of dominance and subordination are defined and contested (Jackson 1991: 219).

Recent analyses of cultural politics have collapsed conventional distinctions between the 'cultural' and the 'political' or 'the economic.' It is contended that one cannot divorce the 'cultural' aspects of development processes from their apparently 'political' and 'economic' dimensions, but neither can the political economy of development be understood without an analysis of its cultural politics. Elements of the natural and built environment commonly take on symbolic as well as material values, which may have important ramifications for development programs, especially concerning local input and participation. Jackson (1991: 226) states:

> Even apparently simple 'economic' resources within the same 'locality,' such as an urban park or an abandoned building, are subject to diverse readings by different groups in different material circumstances: a source of recreation and of danger, an opportunity for reinvestment, or a threat to neighboring properties. Each of these examples demonstrates how 'economic' resources are culturally encoded, their significance depending on such subjective appraisals as much as on any intrinsic material value.

This perspective emphasizes the simultaneity of symbolic and material struggles over resources. It challenges analyses of development to take local cultures seriously, not simply as a quaint epiphenomenon of the structural features of society. It also focuses on the active potential of ordinary people to subvert and reinterpret, in ways that best suit their needs and interests, the dominant meanings attributed to development by the mainstream institutions. For Third World people, this means shaking off the meanings which the dominant development discourse imposes on them. This should open up possibilities for alternative regimes of truth and perception within which new development practices are possible. Increasingly, popular movements are viewed as vital to this process, involving not just economically based struggles over material conditions, but also cultural struggles over alternative meanings of development. New discourses are being developed from the bottom-up based on alternative development meanings and perceptions. They include different types of knowledge, experiences, and ways of doing things. While many popular movements initially developed in response to basic problems of survival, they have often grown to embody a very positive sense of identity for their participants. In Chile, for example, Velasco and Leppe (1989) note that communal kitchens, vegetable gardens, and other popular, community-based development initiatives came to represent more than just efforts to satisfy basic needs; they served broader human needs as well, fostering a sense of community, friendship, self-awareness, participation, and self-empowerment to change one's circumstances. In many ways, then, popular movements have been critical to developing new 'political cultures,' based on a positive sense of community and collective identity that previous top-down development programs only tended to diminish. Escobar (1992: 427) states:

> [T]he new theoreticians see social movements as the search for a new 'political culture,' one in which struggles are less mediated by conventional forms and discourses, and which may even make possible the construction of a new political project based on a different practice of democracy. This practice would be characterized by a more direct and independent style of participation, and involve the politicization of everyday individual and social spheres, the expansion of the realm of the political in general, and the recovery of the social away from the control of the state.

Analyses of popular movements commonly stress the importance of community and sense of place, especially to the development of alternative political cultures. In many countries, the 'national' culture is a rather artificial construct in comparison to local/regional cultures (Hettne 1990; Somjee 1991). The strength of localized popular movements is often attributable to a collective sense of place and community identity, which participants share and which helps to transcend class, gender, caste, and other social divisions (Routledge 1992). Communities are spatially constituted social structures and centers of collective consciousness and socio-spatial identities. In spite of differences, community members often share common interests in defining and implementing locally based development alternatives. These include an interest in preserving and expanding local knowledge and culture, assuming a critical stance with respect to externally imposed development strategies and projects, and the defence and promotion of local popular institutions and organizations. People's view of the relationship between the 'local' and the 'global' is frequently affected by the rise of localized popular movements, which in many cases contribute to a realization that local contexts offer important ideas and methods for understanding more global issues. At the same time, these movements typically stress the importance of understanding, within the context of local struggles, how the everyday practices of local people are linked to overarching global processes. As Escobar (1992: 423) notes, 'a new dialectic of micro-practice and macro-thinking seems to be emerging,' which is primarily being advanced by activists engaged in local struggles through participation in popular movements.

Issues of Knowledge and Empowerment

Much of the attention that NGOs and popular movements have received in recent years has focused on questions of empowerment. Increasingly, it is believed that the alleviation of widespread inequalities and poverty cannot be achieved simply via traditional, top-down redistributive mechanisms, such as subsidies and income transfers, but that it requires the empowerment of the poor. However, this is a radical task which is usually inimical to the interests of the centralized state and allied economic forces. In most highly polarized Third World societies, reducing inequalities and poverty is neither a politically nor an economically neutral activity. Many governments feel a strong ambivalence toward grassroots anti-poverty work, especially if it involves popular education and consciousness raising. For a growing number of anti-poverty workers themselves, though, education and 'conscientization' are believed to be critical for instilling self-confidence and inspiring self-expression among the poor – without which effective human-resource mobilization and participatory decision-making are impossible. As Dube (1988: 88) remarks, the process of conscientization helps the poor to understand the root causes of their problems, which may have quite radical implications:

> Conscientization may be understood as a process of cognitive and evaluative transformation, especially in the poor of the world. It enables the individual to contemplate the environment and the human condition and gain an understanding of the forces that are shaping the contemporary world. Of special interest is the interplay of social, economic, and political currents that result in inequities and injustices in the social order. The individual begins to ask: Why? How? What next? Such reflective exercises, hopefully, will bring home the realization that deprivation and misery are not god-given, nor are they the result of any innate deficiencies in the poor as a group. By and by

the poor would begin to analyze the structures of exploitation and oppression that have brought vast sections of humanity to their present predicament.

Similar to concepts of participation in general, much ambiguity surrounds notions of empowerment. Friedmann (1992: 34), for example, connects political with social empowerment, so that 'political empowerment would seem to require a prior process of social empowerment through which effective participation in politics becomes possible.' Moser (1989: 84) finds that notions of empowerment have frequently been linked to contrasting definitions of participation: 'To date one of the most widely utilized methods of identifying or 'measuring' empowerment in participatory projects has been through the distinction between participation as a means and participation as an end.' In examining the participatory aspects of development projects, Moser believes it is important, first, to identify if and how projects move from a focus on participation as a means (e.g., to increase efficiency, effectiveness, or cost sharing) toward participation as an end (e.g., for empowerment and building human capabilities) and, secondly, to identify what have been the consequences of empowerment in the context of specific projects (ibid.).

From this perspective, grassroots development efforts can be more effective, especially over the longer term, if they move away from a focus on fixed, externally defined goals toward a more flexible, 'enabling' orientation designed to develop the intellectual, moral, managerial, and technical capabilities of local participants (Ghai 1989; Thomas-Slayter 1992). Empowerment is, therefore, commonly regarded as a multifaceted process, involving the pooling of resources to achieve collective strength and countervailing power, and entailing the improvement of manual and technical skills; administrative, managerial, and planning capacities; and analytical and reflective abilities of local people. By including all these facets, Ghai (1989: 218) contends that 'the concept of participation as empowerment comes close to the notion of development as fulfilment of human potentials and capabilities.'

In its most ambitious form as empowerment, participatory development seeks to engender not only practical self-help and self-reliance, but also effective collective decision-making and collective action (Black 1991; de Janvry and Sadoulet 1993). In the end, the sustainability of the empowerment process itself is regarded as at least as important as the immediate completion of particular projects (Moser 1989). Especially important is the creation of local institutions and organizations with participatory methods that permit people to become actively involved at all levels of decision-making. Local participation and empowerment are necessarily based on voluntary association. People only constitute themselves in a truly participatory organization when they perceive that they share common concerns and voluntarily decide to act collectively on them. For this reason, Johnston and Clark (1982) contend that participation should be recognized as a difficult and time-consuming process for the poor – as a considerable investment of their scarce time, energy, and resources. If the poor are to make this investment, participatory organizations must concentrate on meeting the self-defined needs and interests of the poor themselves rather than externally prescribed goals. Moreover, the methods these organizations adopt should follow a 'social learning' approach which builds on people's own initiatives, with outside agencies playing an essentially enabling and supportive role (Friedmann 1992: 160). As Johnston and Clark (1982: 169) comment:

> The sad fact is that analysts, planners, and politicians simply do not know what kind of local organization is actually in the poor's interests. The delusion that sufficient cogitation can overcome this ignorance – that the 'newest direction' will finally be the right direction – may be a greater obstacle than ignorance itself to designing better [development] programs.

In order to really benefit poor and marginalized people, participatory organizations must become sensitive to variations in local conditions, needs, and attitudes. In most Third World areas, culturally and communally specific norms and interests form an essential part of relations of production and social reproduction. Complexity and diversity commonly characterize the livelihood strategies of the poor; some may adopt specialized strategies which rely on a single activity, but most are more versatile and opportunist. As Chambers (1991: 6) notes, different members of poor rural households often engage in different activities at different times of the year: 'They cultivate, herd, undertake casual labor, make things to sell, hawk and trade, hunt and gather a multiplicity of common property resources, and migrate for seasonal work. They bond their labor, beg, borrow and sometimes steal.' Such complexity should serve to caution against reductionist interpretations of the development needs of the poor as defined by outside professionals.

Traditional or indigenous forms of knowledge often offer an important 'window' into the complex and diverse conditions which poor and marginalized people face. Traditional knowledge has frequently been overlooked in the search by outside professionals to find solutions to the development problems of the poor (Bebbington and Farrington 1993; Jain et al. 1986; Taylor and Mackenzie 1992; Wilken 1987). However, increased use of traditional knowledge may make development programs more appropriate to local conditions, provide innovative solutions to certain problems, contribute to a sense of self-worth and collective self-esteem among local people, and enhance popular participation and empowerment. In order to take advantage of these possibilities, development programs should start with the premise that poor people, despite the constraints they face, are knowledgeable and skillful managers of their own environment. If provided with adequate resources, their knowledge and skills place them in an ideal position to devise locally appropriate solutions to their own development problems.

Within development literature, much of the attention that has been given to traditional knowledge has been connected to studies of appropriate technology. For Ovitt (1989: 23), 'appropriate technology represents a direct challenge to manipulation [of the poor] by TNCs or repressive governments since democratic empowerment, grassroots decisionmaking, and economic self-sufficiency lie at [its] roots.' Among the common characteristics of appropriate technology, Ovitt (1989: 26) lists the following:

1 requires low capital costs;
2 uses local materials whenever possible;
3 is understood, controlled, and maintained by the local people who use it;
4 creates employment opportunities locally;
5 involves decentralized, renewable energy sources;
6 is flexible enough to be adapted to local conditions;
7 does not require patents, royalties, import duties, or consulting fees; and
8 assumes collective efforts by the community rather than just individuals.

Appropriate technologies often incorporate traditional knowledge systems, which commonly reflect generations of careful observation of the natural environment, local socioeconomic conditions, and society–nature interrelationships. Wilken (1987), for example, describes a variety of traditional agricultural techniques employed by the peasantry in Mexico and Central America, including horizontal tunnels dug into hillsides to access underground water, silt traps, ant-nest fertilizer, manual irrigation with gourds, runoff harvesting, curved fields, and mulch seed beds. By developing appropriate technologies which use local knowledge and resources, and minimize reliance on other factors, many popular organizations have been able to design relatively inexpensive development projects to meet specific local problems. Such methods are central to the concept of 'socially appropriate technology' (SAT), which recognizes that technology is not neutral and can be used either to make people more dependent and vulnerable or to provide them with a means for empowerment (Baquedano 1989).

In many countries, popular organizations and NGOs have carried out extensive research to design inexpensive and simple technologies which have proved effective in meeting the economic and broader human needs of the poor. In Chile, for example, the Center for Studies in Appropriate Technology in Latin America (CETAL), in coordination with local popular organizations, has developed energy-efficient insulated cooking pots which yield a 50 percent saving in cooking fuel (Baquedano 1989: 129). CETAL has also encouraged the cultivation and use of medicinal herbs to treat common illnesses, and has developed sanitary, manure-producing latrines which provide garden fertilizer and reduce the use of environmentally damaging cesspits. Similarly, the Center of Education and Technology (CET) has cooperated with Chilean popular organizations in developing organic gardening techniques that have been employed in the household and community gardens of poor urban areas to lower food costs and increase nutritional levels (Page 1986: 40). Through a series of training workshops entitled *Somos Capaces* (We Are Capable), the CET and local popular organizations have also been able to introduce greenhouses, hand water pumps, domestic animals, beekeeping, and tree nurseries into many poor communities. These initiatives both reduce the dependence of the poor on the market, and empower them to develop locally appropriate methods to meet their material and non-material needs (Cameron 1994).

Although the appropriate technology literature has usually concentrated on developing traditional or indigenous techniques, attention has recently been given to creating methods to combine various technologies more effectively. This new thinking on appropriate technology does not necessarily preclude the use of outside, advanced technologies by development projects oriented toward the poor. Indeed, it is increasingly recognized that, under certain conditions, advanced externally-developed technologies (e.g., yield-increasing Green Revolution technologies, new bio-technologies) may prove beneficial to the poor (Ahmed 1988; Bebbington and Farrington 1993; Rao 1988). Moreover, as Lele and Adu-Nyako (1992: 104) note, development programs have often ignored the potential benefits of technological advance for the poor: 'Both the integrated rural development projects of the 1970s and the structural adjustment programs of the 1980s overlooked the fundamental importance of science and technology in the generation of appropriate technology to suit the constraints faced by small farmers.'

In order to promote efficiency with equity, however, such technological development must be accompanied by complementary measures designed to assist the poor (e.g., in areas of credit and financing, land reform, infrastructure building, input supply, processing and marketing, dissemination of related knowledge). In addition, an alternative model of technological change should avoid formal 'textbook' solutions to development problems in favor of a flexible approach which makes room for local knowledge and practices. This type of approach is necessarily based on principles of popular participation and empowerment. Poor communities should decide for themselves, through their own local institutions and popular organizations, what mix of traditional and modern technologies best suits their particular needs. If not, technological change will continue to be associated with problems such as the destruction of local cultures, the widening of inequalities, and the further marginalization of the poor and disadvantaged.

Such problems of technological change may be attributed, at least in part, to the common use of Western scientific concepts and methods within development projects. As Escobar (1992: 420) notes, 'Not only has Western science been the major inspiration and legitimation of the failed dream of Development, but it has actually been an instrument of cultural violence on the Third World.' Despite its notable achievements, the Western 'scientific method' has tended to restrict access to knowledge, especially by the poor. At the same time, traditional, aesthetic, and intuitive forms of knowledge have become marginalized. In the end, 'Knowledge itself becomes mystified, losing sight of people's needs' (ibid.). Development projects typically conform to top-down, core–periphery, center-outward biases of knowledge that afford no conceptual space for the ideas and perceptions of the poor and disadvantaged. As a result, in the language of Spivak (1988: 308), 'The subaltern cannot speak.' Third World subalterns or marginalized people are left with no subject position from which to represent themselves.

In order for poor people to express their knowledge and creativity within development projects, favorable conditions have to be created. According to Chambers (1991: 9), these conditions include the establishment of rapport among project participants, whereby outsiders show humility, respect, and interest in learning from local people; the exercise of restraint by development professionals, so as not to over-interpret the views of locals; the employment of intersubjective, participatory research methods; and the utilization of local knowledge, practices, and materials whenever possible. There are many potential paths that development projects might take to create such conditions. To explore them, new approaches and methods for development work are required which are consciously methodological. For rural development work, one such approach that has recently evolved is termed 'participatory rural appraisal' (Chambers 1991), which moves away from the fixed, 'extractive' research procedures of Western science toward more flexible, shared methods of acquiring information. This approach calls on outside development 'experts' to respect the knowledge of the poor and be willing to learn from them. Ready-made, top-down, 'blueprint' methods of development work are rejected in favor of a bottom-up social learning approach which stresses the active participation and empowerment of the poor.

Within this approach, development professionals may continue to play an important role in projects as organizers, facilitators, catalysts, and animators. Frequently, they may also need to engage in 'capacity building,' by providing appropriate conceptual and organizational 'tools' for local people to use.

However, such tools should not conform to the formal methods of Western science, which have often served only to quash local initiative and generate distorted results. Instead, more flexible methods (e.g., use of disciplined observation, guided interviews, and informant panels) ought to be employed that are capable of addressing the diverse and complex realities of the subjects of development themselves. This should enhance not only the informational elements, but also the participatory aspects of development initiatives – leading to the design of projects that are more appropriate and empowering for the poor.

Limitations of Bottom-up Development

Black (1991: 160) states: 'Development, in theory and in practice, is a slave to fashion, and current fashion dictates the promotion of community organization and the involvement of the community in the assessment of needs and the planning of projects.' Nevertheless, while virtually all development agencies currently contend that this is their objective, few have actually put this process into practice. In an analysis of development policies in Africa, for example, Wisner (1988) asserts that a genuine bottom-up, participatory approach to development has yet to emerge. Commonly, the poor have been reduced to passive recipients of often inappropriate goods and services supplied individually rather than on an integrated basis. For alternative strategies to begin achieving their potential, the poor need to be genuinely empowered. This would involve not just piecemeal reforms, but fundamental changes to the status quo and the distribution of power.

As part of this process, closer attention would need to be paid to the meaning of the frequently abused term participation. Although alternative development programs have often been promoted by rhetoric about decentralization and participation, in practice they have generally been tightly controlled by the state and/or outside development agencies. For the state, there is often a basic contradiction between paying lip service to participation for reasons of political expediency, and a real fear that grassroots organizations will generate popular empowerment beyond state control. Consequently, as Moser (1989: 117) notes, '[Alternative] projects of this sort, whose objectives include capacity building, effectiveness, and cost sharing, but which in practice also result in empowerment, tend to be introduced by governments for specific political reasons linked to social and economic transformations at the national level, and to last only as long as those reasons are valid.' Popular, bottom-up agendas have often been forced to compete with professional efforts to redirect them toward top-down, institutionalist goals. Responsibility for group formation has commonly been placed within conventional bureaucratic agencies that have a questionable commitment to many of the qualitative aspects of development policies. The imposition of outside concepts of participation – not only by state agencies, but also by many NGOs – has often undermined indigenous forms of political organization and democratic practice, thereby reproducing paternalistic and authoritarian patterns of domination (Fowler 1991, Fox 1990). As a result, the poor and marginalized have commonly perceived development programs to be outside-imposed, and participation has been reduced to a mainly rhetorical exercise.

Genuine popular participation has frequently been lacking not only in the relations between outside development agencies and local communities, but also in the internal relations of those communities themselves. The distribution of political and economic power at the local level in many countries is such that, if care is not taken, bottom-up development initiatives may result in increasingly skewed resource allocations which further limit the development opportunities of the poor. Commonly, development efforts have paid insufficient attention to local economic, political, and sociocultural structures, especially the ways in which these structures may interact with the organizational forms of projects being introduced. Moreover, little consideration has typically been given to the nature of local institutions and organizations, which may often be undemocratic and exclude particular groups on the basis of class, gender, ethnicity, or other criteria. Fleming (1991: 40–1), for example, notes that women have commonly been excluded from peasant organizations involved in alternative projects, and that informal 'chambers of commerce' comprised of local entrepreneurs have dominated many indigenous NGOs. It appears that many bottom-up development efforts have ignored the possibility that local elite groups might monopolize project benefits or that field-level officers and project leaders might, however inadvertently, align themselves with such groups against the interests of the poor. The tendency within bottom-up development projects has been to conceptualize communities in homogeneous terms. Possible sources of inequality and discrimination within local power structures have been neglected, and it has been assumed that locally controlled projects will be broadly empowering. The structural context within which projects take place has largely gone unexamined. Likewise, issues of how local and community organizations are related to broader socioeconomic and political structures have been ignored. As

Vivian (1994) relates, many projects have suffered from the 'magic bullet' syndrome, in which simple, neat solutions to development problems are sought which fail to take account of the complexities of local structures. However, the poor normally face substantial structural impediments to their improved social and economic well-being. These include an unequal distribution of resources and assets, skewed power relations, and a frequent dependence on local elite groups – even though these elites may well be responsible for the continuing oppression of the poor. In addition to these impediments faced by the poor in general, more specific obstacles (based in structures of gender, ethnicity, religion, or language) may also block the advance of particular marginalized groups. Experience indicates that as long as these structural impediments faced by poor and marginalized groups go unaddressed, bottom-up development projects, despite their frequent good intentions, have little chance of alleviating poverty and inequalities.

Generally, the most basic structural obstacle faced by the poor in Third World societies is the extremely unequal distribution of productive resources and assets. In rural areas, where poverty is usually particularly severe, land and other resources have become increasingly concentrated among elite groups, thereby depriving the bulk of the population of any real opportunity to develop. Many analysts, therefore, contend that broadly based rural development is impossible in most countries without fundamental agrarian reforms, including land redistribution (see, e.g., de Janvry et al. 1989; El-Ghonemy 1990; Griffin 1989; Islam 1992; Lipton 1993; Prosterman et al. 1990; Wisner 1988). Moreover, as the experience of a few countries (especially Taiwan and South Korea) demonstrates, the creation of a more egalitarian pattern of rural development through land reform can provide an essential stimulus not only for agricultural growth, but also for the critical initial stages of industrialization. However, notably absent from most bottom-up development projects, especially those initiated in recent years, has been land reform.

While it should be recognized that conventional, top-down 'confiscatory' land reform is generally politically unfeasible today, this does not preclude possibilities for other, more flexible and innovative types of land reform. Given the problems of conventional land reform, as well as the failure of other rural development initiatives to reduce poverty and inequalities, Lipton (1993: 650) contends that an alternative paradigm or 'new wave' land reform is needed. This would contain three components: (1) a shift toward small farms for 'appropriate' parts of the agricultural sector, with larger-scale (cooperative or private) units in other parts; (2) the promotion of smaller farms through various market-based or tax incentives; and (3) the use of confiscatory reform only if it is controlled by local authorities or NGOs instead of a centralized state bureaucracy. These recommendations recognize the need to adapt land reform, or any other rural development effort, to local conditions, which may vary considerably. The result could well be a dynamic and innovative reform sector with a multiplicity of property arrangements and organizational forms (e.g., small/medium individual farms linked through credit and service cooperatives, 'hybrid' cooperatives combining individual plots with communal landholdings, 'work collectives' composed of individual farmers who pool their labor for certain activities such as harvesting). In Nicaragua, for example, a highly flexible and innovative reform sector of this nature has recently been evolving, with the support of popular organizations such as the National Farmers Union (UNAG) and many NGOs involved in rural development work.

However, most bottom-up development initiatives have neglected larger structural concerns such as land reform, agricultural pricing, and the thrust of macroeconomic policies in general. Projects have remained small and have tended to focus on individual aspects of development. Moreover, they frequently operate in an isolated fashion, outside of the overall policy framework of Third World countries. As a result, efforts to improve the coordination and coherence of various development programs have been hampered, and there is typically little transfer of knowledge from one development body to another, limiting the ability of different agencies to learn from each others' mistakes (Utting 1994). Although there are many examples of localized projects that have improved the lives of a few people here and there, relative to the massive needs of the poor in most countries, such efforts merely represent a drop in the bucket.

Especially during the era of structural adjustments, many bottom-up projects have simply tried to 'fill the gaps' in development programs created by macroeconomic policy reforms. As poverty and other social problems mount, however, there is a growing perception that these initiatives are fighting a losing battle – for every 'gap' filled, many more appear. Neglecting larger structural and macro-policy concerns, alternative projects have tended to focus on isolated problems of survival, ignoring more fundamental issues concerning the systems that generate increasing poverty and inequalities in the first place. Because they have remained isolated and uncoordinated, grassroots projects and movements have been unable to step beyond the micro-level, marked by individual local responses to specific problems, into the macro-realm of effective popular national organizations capable of uniting to confront development issues. As a result, they run the risk of reinforcing tendencies toward the fragmentation of the popular sectors and the creation of societies composed of disconnected apolitical individuals.

In order to avoid such problems, there is growing recognition that local popular movements need to 'scale up' their operations to create micro–macro linkages that can affect the overall framework of development policies (Fleming 1991; Thomas-Slayter 1992; Vivian 1994). Genuine empowerment and participation in development decision-making cannot, in the end, remain dependent solely on local, pressure-group politics or the whims of NGO project funding. Instead, popular democratic structures are needed that can translate decisions from below into needed changes in macro-policy. Designing development policies to serve majority needs and interests necessitates removing barriers to popular participation inherent in existing top-down methods of policy formulation. But simply improving communications and information flows through administrative decentralization is not a sufficient condition for macro-policymakers to integrate bottom-up directives; there also needs to be a political commitment to ensure that this process takes place, and this can only be assured through the constant vigilance of a politically organized grassroots.

In many countries, feelings of mutual suspicion and mistrust characterize relations between grassroots organizations and the state. Popular movements have tended to view the state as an adversary, dominated by elite groups aligned against the interests of the majority. In instances where popular organizations have pursued confrontational strategies, they have often been met with repression. In other cases, popular organizations have rebuffed overtures from the state for increasing cooperation for fear of being coopted by state institutions. For its part, the state has largely ignored bottom-up development efforts, unless they have threatened to provoke popular unrest. The state's attitude toward alternative development initiatives can usually be described as one of benign indifference. Local development projects have been regarded as a relatively inexpensive and convenient means of maintaining social order. But if conscientization and empowerment have become major components of such initiatives, the state has usually intervened to exert its control in a heavy-handed manner.

In recent years, however, both popular organizations and the state have begun to modify their stances toward one another. Many grassroots groups have realized that, given the massive scale of most social problems, progressive development cannot be achieved without state support. Individual development efforts that are poorly coordinated and not integrated with national development planning have often produced unnecessary duplication, considerable confusion and disarray, and many projects operating at cross-purposes from one another (Islam 1990; Thomas-Slayter 1992). In some instances, the state may also offer useful support for popular organizations against reactionary local elites (Sutton 1990). Moreover, more equitable development generally requires a fair measure of resource redistribution, taking from the richer (regions and people) and transferring to the poorer – and this can most effectively be accomplished with state assistance (Sachs 1988). For all of these reasons, an increasing number of popular movements feel the need to influence the state more effectively – through some combination of grassroots political action and lobbying, participation in the electoral process, and, in the end, promoting radical political reforms that will render state structures more responsive to majority needs and interests.

At the same time, an increasing number of Third World governments have come to realize the value of bottom-up development strategies. Such strategies are attractive in that they provide private responses to what have traditionally been considered public responsibilities. If managed properly, increased cooperation with NGOs and popular organizations presents possibilities for making social services more efficient and less costly, without the state simply having to push responsibility for the well-being of the poor onto the informal economy. Local groups are often able to innovate and respond quicker and more appropriately to changing conditions than are more unwieldy state institutions. In some cases, state agencies can use knowledge and technologies developed at the local level by grassroots organizations. Innovations by NGOs and other groups in areas such as participatory research, organizational forms and methods, and technological dispersal can also be incorporated into state agencies. Given all of these advantages, governments might want to treat bottom-up development projects as a type of 'infant industry' which, if nurtured by state support, could help to generate more equitable growth by relatively inexpensive means.

Despite the tendency toward increasing collaboration in recent years, cooperation between the state and popular organizations frequently remains a difficult process. The establishment of closer links between state agencies and popular organizations normally demands a reconciliation of radically different organizational styles and approaches to development, as well as the negotiation of clear areas of jurisdiction. Whether the complex range of relations between the state and popular organizations can be institutionalized in an effective and mutually agreeable manner remains an open question in most countries. On the one hand, popular organizations must increase their visibility and establish their own niche within national development agendas by scaling up their operations and becoming more politically involved, but without losing their autonomy and representativeness. On the other hand, the state must provide necessary resources and a favorable environment for this process, but without becoming overbearing.

In the end, processes of alternative development, if they are to occur in more than a handful of isolated communities, require the support of a strong state. However, a strong state is not top heavy with arrogant and cumbersome bureaucracy dominated by elitist interests; instead, it is agile, responsive, and accountable to the popular majority. It is a socially progressive state that receives its vision and strength from the principles of a popular or 'inclusive' democracy in which a division of power has been negotiated that allows many development problems to be managed by decentralized units of governance in cooperation with the people themselves, organized in their own communities and popular organizations. As Friedmann (1992: 84) remarks:

Inclusive democracy incorporates a fine areal division of powers, it insists on accountability as a central process and secures an open political space for civil encounter and mobilization. Such a democracy, which includes all potential interests and concerns, will assign a significant role to organized civil society, including the very poor, in the making of public decisions at all relevant levels.

9

Women and Gender

The previous chapter explored popular participation and empowerment as one of the new emphases of alternative development. This discussion is extended here into a related area that has also received much belated attention: women and gender relations. Until recently, the role of women in development was all but invisible. New approaches, however, have made considerable progress in addressing this problem. Such efforts have been greatly assisted by the emergence of women's organizations and movements in many countries.

Although much remains to be done, recent theoretical and practical advances have been impressive in areas related to the role of women and gender relations within processes of both production and social reproduction.

Women, Gender Relations, and Development

Until quite recently, women and gender relations have been virtually ignored within the theory and practice of development.[1] Moreover, any recent change has taken place only very slowly and unevenly. Robertson (1988: 181) lists several ways that women continue to be systematically excluded from various aspects of development: from trade unions, peasant organizations, and other groups (frequently using the excuse of women's incapacity and frailty); from managerial and administrative opportunities in large corporations (because of lack of education, fear of competition, and notions of women's role in society as essentially domestic); from technical and managerial positions within development agencies, such as USAID, FAO, and the

---

[1] Elson (1991: 1) identifies gender relations as 'the socially determined relations that differentiate male and female situations.' This definition stresses the social, rather than biological, determination of gender relations.

UN; from many large-scale development projects and programs of national scope; from statistics used by planners to develop policy (e.g., women's agricultural labor is normally omitted from GNP calculations because it is unpaid); from many economically focused or income-generating projects (e.g., the 'ghettoization' of women into small, home-economics projects); and from control over significant resources, such as land and labor, which would help women fulfill their primary economic responsibilities.

Looking back at the modernization approach, Parpart (1993: 447) finds that, if Third World women were considered at all, they were typically regarded as an impediment to modernity and development: '[Neo]colonial discourse represented Third World women as exotic specimens, as oppressed victims, as sex objects, or as the most ignorant and backward members of backward societies.' Consequently, during the first two postwar decades, development theory and practice ignored women on the assumption that they would eventually be forced to adapt to modernization. Development plans were designed according to the premise that productive work was performed by men. Women as workers, owners, or entrepreneurs were virtually ignored (ibid.).

Not only mainstream development approaches such as modernization, but also more radical frameworks have tended to neglect women and gender relations (e.g., Chinchilla 1992; Robertson 1988). Many Marxist approaches, for example, have been accused of being inherently economistic, reductionist, and gender-blind. Women's work has been made largely invisible because much of it is unpaid (or paid for only indirectly through male intermediaries as household heads) and is viewed simply as a form of domestic labor without economic value. An essentially gender-blind class analysis in which women merit little, if any, specific attention has often driven studies of social stratification. Moreover, both Marxists and non-Marxists have frequently assumed that the class position of women can be subsumed under that of related men. Studies of rural inequalities have, until recently, almost always focused on class-based inequalities – as if inequalities based on gender were inconsequential (Ahmed 1987). For an inordinately long time, rural development studies neglected the central role of gender relations in systematically denying Third World women access to land and tenancy rights, training programs and extension services, rural credit, modern farming inputs, and other necessities to improve their position.

Drawing some parallels to notions of 'urban bias' (e.g., Lipton 1977) in development programs, Elson (1991) has introduced the concept of 'male bias.' While male bias commonly intersects with other class, ethnic, or regional biases, its proximate causes may be quite distinct. Male bias 'operates in favor of men as a gender, and against women as a gender, not that all men are biased against women' (p. 3). Among its immediate causes are conscious and unconscious prejudice and discrimination, socioeconomic structures that rationalize male bias, and theoretical models which neutralize categories (e.g., farmers, workers) and interpersonal relations (e.g., in households), thereby rendering gender relations invisible and conflict-free. As these categories and assumptions become enshrined in development policies, women's voices and concerns are silenced. Ultimately, these types of proximate causes of male bias are attributable to deep-seated underlying structures by which production and social reproduction are interrelated in particular societies (p. 13).

Elson and many other analysts (e.g., Afshar and Dennis 1992; Geisler 1993; Mackenzie 1993; Parpart 1993; Sollis and Moser 1991; Townsend and Bain de Corcuera 1993) contend that male bias is not just a past phenomenon, but continues to run through current development initiatives, especially structural adjustment programs. In many poorer, rural countries, the focus of SAPs on large-scale agroexport production has negatively affected smaller women farmers, whose ability to ensure household reproduction through food production has been undercut by changes in the relative terms of trade for food (Mackenzie 1993: 78–9). As a result, many of these women have been forced into seasonal and part-time job markets, usually on quite unfavorable terms (Townsend and Bain de Corcuera 1993: 46).

Moreover, women in general, and poor women in particular, appear to have borne the brunt of many hardships that austerity measures accompanying SAPs commonly induce. Frequently, women have intensified the use of their unpaid labor in response to such hardships. Although it remains generally unrecognized in the economists' models, this is one of the major hidden costs of adjustment (Afshar and Dennis 1992). The macroeconomic theory of SAPs assumes that the maintenance and reproduction of human resources, which unpaid women perform, will continue regardless of the way resources are allocated. It further assumes that changes in income, food prices, and public expenditures accompanying SAPs will affect all household members in the same way because of equal intra-household resource distribution. However, such assumptions have produced growing concerns, especially in two interrelated areas: (1) that SAPs are forcing poor women to work longer and harder, within both the market and household; and (2) that women's labor is not infinitely elastic – a breaking point may be reached where women's capacity to reproduce and maintain human resources may collapse (Elson 1991; Sollis and Moser 1991).

The special needs and interests of women have commonly been ignored not only by macroeconomic programs such as structural adjustments, but also by more specific development projects. By failing to be sensitive and responsive to the differential gender effects of policies, policymakers have often been guilty of compounding the already severe hardships many Third World women experience. Since in nearly every society women are at a disadvantage relative to men (e.g., in terms of income, assets, education, political influence), 'gender-neutral' policies have further marginalized women by weakening their socioeconomic position (Loufti 1987: 112). Because most gender-neutral policies tend to include and benefit only men, there is a clear need to promote measures specifically designed to enhance women's participation at all stages, from planning to implementation, in development initiatives.

Despite much rhetoric about the importance of women in development, organizations and agencies have often allocated only minimal resources toward the needs of women. For example, during the UN Decade for Women (1975–85), only 3.5 percent of the projects of various UN development agencies, representing 0.2 percent of their budget, targeted women. By the late 1980s, the budget of UNIFEM (UN Development Fund for Women) was a mere $5 million out of the UN's $700 million budget (Wipper 1988: 418). Women's projects have normally remained small, short-term, and peripheral to the main thrust of national economic development plans. Moreover, because they have often been formulated by planners who still view Third World women as housewives and who ignore the fact that the large majority are also involved in productive activities, many women's development initiatives are essentially welfare-oriented (e.g., handicrafts, home-economics) projects, ignoring the tremendous diversity of Third World women's needs and interests.

Many analysts have particularly noted the diverse and important, though largely unrecognized, contributions that women make to Third World agriculture. Studies have found that women represent 60–80 percent of the agricultural labor force in sub-Saharan Africa, 33–50 percent in South Asia, and more than 40 percent in Latin America (e.g., Ahmed 1987; Bandarage 1984; Penna et al. 1990; Robertson 1988; Stromquist 1992). Within sub-Saharan Africa, some 80 percent of economically active women work in agriculture, rising to 95 percent in Burundi, Mozambique, and Rwanda (James 1992: 38). In many African countries, women contribute about 70–80 percent in food production, 50 percent in animal husbandry, and nearly 100 percent in food processing (Wipper 1988: 418). According to 1984 FAO figures for rural Africa, women head approximately 30 percent of all rural households, contribute about 80 percent of agricultural labor, produce some 60 percent of the food consumed by rural households, and generate a third or more of all household income, mainly through small-scale agro-industry, trading, craft work, and casual labor (in Jiggins 1989: 953). In spite of this, Robertson (1988: 189) notes that a number of major African studies in the 1980s 'defined farmers as male and dealt only marginally with women as wives, if at all.'

The important role women play in Third World agriculture must be seen as remarkable, given the serious constraints that they normally face.

Ahmad (1984: 73) finds that rural women are generally subject to three major constraints: (1) limited access to land and related resources; (2) lack of control over their own labor and the fruits of this labor; and (3) lack of mobility due to family responsibilities or to social and cultural restrictions. Since land is usually the most important means of production in rural societies, lack of access to land has particularly affected women. In many Third World areas, following the enactment of formal land laws and modifications of land rights during the colonial period, women can obtain access to land, if at all, only through their husbands or fathers, instead of holding direct rights. Thus they are at a considerable disadvantage concerning the terms and conditions by which they provide their labor, irrespective of whether they are unpaid family workers, agricultural laborers, or workers on large commercial farms (ibid.: 77), and are also frequently deprived of access to other vital agricultural resources, such as rural credit, technological improvements and extension services, and modern farm inputs (e.g., Penna et al. 1990, James 1992). Such constraints not only disadvantage rural women, but contribute to the overall food crisis afflicting many countries, particularly in sub-Saharan Africa (Keller and Mbewe 1991, Nyaga 1986).

Despite their enormous potential to accelerate Third World agricultural development, women have been systematically excluded from both mainstream and alternative rural development efforts. Mainstream agricultural modernization schemes, such as Green Revolution projects, have often had particularly negative consequences for rural women – raising unemployment and denying them technological improvements (Agarwal 1989; Ahmed 1987; Cebotarev 1988; Gladwin and McMillan 1989; Zarkovic 1988). Women have also commonly been excluded from the benefits of alternative development initiatives, such as agrarian reforms, resettlement schemes, and integrated rural/regional development projects (Allison 1985; Deere 1985; Fleming 1991; Mayoux 1992; Moser 1989). In a comparative study of 13 Latin American agrarian reforms, Deere (1985: 1037) finds that most have only directly benefited men, as male household heads are the designated reform beneficiaries. Only in a few cases, where the incorporation of women into the reform process was an explicit objective of state policy, did significant numbers of women participate directly. Many reforms sponsored by USAID during the Alliance for Progress period selected potential beneficiaries on the basis of a point system which disadvantaged rural women on educational and other grounds (ibid.: 1042). Regarding donor-sponsored development projects, Sebstad (1989: 950) observes: 'Many – perhaps most – donors give lip service to the importance of women's participation in program conception and implementation, and to the need to be responsive to women's own definition of the problems they face; [but] true participation remains a difficult and elusive goal.' For Moser (1989: 85), this problem is rooted in the failure of most projects to develop disaggregated policies on the basis of gender which recognize the specific needs of women:

> The role that women play in community participation is very rarely mentioned at the policy level, despite the constant comments from those working at the grassroots implementation level that 'without the women the project would never have worked.' The fact that disaggregation on the basis of gender – which recognizes that men and women play different roles in society and consequently in community participation – is not considered of importance in many projects, often has serious implications.

The UN Decade for Women and the 'Women in Development' Approach

Events accompanying the UN Decade for Women (1975–85) originally provided much of the impetus for the rising interest in the role of women in development. A number of analysts have stressed the importance of the Decade for Women in promoting and legitimizing women's movements at the national and international scale (e.g., Hahner 1985; Jaquette 1987). Loufti (1987: 114) states: 'There is hardly a country today that has not formally acknowledged the importance of raising the status of women or [has] failed to adopt the Program of Action of the United Nations Decade for Women.' Geisler (1993: 1967) remarks that the Decade for Women prompted a rethinking of development policies which 'started to conceptualize women not only as carers of the family but also as agents in the productive process' and began to identify 'women's marginalization in the development process . . . as the cause of their deteriorating status.'

The UN organized three particularly influential women's conferences during the Decade for Women. The first two were held in Mexico City and Copenhagen, and were followed by the 'End of the Decade' Conference in Nairobi in 1985.[2] Three potent documents emerged after much analysis and debate over how to develop concrete strategies to ameliorate women's economic, political, and sociocultural marginalization. These documents were the 'Plan of Action', the 'Program of Action', and 'Forward-Looking Strategies'. Essentially, four major goals were delineated for future attention: legal equality, economic empowerment, increased control by women over their own bodies and the prevention of violence against women,

---

[2]   In addition to being important in their own right, these conferences also provided a springboard for South–South linkages among women, including the creation of an international organization, Development Alternatives with Women for a New Era (DAWN), which grew out of discussions before the Nairobi meetings. Subsequently, DAWN has continued to organize and deliberate on development issues, and has published a number of documents concerning strategies for reducing inequalities between genders, classes, and nations (Parpart 1993: 449).

and global peace (House-Midamba 1990: 38). These documents stressed linkages between socioeconomic and political conditions, and called for the full participation and emancipation of women. They stressed the need for alternative strategies which, in contrast to previous development models, would not adversely affect women; they recognized that some governments had recently taken positive steps to promote women's advancement, but stressed that these efforts had thus far been woefully inadequate (Penna et al. 1990: 64).

Closely associated with the UN Decade for Women was the rise of the 'Women in Development' (WID) approach – an 'integrative' or 'reformist' framework for overcoming women's development problems (Bandarage 1984; Moser 1989; Parpart 1993). As exemplified by the pioneering work of Boserup (1970), the WID framework offers a blend of modernization theory and liberal feminism – it assumes that Third World women can be liberated via their integration into the modern sectors of developing economies. Third World underdevelopment is essentially attributed to the continuing presence of traditional values and social structures. However, in contrast to previous modernization studies, the WID approach stresses that the benefits of Western-style development have accrued only to men; its central objective, then, is to spread such benefits to women. Like liberal feminists in the North, the WID framework links women's advancement with improved access to all aspects of educational and employment structures. Moreover, it contends that this goal can best be obtained through reformist legal measures and changes in attitudes – by within-the-system solutions which may be accomplished without fundamental structural change.

As might be expected, this reformist WID philosophy found a resonance with the approaches of many Northern-based development institutions, including UN agencies, the World Bank, bilateral aid organizations, and NGOs. Moser (1989: 1799) notes that the WID approach became especially associated with the work of USAID, based on 'the underlying rationale that women are an untapped resource that can provide an economic contribution to development.' WID advocates were instrumental in the passage of the 1973 Percy Amendment to the US Foreign Assistance Act which enshrined the principle that US development assistance should strive to improve the status of Third World women by integrating them into the development process (Parpart 1993: 448). For USAID and other Northern-based institutions, the WID approach seemed to offer a method for promoting women's advancement, and thus stimulating overall development, via Western-style reformist measures that obviated the need for more radical structural change.

However, for many analysts, this is also the most serious limitation of the WID approach – its failure to consider the possibility that women's advance might, under many circumstances, require structural change (e.g., Bandarage

1984; Kandiyoti 1990; Moser 1989; Parpart 1993). This limitation reflects both the material interests and the ideological boundaries of WID's principal progenitors: modernization theory and liberal feminism. Development for Third World women has essentially been reduced to becoming more Western, more modern. Women's poverty and marginalization are seen to be aberrations within an otherwise just and equitable social system, which can be corrected through legislative reforms, attitudinal changes, and interventionist projects designed to integrate women more fully into processes of modernization.

However, this approach ignores the fact that exploitative social relations may mark the very structures into which women are supposed to become integrated. Hierarchies based in gender, class, ethnic, and other social relations are never challenged. As a result, the WID approach tends to obfuscate a number of troublesome, but important, issues. Complex structural problems are turned into purely technical or legal questions that are amenable to simple reformist solutions. As Kandiyoti (1990: 14) remarks, this may lead to quite contradictory effects:

> Behind the rather bland and uniform sounding recommendations to equip and empower poor Third World women, there may lie a wide range of frankly contradictory objectives from simply making women more efficient managers of poverty, to using their claims and organizations as a political vehicle for far-reaching redistributive measures, both within and across nations. With very few exceptions, policy documents on WID effectively avoid and obscure these troublesome issues by presenting assistance to women as a technical rather than a political issue.

Capitalism, Patriarchy, and the Radical Feminist Perspective

While the WID approach argues that women, in order to overcome their historical marginalization, need to be better integrated into modern development processes, the radical feminist perspective contends that Third World women have been subjected to increasing exploitation precisely because of such integration. For radical feminists, a series of dialectically related social structures, particularly those of capitalism and patriarchy, have conditioned women's integration into development (e.g., Babb 1990; Bandarage 1984; Faulkner and Lawson 1991; Lim 1983; Porpora et al. 1989; Robertson 1988; Wilson 1985).[3] Analytically, it may be possible to separate class relations and capitalism from gender relations and patriarchy, but in reality they are inextricably interconnected. Patriarchy is virtually everywhere shaped by its relationship to the capitalist world system, just as capitalism almost

---

3	Patriarchy may be understood as a form of gender relations in which there is a systematic dominance of men over women.

always is shaped by its relationship to the patriarchal system. Historically, patriarchy may have been expressed in a variety of different forms among Third World countries. But radical feminists assert that a key common element has recently become women's increasing subordination within the capitalist labor market, which has deepened their economic dependence on and vulnerability to men. Thus, in order to understand contemporary forms of women's oppression and marginalization, radical feminists focus on the articulation or mutual accommodation of capitalism and patriarchy, and the resultant gender division of the workforce.

According to Lim (1983), structures of dependent capitalism and patriarchy come together in Third World countries to subject women to 'triple exploitation' based on capitalism, imperialism, and patriarchy. Like all wage laborers, women workers are exploited through the extraction of surplus by capital–labor relations established in the capitalist workplace. In addition, Third World workers are subjected to imperialist exploitation, which refers to the 'differential in wages paid to workers in developed and developing countries for the same work and output' (ibid.: 80). Such imperialist exploitation is readily apparent in the textile industry, in which Southern workers typically earn only a fraction of the wages paid to their Northern counterparts. Finally, women are subjected to patriarchal exploitation based on their subordinate position in society, making them susceptible to poorer wages and working conditions than men in comparable jobs. Because Third World women are uniquely subjected to all three forms of exploitation, their plight is regarded as particularly onerous.

Radical feminists contend that such exploitation has commonly been manifested in the creation of a 'gender division of labor' (e.g., Babb 1990; Baylies and Wright 1993; Faulkner and Lawson 1991; Phillips 1990). This concept has been used to analyze both the 'double burden' suffered by women who work long hours for little compensation in and out of the home, and the relegation of women to dead-end, low-paying jobs in the workplace. Although the gender division of labor may change according to variations in relations of production, Phillips (1990: 94) notes that it is always an 'asymmetric, hierarchical and exploitative relationship, and not just a simple division of tasks between equal partners.' Mackintosh (1981: 13) describes the rise of a typical gender division of labor in the workplace and its consequences for women:

> Once men and women workers have been created as two relatively non-competing groups within the labor force, with men relatively privileged in terms of wages and work conditions, then there exists a material division between men and women which can be exploited by management, and which reinforces women's social and economic subordination in the new economic sphere.

Many radical feminists argue that the gender division of labor widened considerably with the spread of capitalist relations throughout Third World countries during the (neo)colonial era (e.g., Allison 1985; Bandarage 1984; Hellman 1992; Porpora et al. 1989; Robertson 1988; Stephen 1992; Wilson 1985). This is particularly evident in rural areas where the concentration of land and other productive assets under the capitalist agroexport model has greatly exaggerated class and gender differences. While it should be recognized that women's position in rural development can only be analyzed within specific contexts and does not represent a 'neat package,' researchers have uncovered some common patterns. Land concentrations and the commercialization of agriculture have frequently imposed far heavier work loads on women, who have been forced into poorly paid jobs in cash-crop production, while continuing to carry the full burden of domestic labor. As agricultural differentiation has accelerated, women have also typically been relegated to the non-modern, subsistence sector producing domestic foodstuffs. With the increasing use of seasonal, semiproletarian labor in the modern agroexport sector, female production of subsistence foodstuffs has allowed agroexport capital to ratchet down wages to levels insufficient for familial social reproduction. This represents a considerable 'comparative advantage' for capitalist agroexport producers resulting from female exploitation.[4]

In addition to agricultural development, studies by radical feminists have also uncovered a gender division of labor within Third World industry, particularly in the assembly-type operations of transnational corporations (e.g., Babb 1990; Bandarage 1984; Baylies and Wright 1993; Robertson 1988; Sebstad 1989). Much of this research has focused on the impact of the new international division of labor (NIDL) on Third World women and gender relations. Rising women's participation in the industrial labor force has been attributed to the convergence of three macroeconomic trends: the emphasis on export-led industrialization, relying heavily on the use of low-wage female labor; the pursuit of cost-cutting (and, by implication, low-wage) strategies in development processes; and the deregulation of labor markets, especially with the advent of neoliberalism (Sebstad 1989: 938). According to Standing (1989), no Third World country has successfully pursued a development strategy based on export-led industrialization without relying on a huge expansion of female labor. As we saw in chapter 3, this is particularly true of the EOI strategies of the Asian NICs.

In most cases, female labor has been concentrated in industrial homework (i.e., the 'putting-out' system) or in assembly-type factory operations,

---

4   For a more detailed analysis of the common use of semiproletarian labor based on the articulation of peasant and capitalist sectors turn to the section on the agroexport model in chapter 2.

both of which are commonly controlled by TNCs. Frequently, hazardous and unhealthy working conditions and extremely long work hours, often compounded by shift work, extract high costs from female workers, particularly young women, who form a 'temporary' proletariat in the period of their lives between school and marriage. According to Greenhalgh (1985), this illustrates the interlocking and mutually supportive nature of capitalist industrial institutions and traditional patriarchal structures. Industrial capitalism takes advantage of the gender hierarchies within families by using women's lower educational levels, familial obligations, and temporary labor-force status to offer them dead-end, low-paying jobs that no other social group will fill. These discriminatory features of the industrial labor market act, in turn, to reinforce the subordinate status of women in the family (e.g., by stressing young women's income-earning obligations rather than education).

From this perspective, export-led industrialization in many Third World countries has only been made possible by the exploitation of women according to a gender division of labor based on the interlocking structures of patriarchy and peripheral capitalism.

Gender, Class, and Models of Household Differentiation

A growing number of analysts contend that the top-down, economic thrust of development theories, on both Right and Left, has systematically neglected household and family structures to the detriment of women (e.g., Cebotarev 1988, 1991; Evans 1991; Geisler 1993; Moser 1989; Moser and Peake 1987; Taplin 1989; Wolf 1990). For the most part, development theories have assumed that the Third World household consists of a nuclear family (husband, wife, and children). In addition, it is assumed that, within the household, there is a clear division of labor with a male 'breadwinner' and a female 'homemaker.' If household structures are considered at all, the concern is usually for overall stability rather than the individual human rights of family members (especially women). Even analyses of Third World social relations have seldom been concerned with the internal structure of households. Particularly on the Left, this has commonly resulted in a focus on production and class issues to the neglect of the reproductive sphere and gender considerations. Moreover, work is usually defined only in terms of exchange value. Work for use-value that Third World women typically perform (e.g., child-rearing, subsistence food production, fuel gathering and other domestic tasks) is thereby made invisible. For Duvvury (1989: WS97) this is a consequence of 'a cultural/ideological system which views man as the primary breadwinner.'

Cebotarev (1988) argues that, by supporting and institutionalizing patriarchal family structures, development processes have militated against the attainment of equal rights by women. The family infringes women's rights in a number of issues. Some are cultural–ideological, such as the unquestioned male authority in the family or the need to 'protect' women from the outside world. Others are based in the economic organization of the family, in which control over property and income is vested in the male household head, giving his power an authentic material base. Differential access to various male-dominated institutions and networks also contributes to the asymmetry of rights and freedoms. Frequently, more subtle and hidden ways of constraining women's lives are rooted in familial divisions of responsibilities and labor. Women's contributions to production and reproduction are often neither recognized nor remunerated properly, remaining 'invisible' to development planners and other outsiders.

For all these reasons, the household has become the primary site of women's oppression within many studies of gender relations. It is argued that the orthodox 'unidimensional' concept of households submerges much information that is vital to women's development concerning variations in household composition (e.g., by gender, age, kinship) and intra- and inter-household resource allocation and distribution (Evans 1991). Moreover, analysts increasingly see the conventional notion of households as 'a disguise for male power and female subordination' (Kabeer and Joekes 1991: 2). At least to some degree, all people's welfare and labor-allocation decisions are determined at the household level. The household represents a critical link between changes at the macro- or meso-scale and changes in personal welfare; it links together various development policies and individual behavior. The behavior, economic and otherwise, of household members is fluid and dynamic; and must be analyzed carefully rather than assumed.

For example, it cannot be assumed that policies that benefit male household heads will necessarily benefit all household members, who may not only be differentially affected by policies, but may also operate according to different, and possibly conflicting, goals. Such differences, especially within poorer households, are often reflected in a mix of individual and household survival strategies. Such strategies have become the focus of the 'livelihoods systems' approach, which stresses the tremendous diversity of reproductive and productive activities in which Third World households engage. Normally, these strategies have evolved over a considerable time, and seek to mobilize all available resources and opportunities. The strategies of the poorest households generally take on a survival mode; once survival is ensured, they will often shift to security needs, and lastly to savings and growth (Sebstad 1989: 941). Therefore, households and household members may adopt various mixes of livelihood strategies, including 'labor market involvement, savings, accumulation, and investment; borrowing; innovation and adaptation of different technologies for production; social networkings;

changes in consumption patterns; and income, labor, and asset pooling' (ibid.).

To examine variations in household decision-making, some analysts have begun to use a bargaining or 'cooperative conflict' model (Sen 1987; Elson 1991; Kabeer and Joekes 1991; Jackson 1993; Wilson 1991). Household decision-making is seen as a bargaining process between parties whose bargaining power depends on their position both in the household itself and within society at large. Cooperation occurs, and the household persists, as long as this is in the interests of its members. In cases of conflicting interests, decision-making outcomes reflect the differential bargaining power of household members. Women's actions, therefore, often depend on their bargaining position within the household. Moreover, as a site of cooperative conflict, household decision-making is unlikely to result in equal benefits for all.

The emerging livelihoods systems and cooperative conflict models stress that the household is a variable structure with separable, often competing interests, rights, and responsibilities among its members. The household is as much a differentiated unit as are other societal structures (e.g., labor markets) that are segmented by relations of gender, class, age, and so on. In addition, households both influence and are influenced by broader social processes. Until quite recently, there have been few development studies that have examined how households articulate with these larger processes (e.g., concerning the labor-force and production). On the one hand, overarching economic and power relations partially determine the resource-generating activities of households. On the other hand, the households themselves partially determine the choices of which activities to pursue, within the setting of local cultural–ideological relations. Not only intra-household but also inter-household linkages (e.g., cooperatives, mutual aid societies, kinship groups, social movements) may profoundly affect household activities. In order to understand such activities, it is necessary to analyze the complex, and often overlapping, levels of social relations which affect households and their members. As Cebotarev (1988: 196) points out, such social relations may influence women's lives in a number of interrelated areas, including 'the degree of access to and control over material power and authority base . . . [and] also knowledge, ideological–cultural, affective–emotional and personal bases of power.'

Analyses of social relations affecting households may uncover diverse causes behind the continuing exploitation and oppression of many Third World women. Especially within poor households, women are often burdened with much of the responsibility for family subsistence, as well as being important economic providers. However, their ability to fulfill these obligations is frequently limited by gender, class, and other constraints, with resulting inequities in access to critical resources: intra-family differences in the distribution of basic necessities, such as food; the disadvantaged position of women in many labor markets; high rates of illiteracy among women as a consequence of gender differences in schooling; and the lack of access for women to key means of production, such as land, credit, and advanced technologies. Particularly in many rural areas, male out-migration has led to a surge in female-headed households in which women must shoulder virtually the full responsibility for familial reproduction. At the same time, recently introduced neoliberal measures have produced particular hardships for poor women and their dependants (e.g., through the privatization of previously common property resources, rising food prices, deteriorating social services, credit restrictions).

The resulting social reproduction crisis of many poor households, coupled with the multiple sources of women's exploitation, has led to widely divergent survival strategies (Stephen 1992; Weil 1988). Household social relations strongly influence the forms these strategies take, their use being, in large part, a function of internal processes of decision-making, which are themselves influenced by the 'cooperative conflicts' within households (Raghuram and Momsen 1993; Sen 1987). When women are especially disadvantaged by household power relations, survival strategies have often led to the formation of intricate social networks among women from different households, based on reciprocal arrangements of informal mutual aid among relatives and neighbors (e.g., financial aid, child-rearing). Such arrangements frequently arise from women's daily social interaction within the community, focusing on their multiple responsibilities for social reproduction (e.g., maintaining kinship linkages, carrying out sociocultural obligations in the community) (Safa and Butler 1992). Increasingly these networks are evolving into a form of community-based social organization which not only assists women in their immediate household tasks, but also provides a forum for political organizing around basic needs and other critical issues for poor women.

Third World Women's Movements

Since the early 1970s, there has been a dramatic upsurge in the mobilization of Third World women as participants in various movements (see, e.g., Agarwal 1989; Chinchilla 1992; Hellman 1992; Jahan 1987; Kishwar 1988; Stephen 1992). While, in most countries, women's collective action is not new, it has recently been expanded in order to open new opportunities for women, devise cooperative methods by which to overcome common problems, and incorporate a variety of neglected issues into the development agenda. The establishment of women's groups within larger organizations has allowed women with different backgrounds and political skills to voice their ideas and problems in a more effective manner. In many cases, women's participation within popular organizations has added to their importance, especially in their ability to adopt practical new methods with which to challenge existing social, economic, and political orders. Autonomous women's groups have also provided a space for women who wish to extend their political participation beyond merely supporting male activists. As Nyaga (1986: 220) notes, 'The[se] groups have become, for many women, a source of identity, pride and, most importantly, moral support and motivation in their endeavor to improve the quality of their lives and that of their families.'

Although women's organizations reflect the characteristic heterogeneity of Third World social movements, three types of women's groups have become particularly prevalent. First, trade-union types of organizations have been formed to help poor, self-employed women who have historically been denied access to credit and other productive resources in many countries. Perhaps the most well known of such groups is the Self-Employed Women's Association (SEWA), which was created in 1972 in the state of Gujerat, India, and has organized more than 40,000 poor women working in the informal sector (Agarwal 1989; Bhatt 1989; Ghai 1989). SEWA is a trade union designed to empower poor women by making them more visible and providing them with resources, drawing inspiration from the Gandhian philosophy of self-help and mutual aid. With assistance from the ILO, it has created a Co-op Bank which provides credit to local women's cooperatives in both urban and rural areas. SEWA also offers training courses in a wide variety of skills and has sought to address some of the urgent social problems of its members through initiatives such as a maternal protection scheme, the extension of special benefits to widows, child-care provisions, and training programs for midwives (Ghai 1989: 221). In addition, SEWA often engages in advocacy issues on behalf of its members, who previously had little or no opportunity to voice their concerns. For example, SEWA has worked to secure better wages and working conditions for casual laborers and contract workers through changes to national labor legislation. In instances where employees are being abused, SEWA has also engaged in letter-writing campaigns, has filed complaints with government departments and the police, and has brought some cases to court (Bhatt 1989: 1062).

A second common type of women's organization is the community-based popular movement focused primarily on everyday practical problems. Such groups have become particularly prevalent in the poor neighborhoods of many Third World urban areas and, although men frequently also participate, most organizing is done by women. In Mexico, for example, Barry (1992) reports that women account for about 90 percent of the rank and file of groups belonging to the national association of popular urban movements, CONAMUP (*Coordinadora Nacional de Ayuda Mutual Urbano Popular*). In many poor neighborhoods, these groups have evolved from informal social networks developed by women to assist them as the principal caretakers of their households. Within the framework of their roles as wives and mothers, women have mobilized to make public what once was private by developing collective responses to what used to be understood exclusively as individual domestic problems (Jelin 1990).

In effect, women's collective actions have steadily narrowed traditional distinctions between the public and private areas of social reproduction in many countries (Concoran-Nantes 1993). Domestic abuse, food shortages, deficiencies in housing, inadequate schools and health care, and a host of other problems have provided the initial impetus for broader women's mobilization. As Jelin (1990: 189–90) remarks, collective organization around such issues has often opened up 'a new space' for women, 'enabling them to partake in social confrontation, decision making, and supervision as social and political subjects.' Individual domestic problems have become socialized, as women come together to seek collective solutions to their household needs. According to Hellman (1992: 190), this process has often profoundly changed women's social and political identities, and led to more activism and militancy:

> Political activism thus becomes an extension of women's domestic role, and participation comes to be seen as a responsible mother's duty. As a new identity of 'political mother' develops, the private domain is also altered. Women spend more time outside the home, and female competence is no longer defined solely on the basis of women's performance of domestic tasks. Moreover, . . . the activities of the neighborhood groups shatter the passive image of women and transform passivity into combativeness.

A third common type of women's organization, especially in Third World rural areas, is focused on issues of environmental degradation. In many rural areas, women are forced to shoulder the major burden of social and economic distress caused by environmental deterioration. The access of many women to resources needed for familial reproduction is declining, while their workload is increasing. Fuel for cooking and heating, wild foods, fodder, and biomass raw materials for artisanal industries are becoming increasingly scarce in many regions. Such shortages frequently have serious repercussions on the family income, work patterns, and nutrition and health of poor women and their dependants. However, as Cecelski (1987: 63) notes, rural development policies have generally failed to come to grips with the links between rural women's work, energy, and the environment. Moreover, many poor rural women have failed to make these connections themselves as their overriding, immediate concern is to ensure family survival.

In a growing number of countries, though, women's rising participation in local environmental movements has begun to change this situation. Among the best known of such movements are several in India, including the Chipko movement against commercial logging in Uttar Pradesh; the Baliapal movement against the creation of a military testing range in Orissa; and the *Nirmada Bachao Andolan* (Save the Nirmada Movement) against dams on the Nirmada River in Gujerat, Maharashtra, and Madhya Pradesh (Agarwal 1989; Routledge 1992). Within these movements, women have been at the forefront of protest, and have often succeeded in expanding the struggle beyond just environmental issues to include social problems based in gender, class, and ethnic inequalities. As Agarwal (1989: WS60) remarks, women have commonly brought to these movements a holistic ecological perspective which stresses interrelationships and interdependencies between the various components of nature and society that are vital to overall systemic well-being and sustainability.

Society–nature interrelationships have also provided the focus for the recent development of the ecological feminist or 'eco-feminist' perspective (e.g., Merchant 1980, 1992; Shiva 1988, 1991). Ecofeminism seeks to establish an elemental relationship between women and nature, and owes much of its popularity to the fact that environmental and women's movements became prominent at the same time in many parts of the world (Rodda 1993). According to Joekes (1994: 137), 'The ecofeminist school derives from a philosophy of feminism grounded in women's affinity with the forces of nature, as opposed to men's urge to control and manipulate the natural world through application of the scientific method.' Ecofeminists contend that environmental and women's movements are vitally linked, since both are engaged in a struggle that challenges the fundamental categories of Western patriarchy – its concepts of nature and women, science, and development. It is asserted that Western patriarchal projects and modernization schemes have resulted in violence against both nature and women in Third World countries. Environmental problems associated with Western modernization (e.g., deforestation, erosion, reduced genetic diversity) also appear frequently as problems for women (e.g., decreased access to fuel, displacement from food production). For Shiva (1988), both nature (*pakriti* in Hindu thought) and women manifest the creative, regenerative power of the universe. Therefore, ecological recovery from the ravages of Western modernization requires rediscovering the 'feminine principle' by which many traditional societies used to operate (i.e., tolerance for nature's diversity, respect for indigenous knowledge systems).

Ecofeminism draws much of its strength from the assumption that all women, as part of a common 'sisterhood' in oppression, have like interests, and that these coincide with nature's needs for sustainability. However, such generalizations have increasingly come under question. Recently analyses of women and women's movements suggest that the term 'woman' escapes all sorts of easy generalizations which fail to consider the effects of class, ethnic, religious, and other social relations (e.g., Chinchilla 1992; Eckstein 1989; Joekes 1994; Phillips 1990; S. Smith 1990). Social relations prescribe different sets of interests and identities, and different powers of control over resources – not only between but also within genders (similar to classes and other social categories). Even in countries in which all women may be considered second-class citizens, the effects of gender inequality are differentially experienced by women of various classes, ethnic groups, and so on. In many societies, for example, poor black women's identities may be constituted quite differently from those of affluent white women.

In order to create effective development initiatives for poor women, it is essential to have a clear idea of their identities and interests, their resources and capabilities, and the constraints to which they are subjected (especially those amenable to change). Attempts to act on the basis of piecemeal information or simple generalizations that fail to reflect the diversity and complexity of poor women's lives have often been damaging. Presumptions by educated outsiders, whether male or female, about the needs and interests of poor Third World women may frequently be inaccurate, and may produce inappropriate ideas and methods, thereby discouraging grassroots participation. Women's participation in organizations and projects is structured by their socioeconomic roots. Commonly, poor women face a variety of constraints to their ability to organize (e.g., illiteracy, lack of energy and time due to overwork, cultural restrictions on their freedom of movement, ideological limitations in their social roles as women, their focus on immediate problems of survival rather than broader philosophical issues). If these constraints are not addressed in a realistic manner, by devising ideas and methods appropriate to the differential effects of social relations on various women, development initiatives have little chance of enhancing women's participation and empowerment.

Women's Consciousness, Participation, and Empowerment[5]

The empowerment of women has recently emerged as an important goal of most development strategies. However, as with many concepts, the meaning of empowerment is subject to quite different interpretations. For many mainstream analysts who advocate women's integration into the modern capitalist economy, empowerment is translated into increasing women's participation in the labor force. But other analysts point out that merely increasing women's labor participation may not lead to

[5]   This section was produced in collaboration with graduate student Kerry Preibisch, from her 1994 unpublished manuscript.

their empowerment if the structures into which they are integrated are exploitative and oppressive (e.g., Elson 1991; Fleming 1991; Geisler 1993; Parpart 1993). This latter perspective focuses on issues of power as they relate to gender, and calls for the development of strategies that empower women by challenging existing social, economic, and political structures. Creating more equitable and participatory structures in which women can gain control over their own lives is viewed as the primary objective of the process of empowerment. Empowerment becomes a process 'whereby women [are] able to organize themselves to increase their own self-reliance, to assert their independent right to make choices, and to control resources which will assist in challenging and eliminating their own subordination' (Keller and Mbewe 1991: 76).

For most alternative analysts, women's empowerment is a participatory process which proceeds through bottom-up organization from sites such as the household, workplace, and community. Moser (1989: 110) notes, 'Historically it has been shown that the capacity to confront the nature of gender inequality and women's emancipation can only be fulfilled by the bottom-up struggle of women's organization.' Through local, grassroots participation, women are able to develop appropriate ways to deal with their concerns and problems on their own terms. Attempts to impose outside ideas and methods on local women's groups will only lead to inappropriate development initiatives that destroy women's self-confidence and self-reliance (Elson 1991; Parpart 1993). Although outsiders may sometimes provide needed resources or play a facilitating role in the process, genuine women's empowerment has historically been based on grassroots initiatives that are designed to meet the specific needs and interests of local women themselves.

Women's local interests and identities strongly affect the ways that they organize and mobilize politically. Their traditional roles in Third World societies have frequently led to the development of extensive social networks deeply rooted in community life, often enabling women to mobilize collectively (Logan 1990). They are normally maintained through social interaction linked to women's domestic tasks (e.g., meeting at the community well or marketplace, collecting wood for fuel). The interpersonal skills that mediate these social networks have been deftly refined by women as part of their gender-role socialization. As a result, women often exercise interpersonal power at the local level that can have enormous social significance (Jelin 1990). Women's common exclusion from the mainstream political arena has frequently compelled them to devise alternative political strategies (e.g., gossip, persuasion, and consensus building among community or kinship groups). As Stephen (1989: 18) notes in her work on Latin American women's rural mobilizations, such alternative strategies are often effective:

As they were excluded . . . women also developed strategies for participating in local politics through kin networks and cultural institutions. The political skills they developed in these arenas are not built on 'public' speaking and maintaining control over large assemblies. Instead, their skills are related to listening, consensus building, and persuasive discourse with the women to whom they are the closest. They in turn used the women they are connected to through kinship and *compadrazgo* to build a group of political supporters.

Research on women's popular mobilizations in Latin America shows that women often do not conceive this activity as political because it is viewed as distinct from conventional politics (Martin 1989; Radcliffe and Westwood 1993). In a case study of a Oaxacan village in Mexico, Martin (1989: 476) states that, through a history of crooked patron–client relationships with representatives of the state, 'the word politics became synonymous with corruption.' Women commonly associate the arenas of male-dominated politics (e.g., political parties, trade unions, rural cooperatives) with excessive competition, personal interest, manipulation, dishonesty, and corruption. According to Moser (1987), as a result of gender-role socialization, men often pursue leadership positions for personal advancement, while women normally demonstrate a deeper commitment to community goals.

In many Third World areas, women's involvement in community organizations has begun to transform household social relations. Women have been able to overcome their isolation and loneliness by stepping out of the confinement of the home. Through collective grassroots organization, women have begun to enter the political arena, a sphere from which they have traditionally been excluded. Spaces have been created in which women can acquire leadership skills, self-confidence, and self-esteem (Jelin 1990). The existence of these 'safe spaces' is thought to be especially important for many previously isolated and disadvantaged women to discover their identities, become conscious of their situation, explore hitherto forbidden topics, give mutual support, and devise new forms of political struggle and definitions of what it means to 'do politics' (Chinchilla 1992: 47).

In many countries, these local, community-based spaces have expanded through the formation of bottom-up coalitions which unite women across sectors and regions. In Mexico, for example, there are links between many women's organizations, such as the CONAMUP (*Coordinadora Nacional del Movimiento Urbano Popular*), the CNPA (*Coordinadora Nacional 'Plan de Ayala'*), the '19 de Septiembre' syndicate (Women's Garment Workers Union), the *Organización de Trabajadores Domésticas*, and many feminist organizations working at the grassroots level. Such linkages have been further strengthened through national and international women's *encuentros* (meetings). Examples include the *Comisión Organizadora del Encuentro de Mujeres de los Sectores Populares de Mexico* in 1987, which united women from organizations in nine Latin American countries to discuss women's roles in popular movements, and *Las Mujeres Tenemos La Palabra* in 1988, which brought Mexican women activists together on a national scale (Stephen 1989).[6]

By means of collective organization, previously isolated women are given opportunities to interact, and to develop a consciousness of their shared experiences as women. Through increased awareness of their commonalities, women often acquire the confidence necessary to make changes in their own relationships (e.g., 'democratization' of the household). In many polarized societies, this process has been accelerated by participation in autonomous 'women only' groups, which create space for women to develop the skills, consciousness, and self-confidence needed to act more effectively in mixed-gender situations (Concoran-Nantes 1993; Stephen 1989). A quote in a study by Pires de Rio Caldiera (1990: 65) of women's urban movements in Brazil is illustrative of this process:

> It seems to me that men prefer to see women at home washing clothes, cooking, I think that's what it is. If he goes out that's all right but if the woman goes, she'll see what happens. Women must do something they like; I like it so I do it . . . You make friends in the local groups, you make contacts and it opens women up, little by little things become clearer; let's see if we can overcome this fear we have of our husbands.

In regions such as Latin America, Chinchilla (1992: 41) notes that women's participation in local, community-based movements has often been derived 'from an attempt to fulfill, rather than subvert, the traditional gender division of labor (mothers entering the public sphere to save the lives of their children, housewives turning to collective action to provide for the survival of their families, etc.).' According to Kaplan (1982), women's participation which reflects traditional gender roles (e.g., mother, house-wife, nurturer) has frequently generated 'female consciousness.' Female consciousness arises from the socially constructed gender division of labor that assigns women the task of preserving life and familial reproduction. However, by 'accepting this task, women with female consciousness demand the rights that their obligations entail. The collective drive to secure those rights that result from the division of labor sometimes has revolutionary consequences insofar as it politicizes the networks of everyday life' (ibid.: 545).

6   The above women's organizations and meetings in Mexico, respectively, are: the National Council of the Urban Popular Movement (CONAMUP); Plan de Ayala National Council (CNPA), which takes its name from Emiliano Zapata's 1911 plan for land redistribution; the Organization of Domestic Workers; the Organizing Commission of the Meeting of Women of the Popular Sectors of Mexico; and We Women Have the Word.

In many cases, it appears that women's collective actions are redefining the previously private reproductive sphere according to the age-old feminist phrase: 'the personal is political.' Logan (1990) reports that, in many large Mexican cities, women's local participation has developed an 'activist motherhood.' As well as being a motivation for action, motherhood has become a strategy for political mobilization. In Guadalajara, for example, poor women staged a particularly effective demonstration when they marched their children to the city center and bathed them in a huge decorative fountain outside city hall, making 'their point symbolically that they did not have enough water in their communities to carry out their ascribed gender roles as mothers. By using this strategy, they reminded the state that . . . it had chosen aesthetics over human needs' (ibid.: 155). It should be recognized that women are also increasingly assuming non-traditional roles which challenge the dominant ideology of women's proper place in many societies. At the same time, however, the centrality of motherhood in many Third World women's lives, and the cultural veneration that commonly surrounds this role, should not be under-estimated as an important element of women's political mobilization in most countries.

The experience that women have gained from collective action over reproductive issues has often created fertile ground for empowerment based on links between gender-specific consciousness (women's 'practical' interests), feminist consciousness (women's 'strategic' interests), and social consciousness (of class, social sector, nation, and so on) (Chinchilla 1992: 41). Analyses of Third World women's consciousness and popular movements commonly distinguish between 'practical' and 'strategic' gender interests (e.g., Molyneux 1985; Moser 1987, 1989). In many societies, practical gender interests revolve around women's reproductive tasks arising from the gender division of labor. Such interests are usually widely accepted by both women and men as uncontroversial (Wilson 1991). But, while they may provide important organizational bases, these practical interests 'do not challenge the prevailing forms of subordination of women, even though they directly arise out of them' (Molyneux 1985: 233).

By contrast, strategic gender interests challenge existing social structures (e.g., gender divisions of labor, discriminatory systems of property rights) which perpetuate gender inequalities. Because they normally seek to alter the balance of power, strategies based on such interests are usually more long-term, radical, and controversial – and will often be resisted by men (and sometimes women). The need for structural change to meet strategic gender interests (e.g., women's emancipation, gender equality) derives 'from the analysis of women's subordination and from the formulation of an alternative, more satisfactory set of arrangements to those which exist' (ibid.: 232). However, while strategic gender interests normally concern broad structural issues, they are also necessarily intertwined with women's more immediate practical interests. These links are especially important for stimulating grassroots participation through which many women begin to develop the feminist consciousness necessary to move beyond their practical, everyday concerns to more strategic, longer-term issues. As Schirmer (1993: 61) remarks, it is important to note:

> . . . how women who begin with the pragmatic and strategic struggle for human rights (for example, the right to life centered around family survival, exclusive of the conscious demands for the rights of women) learn of the need to connect these struggles to the pragmatic and strategic struggles of women's rights; that is, how to make politics at once political and personal, at once strategic and pragmatic – and then coming to realize that they may have been 'feminists' all along!

Efforts to link strategic with practical gender interests have been associated with the rise of 'popular feminism' in many Third World countries. As defined by Logan (1990: 159), popular feminism allows 'women to continue to work collectively for improvements in the lives of the entire community as opposed to improvements in the lives of individuals. A feminism so defined would also permit low-income women to address issues that symbolize their subordination as women.' Thus, popular feminism entails collective action to bring about change within societal structures historically linked to women's subordination. Given varying conditions within and among Third World countries, however, popular feminism defies any single determination or delineation; in many areas it may differ significantly from First World feminist struggles in both issues and methods. In fact, it might be more appropriate to speak of the rise of a range of new 'feminisms,' each of which may differ according to variables such as ethnicity, class, religion, and so on.

While new ways are continually being found to overcome gender inequalities and oppression, many poor women are also hesitant about prioritizing such issues because 'they identify their principal oppression as being poor, and thus in solidarity with [poor] men' (Craske 1993: 132). Especially among indigenous women, questions of ethnic survival may also figure in priorities.

To quote an indigenous woman from the IXQUIC group in Guatemala: 'It is important for Western women to understand why we don't have the same demands [as them] . . . what could we ask for now? In many senses for Guatemalan women it would mean equal repression and we already have that' (in Westwood and Radcliffe 1993: 8–9). According to Schirmer (1993: 63–4), the different interests and identities of Third World women should be factored into the theorization of women's movements, making feminism a necessarily variable and contested domain:

[S]houldn't we want to know more about how women make sense of their conflicts, how they think about what they are doing and who they see as having the power and ability of making sense of these actions, and how they come to gain a positive self-image? Shouldn't we assume that both 'mothering' and 'feminism' are contested domains.

This perspective stresses 'how women make sense of, and respond creatively to, the changing conditions in which they live' (Bondi 1990: 439). It views women as active participants in various aspects of development, who are increasingly involved in struggles for change. As Jelin (1986: 3) points out, women's participation in popular movements has generated a 'new politics' in many countries based on 'a new way of relating what is political and what is social, the public world and private life, in which daily social practices are linked and interact directly with the ideological and the political-institutional.' This politics of everyday life is playing a vital role in defining new alternative political projects dedicated to breaking down hierarchical structures and spreading the benefits of development more evenly. Women have become central actors in many grassroots initiatives for change, including labor organizations, community-based groups, and environmental movements, belying the traditional image of Third World women as uniformly powerless and helpless victims who passively accept their subordination and marginalization (Agarwal 1989; Moser 1989; Parpart 1993; Taplin 1989). Instead, recent studies have begun to stress the multiple realities of Third World women and the many innovative ways in which they have reinvigorated popular struggles for change in many countries.

Continuing Contradictions

The recent attention that women and gender relations have received in development efforts has resulted in considerable theoretical and practical advance.

Whereas women were scarcely mentioned in development work only two decades ago, today there is seldom a development project that does not directly address women's issues (at least rhetorically). Nevertheless, a number of contradictions need to be overcome if further progress is to made toward meeting women's needs and interests in development. Theoretical, analyses of women and development especially need to move away from the essentialist positions that have characterized much recent feminist scholarship. Essentialism has been a particularly severe problem within the ecofeminist perspective, which proposes a biological affinity between women and nature (Jackson 1993; Joekes 1994). This position denies the centrality of social relations in determining forms of interaction between various social groups and nature. There cannot be any special relationship between all women and nature, because women are not a unitary category. While some women may have a special relationship with the environment, so too may some men. Moreover, rather than being inherently conservationist, some women may be agents of environmental destruction. Attempts to construct a dichotomy between women as environmental custodians and men as environmental destroyers prevent more useful contextually specific analysis.

Essentialism has marked not only the ecofeminist perspective, but much feminist scholarship on women and gender relations in general. A narrow and inflexible view of social relations often results, which displays a limited appreciation of the complexities of interactions among and within social sectors, including genders themselves. Many analysts have assumed that all women have similar interests based on their common gender position. However, this stance is increasingly questioned, especially by women of color, who argue that various women's identities and interests may be constituted quite differently according to the intersection of ethnic, class, and other social relations with those of gender. This has led many feminist authors to criticize the movement's traditional preoccupation with patriarchy to the exclusion of other sources of oppression based, for example, in class and ethnicity (S. Smith 1990: 264). In some instances, upper-class women may join with other elite groups in the oppression of lower-class women. House-Midamba (1990), for example, presents evidence of this among elite women elected to political institutions in Kenya, who tend to be 'pro-status quo' and fail to represent the gender needs and interests of poor and disadvantaged women. Similarly, LeBeuf (1991) reports that women who belong to or have ties with royal lineages in African societies enjoy rights and privileges that separate them from other women, resulting in behavior that closely resembles that of males in power positions. In order to analyze such cases, researchers need to avoid separating indicators such as gender, class, and ethnicity in favor of concepts that treat them conjointly.

Eurocentrism often compounds problems of essentialism in studies of Third World women, especially those conducted by First World feminists. Feminist theory, when applied to Third World women, is normally driven by opportunities to enlarge liberal feminist knowledge, rather than to explore the variety of modes of being female (Parpart 1993). Frequently, broad generalizations result which neglect differences among Third World women rooted in their concrete, lived experiences in various cultures and societies. A Northern feminist agenda is imposed which may not correspond to the needs and interests of many women in the South. As Moser (1989: 1811) relates, a major criticism by Third World women of Western-exported feminism is that '[such] feminism to a woman who has no water, no food, and no home is nonsense.'

Moreover, studies often present 'Third World women . . . as uniformly poor, powerless and vulnerable, while Western women are the referent point for modern, educated, sexually liberated womanhood' (Parpart 1993: 444). According to Mohanty (1988: 62), research by Western feminists has tended to 'colonize the material and historical heterogeneities of the lives of women in the Third World, thereby producing/re-presenting a composite, singular "Third World woman" – an image which appears arbitrarily constructed but nevertheless carries with it the authorizing signature of Western humanist discourse.' Such research not only distorts the multiple realities of Third World women, but also reduces possibilities for cooperative ventures between Western (usually white) feminists and women of color around the world. In order to promote such cooperation, Lazreg (1988: 98) contends that studies should recognize Third World women's lives 'as meaningful, coherent, and understandable instead of being infused "by us" with doom and sorrow.' While the lives of Third World women may often be subject to serious limitations, the outsider's perception of those constraints may itself become a significant obstacle to future progress for these women (Hirschmann 1991: 1691).

To make studies more sensitive to Third World women's multiple identities and interests, some authors are calling for a postmodern feminist perspective which would combine 'a postmodernist incredulity toward metanarratives with the social-critical power of feminism' (Fraser and Nicholson 1990: 34). This would reveal important differences and ambiguities in the lives of women without sacrificing the quest for a 'broader, richer, and more complex, and multilayered feminist solidarity, the sort of solidarity which is essential for overcoming the oppression of women in its endless variety and monotonous similarity' (ibid.: 35). It is contended that feminist theory has too often simply applied the reality of white, Western, middle-class women to women of all classes, ethnic groups, and areas of the world, ignoring possibilities for differences among women themselves (Parpart 1993: 443). By contrast, an attention to difference based in postmodernist feminist thought would remind Westerners that Third World women cannot be lumped into one undifferentiated category. It would also call into question received ideas about Western-style modernization, while offering new insights into the variegated construction of meaning and knowledge by Third World women through their particular experiences. This, in turn, might help to uncover the presently obscured existence of many highly skilled, but poor and marginalized Third World women who have much to offer the development process if given a chance to participate.

The failure to properly account for the diverse realities of Third World women has had a profoundly negative impact not only on the theory, but also on the practice of development. It is clear that development planners and other 'experts' have much to learn from Third World women. But as long as the belief persists that development expertise can only disseminate in a top-down manner from the North, self-confidence and self-reliance in the South will continue to be undermined – negating any possibility that development processes might be empowering for women and other presently marginalized groups. Parpart (1993: 453) contends that much of the discourse on women and development legitimizes the assertion that Western expertise is essential to the progress of Third World women; therefore, it is not surprising that development projects generally stress the acquisition of Western technical skills and equipment, while concerns for local knowledge, participation, and empowerment are elided.

Development programs and projects based on inadequate knowledge of Third World women's lives and attitudes have consistently failed to produce the desired results (Moser 1989; Parpart 1993). Frequently, women's development projects are planned from above, with little or no grassroots involvement, and are formulated around the myth that Third World women are principally housewives, ignoring the fact that the large majority also participate extensively in production (Ahmad 1984; Cecelski 1987; Moser 1989). These projects are normally welfare-oriented, segregated from mainstream development efforts, and focus on 'women's work' such as home economics and volunteer community activities (Randolph and Sanders 1988; Wickramasinghe 1993). For Geisler (1993: 1976), such projects have simply 'conferred to women the doubtful "power" of supplying the services and the labor that states are not willing or able to finance.' The need for most women to gain better access to essential resources for income generation (e.g., land, credit) is commonly ignored, although lack of such access is a major contributing factor to women's impoverishment and exploitation. Development initiatives often approach women in a top-down individualistic manner, rather than designing bottom-up collective solutions that, by promoting self-reliance and empowerment, might give many isolated women the support needed to overcome their problems.

There are some innovative development projects under way that address the felt needs of Third World women themselves and that these women control through their own organizations and movements. However, most development initiatives aimed at the 'people' are managed in a manner that tends to include and benefit only men. At the same time, it should be evident that any sustained improvement in the well-being of the popular majority in the South is inconceivable without the active participation of women, who are responsible for a substantial portion of production as well as most social reproductive activities. There is, therefore, a clear need to enhance the participation of women in development work at all stages from planning to implementation.

An important part of this process involves creating 'gender-training' opportunities to permit development workers at all levels to be aware of the special needs and interests of Third World women (Keller and Mbewe 1991). However, experience shows that any gender-sensitive development strategy must also focus on the household and community. Through bottom-up, community-based organization, households and household members can become conscious of their own social relations, values, and interests – and learn to place these within the larger socioeconomic context. Cebotarev (1991: 429) notes,

> The crucial aspect of this strategy is to assist the family to transform, in a participatory manner, the commonly used 'win–lose' framework of gender relations interpretation into a 'win–win' one . . . [which] recognizes that greater options and possibilities for all family members will benefit not only the individual member but also the family and community as a whole.

At the same time, one should not think significant progress can be made in improving women's lives without basic changes which address the structural forces that create and sustain women's exploitation and oppression. Attention must be given to the broad material and ideological aspects of women's subordination. Assistance to Third World women has often taken the form of stop-gap measures to tackle some of the most visible symptoms of underdevelopment. It seems highly unlikely that such assistance can have much lasting benefit unless it is accompanied by a concerted attack on the mechanisms that reproduce and intensify underdevelopment in the first place (e.g., the distribution of land and other productive assets, wages and credit policies). This, in turn, calls for involvement by Third World states and international development institutions. Various efforts to support Third World women can be most effective and sustainable if they are coordinated with overall development policies. However, given the elitist and patriarchal nature of national and international power structures, it is improbable that the interests of women and other disadvantaged groups will soon be promoted from the top down. Significant change in the lives of Third World women, then, awaits the empowerment of women themselves – which can only take place through prolonged political struggle in cooperation with other like-minded social forces. Although some recent progress in this area has been made through the actions of various popular movements, in most Third World countries these struggles have barely begun.

10

Environment and Sustainability

The previous two chapters dealt with two areas of development which have only recently achieved the visibility they deserve, largely via the attention they have been given in the alternative development tradition. This chapter analyzes a third such area: environmental relations and sustainable development. Worsening environmental problems have prompted a rethinking of the development agenda linking issues of sustainability to more traditional concerns for growth and equity. As a result, a variety of notions of sustainable development have been introduced which have quickly made an impact on development debates. Moreover, as with women's issues, the recent rise of various movements devoted to environmental concerns has greatly influenced alternative theories and practices. Frameworks for sustainable development are moving away from their prior technical fixation toward a more holistic focus which stresses the contextual specificity of environmental problems and includes a people-oriented agenda based on the needs and rights of local people. Although work on this new approach has barely begun and contains numerous contradictions and shortcomings, it also promises to profoundly influence future directions of development.

Environmental Relations and Sustainable Development

Postwar development has dramatically altered the environment and society–nature relations in virtually all parts of the South. Moreover, the effects of rapid environmental change are being increasingly felt not only at the local and regional levels, but also on a global scale. Nevertheless, concern for environmental change has until recently been confined to the margins; progress, usually measured in terms of aggregate levels of production and consumption, has been the accepted measure of development. This Western vision of progress and modernization has commonly produced a pattern of development that ignores traditional society–nature relations within

Third World countries and inadequately addresses issues of social equity, ecological balance, and overall sustainability.

From modernization to neoliberalism, mainstream development initiatives have applied 'nature-conquering' technologies on a massive scale, with devastating effects on many ecologically vulnerable Third World areas (Ghosh 1990; Leonard 1989). According to Black (1991: 120), tropical ecosystems around the world are being destroyed at a rate of 25 million acres per year. The relentless spread of large-scale infrastructure projects has caused especially widespread damage to fragile environments and indigenous cultures in many countries. These projects (roads and railways, ports, power lines, dams) are typically initiated in the name of development and are normally sponsored by national governments and underwritten by loans from the World Bank and other international financial institutions. The clearest beneficiaries of such projects have been elite groups linked to transnational capitals, whose interests have been furthered through concerted lobbying efforts within national and international development agencies. At the same time, the poor and disadvantaged (especially indigenous cultures) have usually been excluded from the decision-making process and have borne the brunt of the costs of these projects (Black 1991; Cheru 1992; Colchester 1994). In India alone, more than 11.5 million people have been displaced without rehabilitation over the past three decades by large-scale projects, especially dams (Gadgil and Guha 1994: 108). Not only has irreparable damage been caused to many local environments and cultures, but the macroeconomic benefits of such projects have often been questionable. In the context of the Amazon Basin, Kyle and Cunha (1992: 8) report, 'With very few exceptions, development projects so far implemented in the region would not pass a cost–benefit test if it were not for heavy subsidies, immensely distorted prices, private appropriation of public goods, and society's underwriting of external diseconomies.'

While the negative effects of large-scale development projects have become increasingly publicized, many other forms of development are creating massive, if less visible, destruction in the South. Such is the case, for example, with the Green Revolution in agriculture, which has been associated with social inequities, groundwater depletion, soil erosion and degradation, chemical contamination, and the reduction of biogenetic diversity in many countries (Conway and Barbier 1988; Shiva 1991). Among the most serious environmental problems resulting from unsustainable agricultural practices has been the gradual deterioration of the soil base in many Third World areas – something the FAO has termed the 'quiet crisis' (Mackenzie 1993: 71). This and other common forms of environmental damage (e.g., deforestation, desertification, water and air pollution, non-renewable resource depletion, pesticide poisoning and other contamination, the decline or disappearance of many species) collectively entail the steady destruction of much of the South's 'natural capital.' Future generations will pay an especially heavy price for present unsustainable development – the benefits of which are currently being monopolized by an elite minority.

Alternative Concepts of Ecodevelopment and Sustainability

Against this backdrop of accelerating and often irreversible environmental destruction, attention has recently been focused on issues related to the sustainability of development. Perhaps more than any other factor, growing environmental damage may be responsible for prompting a rethinking of development. Evidence of this trend is provided in a recent literature survey by Marien (1992: 736–7), which contains the following statements: 'the environment is fast moving to the top of the world's agenda' (Cleveland (1990); the single most important problem of the coming decades is that of 'making our economic peace with the demands of the environment' (Heilbroner 1990); and 'the rise of environmental consciousness will have effects comparable to the consequences of the last three great Western transformations' (Smil 1989).

Rising concern for the environmental consequences of development can largely be traced back to the rise of alternative development approaches in the 1970s. In 1972, the Club of Rome issued a well publicized report entitled *The Limits to Growth*, which warned that life as we know it faces a sudden apocalyptic end if development practices are not dramatically altered to respect the earth's physical limits to growth (Meadows and Meadows 1972). Although most discussion on this report was fairly critical of its focus on absolute limits, it nevertheless succeeded in placing environmental issues squarely on the development agenda. This process was further advanced at a number of major international meetings, including the Stockholm Conference on the Human Environment in 1972, which led to the establishment of the United Nations Environment Program (UNEP), and the Cocoyoc (Mexico) conference on 'Patterns of Resource Use, Environment and Development Strategies' held in 1974 by UNEP and UNCTAD.

These conferences played an important role in the emergence of an alternative development paradigm during the 1970s, which although it included many other concerns, such as basic needs and self-reliance, also clearly focused on the need to harmonize development with the environment. Environmental considerations were given particular prominence within the alternative concept of 'ecodevelopment,' which was introduced by Maurice Strong at the 1972 Stockholm conference and was further elaborated and popularized by Ignacy Sachs (1974) and related theorists (e.g., Glaeser and Vyasulu 1984).

As it evolved, the ecodevelopment approach did not necessarily seek to halt growth entirely, but called for industrialization and other development processes to be made more compatible with environmental sustainability through means such as the adoption of appropriate technologies, the encouragement of conservationist lifestyles, and the use of bottom-up, participatory planning approaches. According to Glaeser and Vyasulu (1984), ecodevelopment includes the following elements, many of which are common to alternative development frameworks in general: harmonizing consumption patterns and lifestyles to environmental needs; using appropriate technologies and ecologically based productive systems; maintaining low energy profiles and promoting renewable energy bases; limiting depletion of non-renewable resources through recycling and other means; finding more socially and environmentally sustainable uses for existing resources; employing ecological principles to guide land use, settlement, and other development patterns; and utilizing decentralized planning methods to encourage local participation. Areas of overlap between ecodevelopment and alternative development in general are further illustrated by Sachs' definition of ecodevelopment as:

> an approach to development aimed at harmonizing social and economic objectives with ecologically sound management, in a spirit of solidarity with future generations; based on the principle of self-reliance, satisfaction of basic needs, a new symbiosis of man and earth; another kind of qualitative growth, not zero growth, not negative growth. (Quoted in Glaeser and Vyasulu 1984: 25)

In contrast to mainstream development strategies, the ecodevelopment perspective gives development no universal meaning or elements (e.g., capital, labor, investment). Instead, an ecodevelopment strategy consists of specific elements – a particular group of people with their own values and needs, living in a distinct region with culturally specific resources. The goal of ecodevelopment is to improve that particular situation – not to bring about 'development' in terms of GDP growth or some other abstraction (Hettne 1990: 188). Ecodevelopment contends that, in the real world, development does not occur in generalized form. Development can only be of something specific – such as a certain 'ecoregion' with its own special conditions and needs. Sachs (1974: 9) notes,

> [E]codevelopment is a style of development that, in each ecoregion, calls for specific solutions to the particular problems of the region in the light of cultural as well as ecological data and long-term as well as immediate needs. Accordingly, it operates with criteria of progress that are related to each particular case, and adaptation to the environment plays an important role.

As in alternative development in general, the ecodevelopment approach calls on countries in the South to become more self-reliant – to create strategies appropriate to their own ecological and cultural contexts, rather than looking to the North for development solutions. Ecodevelopment asserts that there are no universal models which can be successfully emulated. Development strategies should make use of the resources in a given region in ways that both sustain the ecological system (outer limit) and provide for basic human needs (inner limit). For Hettne (1990: 187), this strong normative component, rooted in many elements common to alternative development thinking (e.g., self-reliance, basic needs, sustainability), gives the ecodevelopment approach a paradigmatic quality. Especially if it is complemented by an analysis of power relations at various scales, ecodevelopment has the potential to mount a radical challenge to mainstream development frameworks, because it questions the very values upon which those frameworks are based.

In addition to the ecodevelopment perspective, a focus on environmental issues has also characterized various recent sustainable development frameworks. The concept of sustainability initially appeared during the 1970s and early 1980s, especially via the work of Lester Brown (1981) and others at the Worldwatch Institute. These theorists stressed that no international economic order could be viable if the natural biological systems that underpin the global economy are not preserved. Four development areas were believed to be particularly problematic from the point of view of sustainability (ibid.):

1   lagging energy transition – inherent sustainability problems plague coal and nuclear energy, while renewable energy sources have only been developed slowly;
2   the destruction of major biological systems (e.g., oceanic fisheries, forests, croplands) by development practices that exceed their 'carrying capacities' – both renewable and non-renewable resource bases are shrinking rapidly;
3   the threat of climate change – the results of uncontrolled development (e.g., deforestation, atmospheric pollution) may generate catastrophic effects on the productivity of various ecosystems; and
4   rising global food insecurity – mainstream development practices (such as the transfer of cropland from subsistence to cash-crop production, the shift from indigenous food crops to imported foods) has diminished food security in many areas.

Following its initial popularization, the concept of sustainability has appeared in a wide range of forms in recent development literature. Different authors have given it a variety of meanings. Especially within agricultural economics and related fields, sustainability has characteristically been given a rather narrow, technical definition which focuses on the ability of ecosystems to maintain levels of production (see Colchester 1994). Within this literature, sustainability has become a serious issue essentially because nature is seen increasingly to be constraining human progress (Redclift 1991). Conventional growth models will incur sharply rising costs if environmental warnings, or 'biospheric imperatives,' are ignored. The solution is either to develop technologies which avoid the worst environmental consequences of development, or to adopt measures to assess environmental losses in a more realistic manner, thereby reducing the danger that such costs will be overlooked by policy-makers (ibid.: 38).

Other authors from a variety of alternative perspectives, including radical ecology, ecofeminism, and deep ecology, take a quite different view toward sustainability. For them, the conventional image of progress contained within mainstream development models is inherently contradictory in ecological terms. Accordingly, both the ends and the means of development ought to be critically re-examined. Merely devising technical, within-the-system solutions to environmental problems (such as better methods of costing environmental losses) is regarded as too narrow and ultimately self-defeating. In the end, it leads to a type of 'environmental managerialism' which, instead of examining the roots of environmental problems within socieconomic structures and society–nature relations, begins with such problems and attempts to resolve them on an ad hoc, piecemeal basis. Redclift (1988: 644–5) objects to this type of environmental managerialism on the following grounds:

1   It is essentially a 'protective and reactive' response which only considers environmental problems after development objectives have been set. These objectives cannot, therefore, be fundamentally altered to take ecological and social factors into consideration.
2   It separates the environmental consequences from the social and economic effects of development. Governments and international agencies have often used environmental management as an instrument of social control. Once this fact is recognized, the important issue is raised of who is to do the 'managing' of the environment.
3   It takes as a given the distributive consequences of market-led development. Instead, the interrelated nature of the environmental and distributive effects of development policies should be recognized. This would imply moving away from present practices of merely offering 'compensation' for environmental damage toward establishing environmental objectives that reduce the poverty and vulnerability of the poor.

4   The techniques of environmental managerialism deflect attention away
     from the context of environmental problems. An alternative analy-
     sis would start with this context, and would include questions con-
     cerning the distribution of wealth and resources within and between
     countries.

Many of these issues are addressed within the broader concept of
sustainability offered by the World Commision on Environment and Development (WCED) (1987), whose research fits more comfortably within the
alternative development tradition than does the environmental managerialist
approach. The work of the WCED culminated in the influential 'Brundtland
Report' and was subsequently incorporated into another document, 'Environmental Perspective to the Year 2000 and Beyond,' which was adopted
by the UN General Assembly. The report represented a significant shift in
thinking concerning sustainable development. It placed the environment and
development on the political agenda, and within a political and economic
framework that previous approaches to sustainability sorely lacked (Adams
1992).

The WCED study acknowledged that sustainable development would
require a radical transformation of contemporary economic structures,
which would alter the way that resources are owned, controlled, and
mobilized. To be sustainable, development must meet the needs of local
people, because if not, many people will be obliged by necessity to take
more from the environment than is advisable.

Particularly in the South, struggles over the environment commonly
involve survival strategies to meet basic needs; the cost of individuals
pursuing their own self-interests is often borne by the group (the 'tragedy
of the commons' argument). There is little point in appealing to idealism or
altruism to protect the environment when households are forced to behave
'selfishly' in their struggle to survive (Redclift 1991: 38). Neither is there
much point in implementing technologically based strategies that fail to
address the common roots of environmental destruction within structural
inequalities and impoverishment. Instead, sustainable development should
be linked with goals of distributional equity and social justice – within and
between countries as well as generations. This underscores its inherently
political nature. As the UN World Commission on Environment and Development (1987: 63, 65) argues, achieving sustainability implies basic political
change in line with an alternative development agenda:

> The pursuit of sustainable development requires a political system that secures
> effective citizen participation in decision making . . . This is best secured by
> decentralizing the management of resources upon which local communities
> depend, and giving these communities an effective say over the use of these

resources. It will also require promoting citizen's initiatives, empowering people's organizations, and strengthening local democracy.

As the work of the WCED made clear, sustainability means more than just ecological and agricultural stability; particularly in the context of polarized Third World countries, it has a strong political element linked to the needs and interests of the people. The concern for meeting popular needs, without which conservation objectives cannot be attained, also inevitably gives sustainability an important livelihood component. This connection has been made most explicit in the concept of 'sustainable livelihood security' developed by Chambers and Conway (1992: 7), who also served on the Advisory Panel on Food Security that contributed to the WCED report:

> [S]ustainable livelihood security [is] an integrating concept . . . Livelihood is defined as adequate stocks and flows of food and cash to meet basic needs. Security refers to secure ownership of, and access to, resources and income-earning activities, including reserves and assets to offset risk, ease shocks and meet contingencies. Sustainable refers to the maintenance or enhancement of resource productivity on a long-term basis.

Sustainable livelihood security focuses on local interaction between food security (and other basic needs) and the environment. If they are provided with the conditions to meet their basic needs, people have a vested interest in acting in an environmentally sustainable manner. However, this requires removing the constraints which presently prevent many poor people from taking a longer-term view toward conserving their resource base. Sustainable livelihood security borrows ideas from both the natural and social sciences, linking basic needs issues with environmental concerns. The concept implies conserving natural resources and other environmental factors that are essential to people's livelihoods. It focuses on the local level, where most environmental problems experienced by the Third World poor tend to be concentrated. However, it also includes a broader analysis of links between environmental sustainability, basic-needs provisions, and issues of power at the national and international levels. For all of these reasons, it represents a significant advance in our ability to conceptualize sustainability within Third World settings.

The Political Ecology Approach

A number of other alternative frameworks (e.g., political ecology, radical and deep ecology, ecofeminism) concerned with issues of development and environmental sustainability have recently evolved. The political ecology approach has made an especially important contribution to Third World environmental analysis. According to Neumann (1992), this contains the following elements: (1) a focus on resource users and the social relations in which they are entwined; (2) linking local relations to their broader socioeconomic and geographical context; and (3) historical analysis to understand the contemporary situation. Like the WCED perspective, the political ecology approach critiques efforts to find scientific–technological solutions to environmental problems, which fail to address the basic politico-economic and socio-ethical dimensions of sustainability. Common environmental problems are seen to be intimately connected to structural processes (e.g., land concentrations accompanying agroexport production) that direct how resources are used. As Redclift (1984: 2) notes, 'so many causes of the environmental crisis are structural with roots in social institutions and economic relationships, that anything other than a political treatment of the environment lacks credibility.' Within polarized Third World societies, most environmental issues are also necessarily political and redistributive. Accordingly, sustainable development cannot be achieved without accompanying measures to deepen democratic participation and popular empowerment.

Analysis of the structural roots of Third World environmental problems has given many political ecology studies a strong global focus. It is argued that issues of sustainable development in today's interconnected world cannot be coherently addressed outside of their North/South context, especially the contradictions imposed by the structural inequalities of global capitalism (Redclift 1987, 1991). Nevertheless, both mainstream and alternative approaches to sustainable development have often failed to account for the influences and constraints of the global political economy (Davies and Leach 1991). This has prevented deeper understanding of environmental change as a fundamentally social process, linked to the overarching forces of global capitalist development. Watts (1983), for example, demonstrates how capitalist penetration in central Africa has dramatically transformed rural social structures and livelihood strategies, with profound environmental consequences. Similarly, Shiva (1991) shows how transnational corporate involvement in the Green Revolution has accelerated rural differentiation and environmental damage (e.g., chemical contamination of soils and groundwater) in India. These examples underscore the need to incorporate a global perspective into studies of Third World environmental problems.

While early political ecology studies tended to emphasize the global forces underlying environmental change, recent research has given more attention to social structures and society–nature relations at the local level (Moore 1993; Peet and Watts 1993). Analyses stress that local environments are constituted by a multi-sided complex of relations, including social relations of production and reproduction, which are only partially subject to outside determination. Because environmental problems always contain an important place-based element, they require solutions which are sensitive to local conditions in both social and ecological terms (Blaikie and Brookfield 1987). Rather than treating Third World people as an undifferentiated mass, studies should recognize that local variations in class, gender, ethnic, and other social structures critically affect society–nature relations. Internal social differentiation commonly has a particularly strong effect on patterns of resource access and control. For example, land concentration accompanying rural polarization under agroexport capitalism often subjects peasants to a 'simple reproduction squeeze' (Bernstein 1979), whereby they are forced to intensify food production in agriculturally marginal and ecologically fragile environments. In this situation, environmental sustainability is frequently sacrificed to the exigencies of day-to-day survival.

Commonly, this type of reproduction squeeze especially affects poor rural women, who are major food producers and assume the bulk of responsibilities for maintaining households in many rural societies (Cecelski 1987, Joekes 1994, Mackenzie 1993). Gender, class, and other social relations prescribe different activities, responsibilities, and powers of control over resources among various groups – with poor women being particularly disadvantaged. Local social relations mediate and influence the differential effects of environmental change on communities as a whole on their individual members. Poor rural women appear to be especially affected by patterns of resource control resulting from skewed power relations, both at the broader societal level and within individual households.

In many rural communities, these women are forced to shoulder the major burden of growing impoverishment and environmental degradation. At the same time that their access to basic resources declines, they are required to assume additional household responsibilities, as male family members migrate in search of work. This often means that poor women are forced to work longer hours collecting fuel and water, growing food, and generating income for household survival. Under these circumstances, the needs of environmental sustainability may understandably be ignored, especially if they entail allocating additional time and work. Therefore, sustainability will remain a distant goal if development efforts neglect the complex web of social relations which presently deny an adequate resource base to many poor women and other disadvantaged groups, thereby preventing them from adopting more environmentally sound practices.

Complementary to its stress on social relations, the political ecology approach has recently begun to pay more attention to cultural and ideological aspects of environmental issues (e.g., Bryant 1992; Gadgil and Guha 1994; Leach and Mearns 1991; Moore 1993; Redclift 1991). Whereas much previous analysis constructed abstract analytical categories to examine the role of social relations in environmental change, recent studies explore links between everyday environmental practices and underlying perceptions, meanings, values, and beliefs among various cultures. Innovative methods are evolving to combine some of the traditional concerns of political economy with new ideas concerning cultural interpretation – yielding an alternative to the structural tendency that previously dominated political ecology. Class, gender, and other social relations clearly shape patterns of resource use; in highly differentiated communities, people do not share equal access to either material or symbolic resources. However, struggles over land and resources are simultaneously struggles over cultural meanings. Local cultural categories for environmental features also shape uses of and struggles over resources. As Moore (1993: 383) notes, 'Historical patterns of access to, control of, and exclusion from resources emanate from and, in turn, mold competing meanings and cultural understandings of rights, property relations, and entitlements.'

From this perspective, sustainable development initiatives must be based on an understanding of different people's perceptions of the environment. Methods must be found to get at the sociocultural roots of environmental knowledge – its origins, motivations, and forms of expression among different people. Until recently, studies of sustainable development in the South largely ignored epistemological issues (i.e., of ways of acquiring knowledge and their integration into conceptual systems) (Redclift 1991: 41). The dominant Northern academic system of acquiring knowledge, through the application of scientific principles, remained unexamined or was assumed to be universal. However, much of the knowledge people outside of mainstream science possess, especially among traditional Third World societies, is encoded in customs, rituals, values, and beliefs linked to the cultural practices of everyday life. Understanding this cultural realm is crucial to comprehending the ways that different people use environmental knowledge, with obvious implications for sustainable development.

Becoming aware of local people's environmental knowledge is important, to permit sustainable development initiatives not only to devise more appropriate means of avoiding environmentally destructive practices, but also to make better use of vast local expertise. It should be acknowledged that, especially in rural areas, local people are often 'experts' about their environments, and that any attempt to foster sustainable development without their active participation is unlikely to succeed. While development planners need to avoid romanticizing about traditional or indigenous groups, whose environmental practices may not be without contradictions, at the same time, the knowledge and opinions of these people should be be taken seriously. However, many outside development professionals ignore or remain unaware of indigenous systems of knowledge and research procedures which may differ from accepted scientific methods (Richards 1985). In the area of pest management, for example, Bentley and Andrews

(1991: 116) contend that agricultural scientists often fail to recognize research and experimentation by local farmers because it does not use statistics, proper blocking treatments, replicable techniques, or control groups. Such experimentation is commonly carried out in little patches of crops which few outside scientists recognize as forms of 'test plots.'

Worse, it is commonly believed that while professionals take a dynamic, long-term view of sustainability, poor rural people live 'hand to mouth' and take a static, short-term view. As Chambers (1991: 7) points out, the opposite is often true; when poor people have secure rights to resources, their behavior normally manifests a longer-term view: 'they create, protect and develop microenvironments, like terraces and structures to capture and concentrate soil, water and nutrients; they plant and protect trees which they will never live to harvest.' Traditional and indigenous cultures typically have a rich and intricate knowledge of their own environments which has been passed down over generations. As it becomes manifested in local institutions and practices, this knowledge is continually evolving, finding innovative solutions for both old and new problems (Amanor 1994). For this reason, Colchester (1994: 82) remarks that the 'social, cultural, and institutional strengths inherent in traditional systems of resource use need to be built on to achieve sustainability and not dismissed as "backward" and "wasteful."'

In many traditional societies, this knowledge is reflected in indigenous technologies, whose appropriateness is context dependent; some techniques may prove unsuitable to changing conditions (Bebbington and Farrington1993). At the same time, however, many of these technologies are flexible, carefully adapted to local particularities, and especially appropriate to the conditions of capital-poor, labor-rich productive systems (Redclift 1987; Richards 1985; Wilken 1987). Typically, such technologies are not universally applicable, but are locally grounded (Sachs and Silk 1990). They also commonly meet many key goals of sustainability because they use less fossil-based inputs, are less dependent on the vagaries of global markets, and are more flexible and diversified than their Western counterparts (e.g., Green Revolution technologies). Such diversity may provide a range of benefits, including meeting a variety of basic needs (e.g., for food, clothing, medicines, building materials); offering heightened resistance to diseases and climatic changes to which certain species may be susceptible; and allowing for more socially and ecologically appropriate development to evolve in fragile ecosystems.

To make effective use of traditional knowledge and technologies, sustainable development initiatives need to adopt methods which encourage local participation at all stages, from research to policy decision-making and implementation. In the end, sustainable development cannot be achieved without the active participation of the popular majority which, in turn, demands attention to power relations and political factors. Political ecology studies have only quite recently begun to give serious treatment to politics (Peet and Watts 1993). Nevertheless, recent research has taken a number of interesting directions, especially through analyses which link various forms of political action to questions of resource access and control. Such research has focused on a range of subjects which express the heterogeneity of Third World environmental movements, including gender struggles centered on environmental issues (e.g., Agarwal 1989; Mackenzie 1991), environmental-livelihood movements (e.g., Broad 1993; Gadgil and Guha 1994), and the emerging 'liberation ecology' approach which unites concerns for nature and social justice (e.g., Martinez-Alier 1990).

Third World Environmental Movements and Popular Empowerment

Recently, environmental movements have formed an important political expression of Third World environmentalism and growing opposition to unsustainable development practices. As with social movements in general, these environmental organizations are largely unregulated by, and distinct from, the state. They normally operate in a bottom-up, participatory manner outside of the sphere of formal party politics. Links may sometimes be formed with other like-minded groups, but coalition-building usually proceeds from the bottom up and is focused on particular issues (e.g., overcoming impoverishment and environmental destruction simultaneously). Increasingly, links are also being formed with outside NGOs and development agencies, which recognize the many advantages that participation by local organizations may bring to development work (e.g., their greater flexibility, knowledge of local conditions, ability to respond quickly to changing circumstances). In many countries, these advantages are proving especially important to the development of innovative, locally appropriate methods for addressing the range of technical, social, and economic factors which together affect the environment (Vivian 1994).

The recent rise of environmental movements has had a dramatic effect. Just a few years ago environmental issues were given little if any attention in the development agendas of most countries. Moreover, development efforts largely disregarded the feelings and desires of local people concerning the social and environmental impact of various projects. However, this situation appears to be rapidly changing – especially in the parts of the South where grassroots social movements are organizing to combat socially inequitable and environmentally harmful development initiatives. Today, formally or informally organized social movements are regularly confronting development agencies to halt environmentally questionable projects in countries such as Bangladesh, Botswana, Brazil, Ecuador, Honduras, India, Mexico,

Malaysia, Philippines, and Thailand (e.g., Agarwal 1989; Colchester 1994; Dore 1992; Gadgil and Guha 1994; Parayil 1992; Utting 1994).

While their bottom-up participatory orientation gives many Third World environmental movements a broad commonality, they also typically have quite diverse membership, ideological orientation, organizational methods, and central issues, as is displayed, for example, by environmental movements in India, which have gained some prominence in the recent development literature. Among the methods these movements practice are direct nonviolent confrontation to block destructive development practices; passive resistance and non-cooperation to undermine inappropriate development projects; promotion of alternative environmental messages through skillful use of the media and tourism; and devising alternative, more appropriate development projects, often with NGO assistance.

The origins of contemporary Indian environmental movements can largely be traced to the Chipko movement in the early 1970s against commercial logging in the Central Himalayas, for which the term 'tree huggers' was coined. This struggle was representative of the focus of much early Indian environmental conflict on forestry issues. Although subsequent Indian movements have formed around a range of other problems (e.g., industrial pollution, unhealthy working conditions, chemical contamination of agricultural areas), struggles over dams have become especially significant (Agarwal 1989; Gadgil and Guha 1994). Many of these protests have involved tribal and other indigenous groups whose traditional lands were threatened by large-scale dams. Women have usually also been in the vanguard of these struggles.

According to Gadgil and Guha (1994: 127–8), Indian environmental movements display three dominant ideological orientations, further contributing to their heterogeneity. The first may be called 'Crusading Gandhian,' which views environmental destruction and social conflict as above all a moral problem which may be analyzed using Gandhian philosophy and traditional Hindu scriptures. This approach calls on India to reject Western models by returning to its cultural roots in precolonial village societies which stressed social and ecological harmony. The second trend is 'Ecological Marxist,' which argues that environmental and social problems are essentially based in patterns of unequal resource access, rather than questions of morals and values. Ecological Marxists in India are perhaps most closely identified with People's Science Movements, the best known of which is the *Kerala Sastra Sahitya Parishad*, which has widened its initial focus of bringing 'science to the people' to include environmental issues. The third tendency is termed 'Appropriate Technology,' which occupies the vast middle ground between the first two approaches and seeks to create practical development alternatives based on a working synthesis of agriculture and industry, large and small units, and Western (modern) and Eastern (traditional) technological systems. Its emphasis is not so much on challenging the 'system,' *pace* the Marxists, or the system's ideological underpinnings, *pace* the Gandhians, but on demonstrating pragmatic socio-technical alternatives to the centralizing and degrading technologies currently in use.

In India and elsewhere, environmental movements have frequently been linked to alternative, bottom-up development efforts. In many Third World cities, for example, urban 'ecodevelopment' initiatives have begun using the latent resources of cities to meet pressing social needs (e.g., Bartone 1991; Drakakis-Smith 1990; Sachs 1988; Stren et al. 1992; Viola 1988). Examples include the reorganization of local public transit systems to conserve energy (e.g., Curitiba, New Delhi), energy recovery from landfills and other waste material (e.g., New Delhi, São Paulo), and urban food production on vacant or idle land (e.g., Harare, Mexico City, Santiago). Likewise, locally managed ecodevelopment projects have been initiated in many Third World rural areas (see, Bebbington and Farrington 1993 Bentley and Andrews 1991; Cheru 1992; Nesmith 1991; Thrupp 1990; Thomas-Slayter 1992). Examples include social forestry to decrease soil erosion and fuelwood scarcity, integrated pest management to reduce chemical contamination of soils and groundwater, and ecotourism based on the creation of nature reserves to preserve fragile ecosystems while providing local employment.

Increasingly, Third World environmental movements are adopting alternative development agendas which link problems of environmental destruction to issues of basic-needs provision, distributional equity and social justice, local self-reliance, and popular empowerment. Throughout the South, the claims being made by such local movements show a striking similarity: the right to secure basic human needs; the right to ownership of their territories and resources; the right to self-determination; and the right to represent themselves through their own institutions. Many development workers have long realized that effective resource management cannot be achieved without the goodwill and cooperation of viable local communities. Accordingly, deepening popular participation and empowerment necessarily go hand in hand with sustainable development. Identifying and matching changing social needs to local resources requires the constant and effective participation of grassroots organizations and movements. Nevertheless, Third World governments and international development agencies have been hesistant to adopt bottom-up approaches for addressing environmental problems. The idea that local popular organizations should actually control resource use and development practices has seldom been seriously entertained.

The environmental crisis confronting many Third World areas is also a crisis of grassroots political participation and empowerment. The former cannot be resolved in isolation from the latter. The promotion of sustainability is by definition, therefore, political. Environmental movements have added a new dimension to notions of democracy and civil society in many countries, often posing a serious ideological challenge to dominant ideas concerning the meaning, content, and pattern of development. However, while there are opportunities for broadening local participation and grassroots coalition building, popular environmental movements must also find methods to 'scale up' their struggle against ecologically and socially destructive development strategies. As Colchester (1994: 93) notes, 'The assertion of indigenous rights and the transfer of resources back to local communities is being, and will continue to be, resisted by those who benefit most from present development strategies.'

Continuing Contradictions of Sustainability

Although environmental issues have received increasing attention from a variety of alternative development approaches, in practice environmental sustainability remains an elusive goal for most Third World countries. Moreover, alternative theories of sustainable development often appear contradictory and utopian. Curiously, relatively little attention has been given to the realm of institution building, which is central to devising methods to achieve sustainable development practices in the real world. Despite the fact that the route is at least as important as the destination in issues such as sustainability, the development literature is only beginning to examine the appropriateness of alternative institutional means to alleviate environmental problems in different Third World contexts. If sustainability is to be more than just a pleasant vision, the theory and practice of development must link environmental problems to the specific socioeconomic, political, and institutional contexts within which they occur. It is especially important to include questions of power in sustainable development initiatives from the very beginning. Efforts to create more appropriate and sustainable resource utilization, particularly in highly polarized societies, cannot be expected to succeed if they fail to address the complex web of power relations within which patterns of resource access and control are largely determined.

Frequently, sustainable development initiatives by NGOs and other groups have attempted to treat environmental problems with simple, neat solutions or 'magic bullets' (Vivian 1994). The focus is commonly on biological and/or technical solutions which neglect the social dimension. Little coordination takes place between various development projects and institutions, which frequently operate at cross-purposes or engage in debilitating 'turf wars' (Thrupp 1990, Utting 1994). Sustainable development efforts are often contradicted by other environmentally destructive policies or are implemented in a manner which ignores socioeconomic concerns related to the rights, needs, and priorities of local people. In many rural areas, for example, environmental initiatives remain uncoordinated or, worse still, are in conflict with programs designed to address food security needs (Davies and Leach 1991). This separation inhibits consideration of the mutual implications of food security and environmental sustainability. Normally, assured access to productive resources is vital both to rural people's food security and to their ability to protect the environment (Cheru 1992; Davies and Leach 1991; Thiesenhusen 1991). It follows that food security entitlements and local participation are requisite starting points for any viable environmental project in poor rural areas.

The interconnected nature of Third World environmental and social problems underscores the need to involve local people via popular institutions and organizations in the formulation and implementation of environmental projects. In most cases, environmental initiatives cannot succeed without bottom-up incentives for the local population to participate. People must feel that they have a personal stake in the benefits offered by sustainable development practices. Too often, environmental projects are imposed by outside development professionals and scientists who are not accountable to the local people. Little consideration is given either to the particular needs and desires of these people or to their varying perceptions concerning the environment, though various groups may perceive the environment and value resources quite differently. Moreover, the social distribution of the costs and benefits of development changes affecting resource use is usually very uneven. In particular, skewed power relations often produce a yawning gap between patterns of vulnerability and responsibility; those who bear most of the costs of environmental change frequently have little control over development decisions that generate such change. To reduce this gap, which is also commonly linked to local non-cooperation and resistance to top-down environmental projects, popular participation and empowerment need to be given a central place in sustainable development initiatives.

At the same time, increased concern for creating participatory institutions at the local level should not come at the expense of neglecting broader political and economic considerations. As is common to alternative development approaches in general, local environmental projects have often encountered problems in 'scaling up' their development efforts (Amanor 1994, Thrupp 1990, Vivian 1994). This has frequently meant that the negative impact of sweeping development models and macro-policies has more than offset hard-earned local achievements. Major constraints to locally sustainable development commonly arise from the wider political economy, especially through the imposition of development models which further marginalize the poor and disadvantaged. As a result, many people are pulled in two directions at once: by their desire to follow sustainable development practices, on the one hand, and by the exigencies of environmentally destructive market forces, on the other. This problem appears to be particularly characteristic of current Third World agricultural strategies, which have accelerated social polarization and environmental destruction in many rural areas. Especially under the demands of SAPs and neoliberalism, the tendency has been to sacrifice social and environmental sustainability in favor of an all-out effort to maximize the production levels and short-term earnings of an agroexport elite.

According to Utting (1994: 232), this has meant that environmental planning in many Third World countries has suffered from a 'lack of macro-coherency' or 'the failure to locate environmental initiatives within a coherent development policy framework.' Policies and projects have been characterized by a narrow sectoral focus which pays insufficient attention to the broader development context and linkages between patterns of capital accumulation, widening inequalities and impoverishment, and rising environmental degradation. In effect, most sustainable development projects have been reduced to trying to minimize the negative social and ecological effects of market-led, growth-first development strategies. To address this contradiction, sustainable development needs to overcome its utopian tendencies, especially by paying more attention to questions of power. Uninformed by analyses of power relations at various (micro, meso, macro) scales, sustainable development approaches, no matter how internally coherent or technically rational, have little chance to achieve their goals. Since sustainable development normally requires complementary socioeconomic change, it cannot be successful if broader development structures block such change.

Therefore, sustainable development efforts must include the power variable from the very beginning by placing environmental issues within their wider socioeconomic context. Growing environmental problems at various scales underscore the need to devise concepts and methods capable of addressing both the global dimension and local/regional particularities of environmental change. This rethinking process should neglect neither social relations nor society–nature relations, which together give environmental problems a necessarily multidimensional nature. Methods must especially be found to deal with the underlying political and economic causes of environmental degradation within current development structures and institutions. It is particularly important to address the roots of many environmental problems in the policies of Northern-based development institutions – both through popular pressure to transform those institutions and by creating alternative structures and mechanisms to enhance North–South and South–South cooperation for sustainable development. Perhaps the most revealing example of the environmentally destructive economic logic that currently drives many global development institutions is offered by the following statement in a leaked internal memo from Larry Summer, a Vice-President of the World Bank:

Health-impairing pollution should be done in the country with the lowest cost, which will be the country with the lowest wages . . . I think the economic logic behind dumping a load of toxic waste in the lowest wage country is impeccable and we should face up to that. (in Dore 1992: 85)

While most Third World environmental problems have a pervasive macro-dimension at the level of the global political economy, they are also subject to important micro-determinants based in local social relations and structures. This means that environmental initiatives will succeed only to the extent that they are site-specific, culturally appropriate, and socially sensitive. In most Third World areas, finding methods to meet the special needs of the poor and disadvantaged is a requisite for creating more sustainable development practices. No amount of research or advice will persuade people to conserve resources if powerful economic forces are driving them in the opposite direction. Sustainable development ultimately depends on the day-to-day actions of local people pursing a variety of strategies aimed at securing their livelihoods. Accordingly, sustainable development initiatives need to reject approaches by which outside concepts, formal models, and ready-made plans are imposed on local communities. Instead, attention should be given to local values, beliefs, customs, practices, and social institutions.

This approach requires a shift in sustainable development away from its common technical fixation on top-down resource management toward a more holistic environmental focus which includes a people-oriented agenda and invites local participation. In order to be appropriate to the people affected, sustainable development practices must be based on more than just ecological soundness. They must put local people's priorities first, by promoting methods that stress dialogue, participation, and learning by doing, emphasizing the inseparability of social and environmental problems from the perspective of those experiencing them. This entails an integrative and socially aware approach to sustainable development which combines environmental protection with the needs and rights of local people. Essentially, a new framework for sustainable development is required – one which searches for appropriate solutions to contextually specific environmental problems, creates a spirit of discovery and enquiry in collaboration with local people, recognizes the validity of traditional environmental knowledge and practices, and deepens popular participation and empowerment along with sustainable development practices.

11

Popular Development

Third World societies are complex and multifaceted, as is the concept of popular development. Many of the key elements of popular development discussed in previous chapters of the book will be summarized here and linked together to allow a clear picture to emerge. Popular development addresses many of the central issues and questions that have continuously occupied postwar development studies, but it also employs new concepts and methods designed to overcome the most serious shortcomings of our current development frameworks. In this way, it seeks to advance the rich alternative development tradition that has evolved in recent years. But, as the term implies, popular development is focused on a central concern of all alternative approaches – creating development appropriate to the needs and interests of the popular majority in Third World countries. This especially requires increased attention to the contextuality of development at various scales, involving a complex and ever-changing interplay of objective conditions and subjective concerns.

Rejection of Grand Theories and Eurocentric Biases

Much of the optimism that characterized development studies in the early postwar period has now given way to disillusionment. As the theoretical and practical shortcomings of the established models of both Right and Left have become apparent, a rethinking of development has begun to take place. This necessitates removing the conceptual blinders and methodological straitjackets of conventional thinking in favor of a broader, more flexible vision of development capable of addressing diverse Third World realities.

From the older Keynesian framework to the newer neoliberal paradigm, postwar development studies have been dominated by 'grand theories.' This has generated increasing tension between the desire to formulate universally valid principles and formal models, and the need to understand the variety of actual experiences and potential alternatives of Third World development. Most development theorists have been preconditioned to look for parallels between the development history of the West and the contemporary situation in non-Western societies. Typically, the rich and diversified development experiences of different societies have been simplified and distorted by formal models and theoretical constructs which reduce development to a few universally valid factors and organizing principles.

However, the postwar experience demonstrates that solutions to development problems must be sought in the contextuality of development, which is a product of particular historical processes. The context of development is constantly changing in scale, over time, and among societies – creating both new obstacles and new opportunities for variations. Accordingly, it is impossible to draw conclusions from the experiences of particular societies at particular times and expect these findings to be necessarily valid for other cases. Development theories and strategies must come to grips with the many pluralisms of societies that produce important variations across both time and space. Development frameworks, especially those designed to contribute toward policy-making, need to devise ideas and methods capable of accommodating this geographical and historical diversity. Tendencies toward grand theorization, therefore, ought to be rejected as inappropriate to the analysis of diversity and change – which make development a necessarily multilinear process subject to divergent constraints and opportunities according to the complex interplay of both objective and subjective factors.

Although one discipline cannot adequately address the multifaceted nature of development processes, 'discipline-centrism' is an ongoing problem in development studies. The development process is artificially fragmented and compartmentalized to fit the areas of specialization, research methods, and theoretical frameworks of individual disciplines. Instead of being constituted as a unified and distinct area of intellectual inquiry, the field of development studies is incorporated, in bits and pieces, into various disciplines. By contrast, interdisciplinary approaches to development have yet to gain much respectability in an intellectual environment which tends to favor more 'scientific' and 'rigorous' research in disciplinary specializations. As a result, many discipline-based development studies have failed to take advantage of ideas and methods developed in other disciplines, as well as from a variety of non-academic sources in the South itself.

Grand theorization and discipline-centrism are frequently also associated with Eurocentrism. Research tends to examine only those phenomena and events that can be conceptually compartmentalized and take on theoretical significance within the accepted version of the West's development experience. Other subjects are typically ignored, and scant attention is afforded to the views, wishes, and values of Third World peoples themselves. In a highly normative and ethnocentric manner, Western values are universalized and linked with progress, while traditional Third World values are denigrated and tied to stagnation and underdevelopment.

The domination of developing countries by Eurocentric development strategies has far-reaching consequences. In broad terms, it arrests indigenous development processes and preempts possibilities for alternative, more self-reliant and broadly based development projects. Indigenous social structures that have traditionally supported majority interests and have served as sources of identity and popular participation are undermined without being replaced by viable alternatives. Alien Western values, relationships, and institutions fuel rising tensions, uncertainties, and feelings of anomie – thereby subverting efforts to bring about needed socioeconomic change. More appropriate development strategies must pay close attention to the historical legacy that created Third World societies, to sociocultural factors and other particularities of those societies, and to overarching structural conditions that make development along classical Western lines highly unlikely.

Bridging the Gap Between Theory and Practice

Grand theories and formal models inevitably reduce people to ideal types, predicting behavior on the basis of a limited number of set variables. In development, this results in a yawning gap between theory and practice. Much theoretical research is of little use to practitioners of development in the field, while practical development projects often lack direction or repeat mistakes for want of a firmer theoretical grounding. Typically, the theoreticians and practitioners of development occupy different worlds and speak different languages. Theoretical models are too formalistic and abstract to be relevant to the everyday world of their would-be beneficiaries. Meanwhile, there is an accumulated wealth of practical knowledge derived from actual experiences and on-the-job training that only poorly informs the theoretical literature. One of the most valuable resources for theory-building ought to be the development experience itself, as it is conceived and practiced by various social groups. As anyone who has worked in development at the local level knows, projects seldom evolve as planned. Greater familiarity with local experiences might produce more useful and applicable concepts, more appropriate methods, and more realistic expectations of the people involved in the actual work of development. Meeting such goals requires a more practical, applied stance and more involvement in policy making, project planning, and implementation.

Because the context of development is constantly changing at a variety of interlocking scales, no fixed theoretical approach is likely to prove satisfactory. Development theories must therefore be flexible and responsive to changing conditions; it is a time for conceptual tentativeness and methodological experimentation rather than set models and rigid methods. Many contemporary Third World problems represent a crisis for development theory simply because it has not provided the appropriate ideas and tools to address them. Progress requires more awareness of the ways in which our theoretical constructs deny negative possibilities, filtering out potentially important factors that cannot be comfortably accommodated. Studies often employ a priori models which rule out subjective factors and separate the observer (researcher) from the observed (research subjects). Broader social, human, and moral issues not amenable to modeling techniques are excluded from analysis. People are treated as objects to be studied rather than as subjects of development in their own right whose knowledge and interpretations of the world might contribute not only to the findings but also to the design of development work.

Much recent criticism of mainstream development approaches has focused on the human factor and the subjective realm of values, meanings, and interpretations. Postwar development studies, in the rush to be more 'scientific' or 'objective,' have adopted methods that are inappropriate to the study of social subjects and that neglect many issues concerning means of acquiring knowledge. Historically constituted values and meanings are either excluded from analysis or treated as simple universals in a way that denies their social construction. Above all, the universalization of values within development frameworks has repudiated the wishes and aspirations of marginalized social groups. These silences underscore the fact that development studies, like all discourses, is intersected by power relations – a connection that undermines positivist notions of objectivity and the distinction between facts and values.

In general, development studies must pay more attention to human complexities and to the dynamic, open-ended, and non-determined nature of social processes. Stress should be placed on local diversity, human creativity, and processes of social change. Research should explore various subjective elements of development alongside overt materialist behavior and objective factors. This entails 'bringing the actors back' into development work in their own economic, political, and sociocultural contexts. Studies should try to interpret others' understanding of the world without preconceived notions and conceptions. Contrary to much development research, it should not be assumed that one factor (typically the economic) is dominant or that social processes conform to some sort of predetermined universal logic.

Increasingly, advocates of alternative development approaches contend that a primary concern for 'humaneness,' including social, ethical, and moral considerations, should replace the abstract, technical focus of the mainstream frameworks. If development is fundamentally about processes of human action and interaction rather than just about goods and services, then it is clear that development theory must deepen its understanding of what it is to be human. Human practices are composed of mutual actions and social relations that reflect important differences in intersubjective values and meanings. Accordingly, one must understand the discourse, or the underlying configuration of interpretations and meanings, before one can understand the social practices. Frequently, social practices which may on the surface appear to be similar may be interpreted quite differently and may take on distinct meanings for different groups of people.

The failure of many development approaches to understand and account for differences in social practices has had a particularly serious impact on local participation. Within most Third World societies, the 'professionalization' of development work under exclusionary Eurocentric frameworks, the control of this knowledge by elites in their own interests, and the devaluation of alternative sources of popular knowledge have all prevented the majority from participating in development decision-making. Local self-confidence has been undermined and grassroots groups have been blocked from acquiring the knowledge and skills needed to analyze and solve problems for themselves. The transformation of people into agents of their own development, which ought to be the focal point of any broadly based democratic development strategy, has been retarded by exclusionary theories and elitist practices. At the same time, theorists are prevented from understanding much about the real world of the popular majority, to which development is ostensibly directed, because these 'experts' are so far removed from that world. Because knowledge is associated with formal Western education and technical training, indigenous concepts and methods are ignored or relegated to a strictly subordinate position. Development studies that do not conform to accepted Western theories and methods are neglected, as are other local sources of new insight, understanding, and creative solutions to development problems.

New Realism Versus Old Dichotomies

Moving beyond outmoded development models that have outlived their historical usefulness necessitates transcending a series of false dichotomies that have polarized much of the postwar development agenda. Examples of these dichotomies include state planning versus the market, inward- versus outward-oriented development, urban industrialization versus rural agriculture, and top-down centralization versus bottom-up decentralization. To transcend these dichotomies, a new approach is required which avoids framing development issues as either/or choices according to preconceived theories. Postwar experience shows that the different sides of each of these dichotomies need not necessarily be mutually exclusive. In fact, given appropriate and properly coordinated policies, they can often be mutually reinforcing. However, the case for or against a particular strategy should largely depend on the historical and geographical conditions, the sociocultural and political institutions, and the specific needs and interests of individual countries.

This new approach stresses pragmatism, flexibility, and the contextuality of development. It requires freeing up our minds and searching for innovative solutions, because the stale, ideologically driven debates to which we have become accustomed have lost their relevance. No development orthodoxy, whether that of market-led neoliberalism or state-centered Keynesianism, can provide blanket solutions to the problems of all countries at all times. Rather, strategies must address the contextuality of development, which is the product of both specific historically determined conditions and the subjective concerns of different groups of people. Failure to understand the special opportunities and constraints presented by such factors renders our formal models and universal strategies irrelevant to the real needs and problems of many Third World locales.

Potentially, one of the most beneficial aspects of the neoliberal agenda is its focus on reducing state waste and inefficiencies. It is undoubtedly healthy to avoid the old assumption that the state can do anything and everything, which unfortunately has marred many Keynesian-inspired strategies. Nevertheless, it cannot be assumed that the market by itself can automatically stimulate development in all countries under all conditions. In the current ideological climate, the neoliberal contention that market failures are trivial, but government failures are enormous, has become a powerful slogan. But as a focus for serious analysis, it is wholly inadequate to understand the many interrelationships between market and government failures that underlie most development problems. Moreover, it exonerates other major actors (e.g., transnational corporations, Third World elites, international financial institutions) from any responsibility for Third World underdevelopment. While this position may serve certain ideological interests, it offers only a simplistic, naive conceptual foundation for finding appropriate solutions to real-world development dilemmas.

Recent evidence from East Asia and other areas shows that the efficient use of market forces does not necessarily preclude state development planning. Selective and carefully coordinated state intervention can alter markets to more efficiently serve a range of development goals. However, a key problem is finding the proper mix of market and state, and then devising a set of institutional and organizational arrangements compatible with this mix. Neither the state nor markets are neutral institutions; both can work for good or ill. Development strategies should consider under what conditions states and markets can work to serve broad development objectives and how to bring about these conditions. Solutions will necessarily be particular to individual countries, will pay close attention to diverse contextual elements that shape development, and will involve more than just promoting economic growth.

In many underdeveloped countries, much of the private sector is undynamic or has been incapacitated by decades of infrastructural neglect. Under these circumstances, sudden market liberalization may result in a precarious vacuum, inviting anti-competitive behavior by the few who have the means to step in. To avoid this scenario, a more balanced approach is needed to questions of liberalization and relations between the state and markets. The main questions for development strategies seem no longer to concern the extent of state intervention and/or the size of the public sector, but concern the comparative advantages of the public and private sectors, how these sectors may complement each other, and how their performance may best be improved. The state should be asked only to do what it can do best and should stay out of other areas. Nevertheless, it can take important measures to promote development of both the private sector and society at large. While there are normally costs involved in state interventions, unfettered markets often exact even higher costs, especially among marginalized groups in polarized societies.

With the rise of neoliberalism, market-led development has increasingly been tied to export promotion and an outward orientation. Moreover, this development agenda is frequently juxtaposed with import substitution and other inward-oriented policies. However, these two approaches may be complementary under certain circumstances. As the East Asian development performance demonstrates, selective measures supporting domestically oriented sectors may be compatible with export promotion and other outward-oriented policies. This may enable countries to make better use of the rational core of comparative advantage theory to enlarge their participation in international markets, while simultaneously providing conditions for a more participatory, internally articulated form of development.

The adoption of a strategic trade policy may be preferable to a free-trade stance for most Third World countries. Free trade may permanently confine underdeveloped countries to a 'trap of static comparative advantage' in which they are unable to diversify away from primary commodities and other low-wage goods. The Asian NICs show that a strategic trade stance may be used to promote exports compatible with broader development goals. Contrary to neoliberal opinion, the efficiency of trade liberalization cannot be established a priori for individual countries with different needs and priorities. Any strategy that does not address the wider aspects of development but focuses solely on liberalization measures runs the risk of generating unforeseen and destabilizing results, as well as missing new potential development sources.

Much of the argument for strategic trade policy rests on the advantages of protecting selected infant industries, at least during their formative period, from well-established foreign competitors. In addition, small/medium producers may require support either to enter global markets or to meet new competitive conditions in liberalized domestic markets. While the old-style 'umbrella' approach to protectionism should be avoided, this does not preclude supporting specific sectors according to particular development goals (e.g., fostering structural change, creating jobs, avoiding peasant impoverishment). Within the manufacturing sector, infant industries may require initial state support to gain a foothold against foreign competitors. This may allow domestic firms to capture economic rents from foreign competitors. It may also create opportunities for significant 'spin-offs' in terms of technological diffusion, demonstration effects, and skills development.

Within the rural sector, small/medium farmers may require assistance to pursue new opportunities in global markets or to avoid being swamped by foreign competition in liberalized domestic markets. Risks preventing peasants and other small producers from entering lucrative export markets may be reduced by well-targeted state programs (e.g., technical assistance, rural credit, diversified processing and marketing channels). Moreover, state support may be used to protect peasant food producers from 'dumping' by highly subsidized Northern exporters.

Various means may be chosen, but active state involvement is a requisite in most Third World countries if export promotion is to avoid reinforcing polarization tendencies that have accompanied previous export-led development strategies. If managed efficiently, the dividends from stimulating more broadly based and socially sustainable growth would far outweigh any short-term costs. There is nothing inherently wrong with development strategies that seek to increase export production, if they are consistent with larger societal goals (e.g., promoting growth with equity, maintaining access to affordable food and other basic needs).

Indeed, most peasants and small/medium producers are not opposed to diversifying into export sectors. However, they wish to do so on terms which allow them to effectively compete with larger producers and minimize risks to acceptable levels. This normally entails putting mechanisms in place to facilitate productive diversification without risking subsistence needs. It also entails the creation of institutions (e.g., cooperatives, producer associations) to enhance producers' technical and marketing expertise and to assist, when necessary, in market diversification. Without these types of measures, the current cycle of export promotion threatens to reinforce the exclusionary, elitist character of traditional export models. Conversely, however, the inclusion within export strategies of policies designed to meet the specific needs of small/medium producers could reverse tendencies toward polarization. It might also help to lay the social foundations for increased political stability, without which any future development strategy is unsustainable.

The social requirements for more balanced and sustainable development are normally closely linked to sectoral and regional considerations. Nevertheless, development strategies have typically been biased toward industrialization in core urban areas at the expense of agriculture and related activities in outlying rural regions. As the experience of the Asian NICs once again shows, however, rapid industrially based growth may be complementary to continuing agricultural development, even when dominated by small/medium producers. In South Korea and Taiwan, land reforms created initial conditions conducive to balanced rural development. Moreover, a variety of additional state policies incorporated the rural sector into the national economy in a way that simultaneously accelerated agricultural and industrial production, while generating widely distributed income increases for both rural and urban households. This helped to avoid the massive rural marginalization that has accompanied the industrialization drives of other Third World countries.

The Asian NICs have been relatively successful in addressing three false dichotomies of development frameworks – state planning versus the market, outward-versus inward-oriented growth, and industrialization versus agricultural development. However, another dichotomy must also be dealt with to bring about popular development – that between top-down, centralization and bottom-up, decentralization. NIC development has been strictly, and often ruthlessly, controlled by a highly centralized, authoritarian state apparatus. The top-down, paternalistic manner by which development projects have been administered has systematically undermined local forms of social organization and democratic practice. Although some variation exists, none of the NICs have made much progress in creating democratic structures that would facilitate meaningful popular participation in development institutions and the political arena.

Postwar development planning has tended to counterpose centralization to decentralization, instead of treating them as integrative or, under some conditions, possibly complementary strategies. Most countries emerged from the Second World War with highly centralized state structures that suited the top-down thrust of Keynesian development planning. Decentralization efforts were often initiated in the 1960s and 1970s, prompted by dissatisfaction with centralized Keynesian planning. By the 1980s, mounting political and administrative problems caused a shift back to more centralized control. In recent years, however, ideas of decentralization and local participation have again gained favor, as notions of centralized government and planning are increasingly challenged, in both theory and practice.

However, just as centralized development failed to meet early postwar expectations, more recent decentralization initiatives have often failed to produce the desired results. In most cases, the actual practice of decentralization has not matched the rhetoric. Few genuine attempts have been made to implement democratic decentralization. If some political decentralization has occurred, power has typically devolved to members of local elite groups and/or centrally recruited civil servants. Popular participation has effectively been neutralized by elaborate mechanisms of centralized supervision and control. Moreover, most decentralized planning programs have been given inadequate human and material resources to fulfill their functions. Without the requisite resources, they remain empty shells.

Decentralization is often viewed mainly as a technical solution to bureaucratic inefficiencies. Typically, when decentralization efforts are analyzed, the measures of successful performance are those of administrative decentralization. Successful programs are those that improve service delivery, reduce waste and delays, and enhance cost recovery. The empowerment of disadvantaged groups is, at best, viewed as a secondary goal. A narrow, economic agenda is established about what matters in decentralization. However, as with all other aspects of development, decentralization is basically a political process. For decentralization to be empowering, it must be linked to broader goals of economic and political democracy; participation cannot merely be functionalized as a technical planning exercise. Fundamentally, decentralization has to do with power: who rules and how that rule is carried out. Moreover, these questions are normally complex and do not lend themselves to universal or unidimensional solutions. Different groups may prevail in different issue arenas or in the same arena at different times. Coalitions and alliances are assembled at various scales, are dissolved, and become reconstructed in new forms.

Viewed from this perspective, neither decentralization nor centralization can be regarded as necessarily supportive of development goals, such as increased popular participation, economic and political democracy, and social justice. There is little point in identifying particular institutional arrangements in advance as more democratic or egalitarian; the appropriate choice has to be made in light of local circumstances. The attainment of certain goals cannot be guaranteed by the institutions themselves, as if they can be separated from the societal setting in which they are constituted. The extent and form of desirable (de)centralization can only be determined in concrete situations. Similarly, the measures of success for (de)centralization can only be specified in particular contexts. Therefore, it is pointless to try to construct a general strategy for (de)centralization, or a universal, formal model of (de)centralized government, or a generic set of evaluative criteria that are applicable in all cases. Neither decentralization nor centralization in and of itself promotes popular development. The choice over which to use must depend on the objectives and the particular setting. What works reasonably well under some conditions may prove disastrous under others.

Balanced and Sustainable Development

Given the growth exigencies of the present global economy, development strategies that disregard macroeconomic balance and allocative efficiency are bound to fail. However, these economic imperatives cannot be allowed to override the broader, long-term requirements of Third World development, such as relatively equitable income distribution, basic-needs provisions, human-resource development, popular participation and democratization, socially and spatially balanced growth, and cultural and environmental sustainability. In most countries, these requirements necessitate new conditions to improve economic opportunities, develop human capabilities, and enhance social cooperation and democratic participation. But the tendency of our mainstream models, especially when framed by structural adjustment programs, is to sacrifice such concerns to the immediate accumulation demands of an elite minority. In so doing, they often neglect issues of general welfare and social cohesion for short-term profits and unsustainable, unbalanced growth. Furthermore, the narrow macroeconomic focus of these models delays more sweeping changes needed for sustainable, majority-based development.

While no development strategy explicitly aims for inegalitarian development, neoliberalism and other mainstream strategies implicitly assume that inequalities, both socioeconomic and regional, are a necessary price for growth. A trade-off is assumed between distribution and growth, especially during early development stages, so that redistributive measures may raise short-term consumption by the poor, but at the cost of reduced investment and diminished prospects for long-term growth. By contrast, alternative development strategies assume that there need be no conflict between redistributive measures and growth-oriented policies. In fact, it is generally asserted that redistribution before growth, and not vice versa, makes more sense, particularly in highly polarized societies. A higher priority is therefore given to immediate redistributive measures, especially those directed toward human-resource development. A direct, targeted approach to alleviating poverty and reducing inequalities is favored rather than waiting for the 'trickle-down' effects of growth to eventually occur.

The escalating social and environmental costs of mainstream development have provided the focus for much recent research in sustainable development. Some studies emphasize the high 'opportunity costs' associated with irreversible environmental damage that may foreclose future development options. Other studies stress the neglect of the essential 'utility-yielding' role of ecosystems and their environmental functions. And a third group of studies has broadened the discussion to issues of 'coevolutionary development,' or interrelationships between social and ecological systems in which feedback mechanisms that previously maintained ecosystems have been progressively shifted to the social system. These sustainability studies challenge development frameworks to re-examine a series of interlocking environmental and social questions that have largely been ignored. It is asserted that development is not well measured by indicators such as GNP which focus solely on economic growth. Instead, attention should be given to issues such as redistributive justice and egalitarian ethics, human-resource development, protection of the environment and species survival, and the diverse interests and desires of marginalized and disadvantaged groups. Strategies that may generate high growth but also produce widespread alienation and distributional disparities ought to be avoided. Likewise, growth-oriented strategies that generate unacceptable levels of environmental destruction should be abandoned in favor of alternative approaches that address needs for ecosystem maintenance and the preservation of biodiversity.

Because most environmental problems contain an important place-based element, they require solutions which are sensitive to local social and ecological conditions. Development approaches should recognize that society–nature relations are critically affected by local variations in class, gender, ethnic, and other social structures. Commonly, social differentiation has a particularly strong effect on patterns of resource use and control. In many rural areas, for example, land concentrations accompanying the rise of agroexport capitalism have subjected peasants to a 'reproduction squeeze,' whereby they are forced to intensify food production in agriculturally marginal and ecologically fragile environments. In this situation, environmental sustainability is frequently sacrificed to the exigencies of day-to-day survival.

This type of reproduction squeeze is often hardest on poor rural women, who are major food producers and assume the bulk of responsibilities for households in many rural societies. Gender, class, and other social relations prescribe different activities, responsibilities, and powers of control over resources among various groups – with poor women being especially disadvantaged. Local social relations mediate and influence the differential effects of environmental change on communities and their individual members. Poor rural women appear to be particularly affected by patterns of resource control resulting from skewed power relations, both at the broader societal level and within individual households.

The interconnected nature of Third World environmental and social problems underscores the need to involve local people via popular institutions and organizations in the formulation and implementation of sustainable development projects. In most cases, such initiatives cannot succeed without incentives for the local population to participate. Local people must feel that they have a personal stake in the benefits offered. Too often, projects are imposed by outsiders who are not locally accountable. Little consideration is given either to the particular needs and desires of local people or to their perceptions of the environment. However, various groups may perceive the environment and value resources quite differently. Moreover, the social distribution of the costs and benefits of changes in development affecting resource use is usually highly uneven. Skewed power relations often produce a yawning gap between patterns of vulnerability and responsibility; those who bear most of the costs of environmental change frequently have little control over decisions that generate such change. To reduce this gap, which normally results in local non-cooperation, sustainable development initiatives need to stress popular participation and empowerment.

This requires a shift away from technical fixations on top-down resource management toward a more holistic environmental focus which includes a people-oriented agenda and invites local participation. To be appropriate to the people affected, sustainable development practices must be based on more than just ecological soundness. They must also put local people's priorities first, by promoting empathetic development methods that stress dialogue, participation, and learning by doing. Emphasis should be placed on the inseparability of social and environmental problems from the perspective of those experiencing them. This entails a more integrative and socially aware approach which combines environmental protection with the needs and rights of local people. Essentially, a new sustainable development framework is required – one which searches for appropriate solutions to contextually specific environmental problems, creates a spirit of discovery and inquiry in collaboration with local people, recognizes the validity of traditional environmental knowledge and practices, and deepens popular participation and empowerment along with sustainable development practices.

In recent years, the advancement of new ideas and practices in sustainable development has been greatly aided by a range of Third World environmental movements. Many of these groups are adopting alternative development agendas which link environmental problems to issues of basic-needs provision, distributional equity and social justice, local self-reliance, and popular empowerment. Throughout the South, the claims being made by local environmental movements show a striking similarity: the right to secure basic human needs; the right to ownership of their territories and resources; the right to self-determination; and the right to represent themselves through their own institutions. Many development workers have long realized that effective resource management cannot be achieved without the goodwill and cooperation of viable local communities. Identifying and matching changing social needs to local resources requires the constant and effective participation of grassroots organizations. Nevertheless, development efforts have hesitated to adopt participatory, bottom-up approaches for addressing environmental problems. The idea that local popular organizations should actually control resource use and development practices has seldom been seriously entertained.

Ultimately, the environmental crisis in the South is also a crisis of political participation and empowerment. The former cannot be resolved in isolation from the latter. The promotion of sustainability is by definition, therefore, political. Environmental movements add a new dimension to notions of democracy and civil society in many countries. They often pose a serious challenge to dominant ideas concerning the meaning, content, and pattern of development. However, while environmental issues may offer opportunities for broadening local participation and grassroots coalition-building, popular movements must also find methods to 'scale-up' their struggle against ecologically and socially destructive development strategies. If not, the assertion of democratic rights and the transfer of resources back to local communities will continue to be thwarted by the powerful outside interests who benefit most from top-down development strategies.

The Indigenization of Development

The need to devise theories and methods more appropriate to the particularities of Third World societies has focused increasing attention on indigenous concepts and practices of development. The indigenization of development thinking has become a central element of attempts to create more comprehensive and relevant approaches. This process of indigenization involves the creation of counter-institutions and practices to promote critical and independent development work. It implies an intellectual emancipation and a fundamental reassessment of the major Western-based development paradigms. It calls for new forms of development based on the knowledge and needs of Third World people themselves rather than the 'expertise' of outsiders. Such development would not necessarily preclude Western concepts and methods, but entails a more realistic view of them as reflecting a specific geographical and historical context. It rejects efforts to remold other people according to ethnocentric 'universal' models and pre-defined standards. Instead, it calls for development workers to be open to differences and learn from them. This means learning about other social groups and cultures, taking an interest in local knowledge and cultural practices as a basis for redefining development approaches, adopting a more critical stance with respect to established theories and methods, and promoting the participation of indigenous popular organizations in all stages of development initiatives.

The indigenization process opens up new possibilities for multiple, polycentric development approaches, each informed and inspired by local traditions and popular creativity. Until now, most Third World countries have had little choice but to conform or react to development blueprints drawn up elsewhere. Recently, however, some Third World areas have begun to experiment with indigenous concepts and methods based on their own development experiences and intellectual traditions. Many of these alternative approaches seek to foster participation and empowerment by creating a sense of self-worth among Third World people through rediscovering and reinterpreting local histories and cultural traditions. Before people can determine their own futures, they must first take back their pasts, which have frequently been lost or discarded by the imposition of alien development models.

The rise of indigenous development approaches has prompted a reexamination of the role of traditional social relations, values, and structures in Third World development. Rather than being dysfunctional or obsolete, indigenous institutions may offer opportunities to make development initiatives more locally appropriate and effective. Following a careful and sensitive reassessment, many of the values and institutions of traditional societies may prove surprisingly conducive – particularly to alternative forms of development that seek to maximize popular participation and empowerment. Traditional societies also often have flexible and efficient support systems and other informal institutional mechanisms that may be adapted for use in development projects. In some rural areas, for example, development initiatives have begun to explore ways to employ traditional savings institutions to mobilize hitherto under-utilized resources to finance local projects.

An important 'window' into the complexity and diversity of traditional societies may be provided by creating mechanisms to elicit traditional or indigenous forms of knowledge, which have frequently been overlooked by outside professionals seeking solutions to Third World development problems. Increased use of indigenous knowledge may make development programs more appropriate, provide innovative solutions to certain problems, contribute to a sense of self-worth and collective self-esteem, and enhance popular participation and empowerment. To take advantage of these possibilities, development programs should start with the premise that local people, despite the constraints they often face, are knowledgeable and skillful managers of their own environments. If provided with adequate resources, their knowledge and skills place them in an ideal position to devise locally appropriate solutions to their own development problems.

Becoming aware of indigenous forms of knowledge is important to permit development initiatives not only to devise more appropriate means of avoiding socially or environmentally destructive practices, but also to make better use of a vast local expertise. While development workers need to avoid romanticizing about traditional or indigenous groups, whose development practices may not be without contradictions, at the same time, the knowledge and opinions of these people ought to be taken seriously. Nevertheless, many outside development professionals ignore or remain unaware of indigenous systems of knowledge and research procedures which may differ from accepted scientific methods. To allow local people to express their knowledge and creativity within development projects, favorable conditions must be created, including the establishment of rapport among project participants, whereby outsiders show humility, respect, and interest in learning from local people; the exercise of restraint by development professionals, so as not to over-interpret the views of locals; the employment of intersubjective, participatory research methods; and the utilization of local knowledge, practices, and materials whenever possible. There are many potential paths that development projects may take to create such conditions. To explore them, new approaches are required which are consciously methodological, moving away from the fixed, 'extractive' research procedures of Western science toward more flexible, shared methods of acquiring information. These approaches call on outside development 'experts' to respect the knowledge of local people and be willing to learn from them. Ready-made, top-down, 'blueprint' methods of development work are rejected in favor of bottom-up social learning approaches which stress the active participation and empowerment of local people.

Within these approaches, development professionals may continue to play an important role in projects as organizers, facilitators, catalysts, and animators. Frequently, they may also engage in 'capacity building,' by offering appropriate conceptual and organizational 'tools' for local people to use. Normally, however, such tools should not conform to the formalistic methods of Western science, which have often served only to quash local initiative and generate distorted results. Instead, more flexible methods ought to be employed that are capable of addressing the diverse and complex realities of the subjects themselves. This should enhance not only the informational elements, but also the participatory aspects of development initiatives – leading to the design of projects that are both more appropriate and empowering for local people.

Space and Place in Development

Increasingly, development projects are exposing a basic tension between mainstream planning theories, with their Northern-based assumptions, and the development realities of the South. As in many other fields, there is a striking tendency in spatial and regional planning to pay homage to influential theories of the past rather than to construct new, more appropriate frameworks. The realities of the South are continually made to fit the theory, rather than testing the latter on the former. As a result, spatial and regional policies often use ideas and practices borrowed from the North that are inappropriate to Southern conditions.

Mainstream planning displays fairly consistent biases toward relatively capital-intensive, large-scale, import-intensive forms of development that are dominated by elite groups and concentrated in the core locations. Such biases have been particularly marked in projects which are controlled by centralized planning decisions and which are heavily dependent on foreign funding. By contrast, more popular, inclusive development is likely to be consistent in most countries with a more dispersed pattern of regional growth in which institutional mechanisms provide real opportunities for the popular majority to participate in decision-making concerning planning. This is particularly true for many poorer countries where economic diversification to serve majority interests must, at least initially, be focused on linkages with primary sectors, and where possibilities exist for cooperatives and other popular institutions to play an important role in this process. Such conditions tend to encourage complementary, rather than competitive, relationships in both socioeconomic terms (among classes, social groups, and economic sectors) and spatial terms (between peripheral rural regions and core urban areas).

These conditions cannot normally be created under conventional top-down planning models. Instead, more bottom-up approaches are needed to incorporate people's participation into planning processes, from problem identification through to plan implementation. However, unlike mainstream planning strategies 'from above,' there is no categorical recipe for alternative strategies 'from below' beyond a few general principles related to removing previous development constraints and expanding democratic participation. This implies a basic shift in planning away from ready-made plans toward devising strategies in accordance with particularities of place. Because this is a complex process involving many intertwined factors, related research will necessarily be interdisciplinary. Gaps between theory and practice as well as between academic disciplines, which have retarded the ability of planners to gain a more comprehensive and insightful understanding of their subject matter, will need to be overcome. Regional development studies will need to integrate micro-level investigations of particular places with macro research on the economy, polity, and society. Understanding the need to integrate a focus on space and place with associated measures designed to alleviate underlying problems such as poverty, unequal resource distribution, and environmental degradation, is critical to the future advance of regional development policy in the South.

Many of the problems of mainstream planning stem from its use of an inappropriate conception of space. Space is unacceptably seen as separable from other aspects of life and development. The historically changing relationship between spatial and social organization is ignored, robbing spatial analysis of much of its utility. Spatial organization is divorced from underlying social relations, and space is implicitly and incorrectly afforded with causal powers. In addition to the conceptual limitations that it imposes, this type of 'spatial separatism' causes regional planning to conflate the spatial with the socioeconomic effects of policies. Measures of spatial efficiency are inappropriately transformed into those of generalized socioeconomic efficiency, while levels of spatial equity are conflated with those of social equity. The conflation of spatial problems with social problems inadvertently often produces development programs that are harmful to the very people they are intended to help. For example, much evidence suggests that programs designed to alleviate rural poverty by promoting market integration and capitalist-oriented growth in peripheral regions have instead accentuated intraregional inequalities, rural differentiation, and peasant marginalization.

It is relatively easy to identify places as being rural/urban or peripheral/core areas. But this is a much more difficult exercise for people – and it is people, instead of places, that are responsible for different types of interaction over space. What is needed is an analysis of how various classes and social groups are differentially affected by rural–urban, core–periphery, and other spatial interactions. This can only be accomplished by moving beyond a focus on space and place per se to address the great diversity of factors (e.g., economic, politico-ideological, sociocultural) that influence commodity flows and other forms of spatial interaction within and among regions. The actual role of various components of spatial structure (e.g., land-use and tenure patterns, rural circulation systems, settlement hierarchies, core–periphery linkages) is difficult to gauge without an analysis of their concrete regional context, taking into account relations of production and social reproduction as well as structures of political decision-making and control.

Concepts of space, place, locality, and region have recently been afforded increasing attention in the social sciences and humanities. Many analysts contend that the nature or character of regions must form an important part of how social processes are conceptualized within those regions.

This, in turn, requires a social theory in which the regional setting is not treated simply as an abstraction or a priori given, but is viewed as the result of social processes that reflect and shape particular ideas about how the world is or should be. From this perspective, regions become more than just backdrops for case studies; they are themselves part of the social dynamic of development. Communities and societies are, in large part, defined territorially, and social identity is commonly tied to territorial affiliation. This can clearly be seen in the rise of community or place-based social movements in many countries.

Rising interest in the role of space and place in development has generated a new research area in geography and related fields called 'locality studies,' which focuses on the ways in which the general dynamics of development are conditioned by local contexts. Communities and regions are viewed as spatially constituted sociocultural structures, as well as centers of collective consciousness and group identities. Regional cultures and identities are formed by social practices and relations that are, at least partly, autonomous from the logic of capital. Moreover, struggles among various groups over resources and power are seldom restricted merely to economic issues, but normally also extend into cultural areas involving divergent lifestyles and meanings.

An important part of the way that localities become constituted takes place through everyday cultural practices and personal histories, which also help to shape processes of regional production, reproduction, and transformation. Concepts of place can therefore not be simply reduced to a specific site or scale. Instead, place is composed of situated episodes of life history which unavoidably have various geographical dimensions: real, imagined, or utopian. As centers of meaning, places are constituted through personal histories; however, they also take on a more permanent and distinct presence over time through collective actions and institutional practices. As a result, social relations take place not only at different scales, but also with different meanings which are normally wrapped up in place-based identities and territoriality. From this point of view, territory is not just part of a functional or physical spatial system, but also takes on a sociocultural, experiential connotation.

This concept of territory stresses subjective as well as more objective elements of spatial and regional development. It moves away from functionalist models which, up to now, have characterized most spatial/regional planning and have negatively affected local participation. Ideally, development planning should be a social learning process which makes room for the various ideas and perspectives of different groups. Therefore, the institutional design of the planning process ought to be open and capable of incorporating contributions from the widest possible variety of local sources. Planning practices should facilitate a two-way learning process

– on the one hand, disseminating outside information and techniques, but, on the other, permitting local ideas, perceptions, and methods to inform decision-making and implementation procedures. Emphasis should be placed on participatory methods designed to elicit the various perceptions, intersubjective meanings, and objectives of local actors concerning development – which may be quite distinct from those of outsiders and without which development projects can seldom be appropriate.

Efforts to broaden local participation underscore the fact that spatial/regional planning, like all aspects of development, is inherently political. Space is, in a very real sense, a politically contested domain. Planning cannot be effective if it attempts to isolate spatial problems from the territorial expressions of power relations. Discussions and debates over spatial strategies should be informed by an analysis of the broader political context within which these strategies occur. This means transcending the conception of spatial planning as a purely technical, apolitical process.

The neglect of politics and power relations is part of a larger conceptual problem: the failure to situate space and place within the broader context of development processes conditioned by particular, historically constituted patterns of social relations. Spatial problems at the local/regional level normally reflect deeper socioeconomic difficulties, which cannot be addressed without an analysis of underlying relations of production and social reproduction. Such analysis reveals the roots of many spatial/regional problems within patterns of class, gender, ethnic, and other social relations.

Moreover, spatial frameworks must grapple with larger questions of power at the national/international level. Consideration needs to be given to issues such as the local/regional impact of state policies, the growing penetration of transnational corporations and institutions, and the overall development path adopted by particular countries. All of these concerns are basic to an understanding of why the poor commonly remain economically and politically marginalized, despite the best efforts of many local development projects.

Questions of Power

In large part, the present crisis afflicting theories and strategies of development stems from their failure to answer a basic question: Whose development? If sustainable, majority-based development is to be more than just a utopian vision, development initiatives must come to grips with this fundamental issue. Accordingly, questions of power need to be included in development frameworks from the very beginning. Changes in forms and paths of development must be supported by complementary changes in power structures. This means devising strategies which are capable of getting at the social roots of many Third World development problems within discriminatory and exploitative societal structures.

Typically, however, neither mainstream nor alternative development initiatives have paid much attention to economic, political, and sociocultural structures. Little consideration has been given to the nature of local institutions and organizations, which may often be undemocratic and exclude particular groups on the basis of class, gender, ethnicity, or other criteria. As a result, many development efforts, including a large proportion of bottom-up projects ostensibly designed to enhance local participation, have become quite exclusionary in practice. Unfortunately, these development initiatives have ignored the fact that project benefits might be monopolized by elite groups or that field-level officers and project leaders might become, however inadvertently, aligned with such groups against the interests of the poor and disadvantaged.

The tendency within most development projects is to view communities in homogeneous terms, thereby limiting possibilities for divergent needs and interests along class, gender, ethnic, or other lines. Possible sources of inequality and discrimination are neglected, and it is assumed that locally controlled projects will be broadly empowering. Issues of how local and community organizations are related to broader socioeconomic and political structures are ignored. All too often, projects suffer from a 'magic bullet' syndrome, in which simple, neat (usually technocratic) solutions to development problems are sought which fail to account for the complexities of local structures. However, the poor normally face substantial structural impediments to their improved social and economic well-being (e.g., skewed power relations, unequal resource distributions). Furthermore, the advance of particular marginalized groups may be blocked by more specific obstacles (based in structures of gender, ethnicity, religion, or language). As long as these structural impediments go unaddressed, development projects, despite their frequent good intentions, have little chance of meeting the needs and interests of the poor and marginalized.

Among the marginalized groups of Third World societies, poor women have been especially neglected by development efforts – not only by macroeconomic programs such as structural adjustments, but also by more specific local projects. By disregarding the differential gender effects of policies, policymakers have often been guilty of compounding the already severe hardships experienced by poor women. Since in nearly every society women are at a disadvantage relative to men (in terms of income, assets, education, political influence), 'gender-neutral' policies have often further marginalized women by weakening their position within particular social groups and economic sectors. Because most gender-neutral policies include and benefit primarily men, there is a clear need to promote measures specifically designed to enhance women's participation at all stages of development efforts, from planning to implementation.

To create more effective development initiatives, it is essential to have a clear idea of women's identities and interests, their resources and capabilities, and the constraints to which they are subjected (especially those that are amenable to change). Attempts to act on the basis of piecemeal information or simple generalizations that fail to reflect the diversity and complexity of women's lives have often been quite damaging. Presumptions by educated outsiders, whether male or female, about the needs and interests of Third World women are often inaccurate, generating inappropriate concepts and methods that discourage participation. Women's participation in various organizations and projects is structured by their socioeconomic roots.

Commonly, women face a variety of constraints to their ability to organize (such as high illiteracy, lack of energy and time due to overwork, cultural and ideological restrictions). If these constraints remain unaddressed, development initiatives have little chance of enhancing women's participation and empowerment.

Empowerment and People-Oriented Development

The failure of contemporary development to meet popular interests underscores the need to devise more people-centered approaches which stress empowerment and participation. Experience shows that problems such as underdevelopment, inequalities, and poverty cannot be solved by top-down strategies such as neoliberalism or Keynesianism, but require a shift to alternative approaches based on popular empowerment. From this perspective, development initiatives can be more effective if they move away from a focus on fixed, externally defined goals toward a more flexible, 'enabling' orientation designed to increase the intellectual, moral, managerial, and technical capabilities of local people. Empowerment becomes a multifaceted process, involving the pooling of resources to achieve collective strength to oppose elitist structures. It entails the improvement by local people of their manual and technical skills; administrative, managerial, and planning capacities; and analytical and reflective abilities. By including all of these elements, processes of empowerment can contribute to new forms of development in which the fulfillment of human potentials and capabilities is especially important.

In its most ambitious form, as empowerment, participatory development seeks to engender not only practical self-help and self-reliance, but also effective collective decision-making and action. In the end, the sustainability of the empowerment process itself should be regarded as at least as important as the immediate completion of particular projects. Especially important is the creation of local participatory institutions and organizations. People only constitute themselves in a truly participatory organization when they perceive that they share common concerns and voluntarily decide to act collectively on them. For this reason, participation should be recognized as a difficult and time-consuming process, especially for poor and marginalized groups. It requires a considerable investment of their scarce time, energy, and resources. If they are to voluntarily make this investment, participatory organizations must concentrate on meeting the self-defined needs and interests of local people themselves rather than externally prescribed goals. It follows that the methods adopted by these organizations should adhere to a 'social learning' approach which builds on people's own initiatives, with outside agencies playing an essentially supportive role.

The recent rise of participatory popular organizations and movements has been linked in many countries to the emergence of new subjectivities. Even if their motives may be largely economic, most members of popular movements do not conceive of their struggle in purely economic or class-based terms. Theoretically and practically, the result has been a rupture in the unified political space of the Left which was previously characterized by a single privileged subject – the working class. This unity has been shattered to make room for the plurality of collective actors that make up the popular sectors. New grassroots conceptions have been advanced which view society as an essentially pluralistic and multi-sided entity which is produced, at least in part, through the construction of collective identities by diverse social actors according to multiple views and interests. These identities are necessarily unstable; they have been formed through complex processes of articulation among various social groups. Furthermore, they have generated a whole new style of pluralistic, bottom-up, non-party political activity, which is slowly revitalizing the Left in many countries by transforming the very nature of what constitutes progressive political practice.

The emergence of new popular subjectivities has often been furthered by popular education and 'conscientization' efforts. These have been especially important for instilling self-confidence and inspiring self-expression among marginalized groups – without which effective human-resource mobilization, participatory decision-making, and genuine empowerment are impossible. Conscientization may be understood as a process of cognitive and evaluative transformation, particularly for the poor and marginalized, which seeks to produce individuals who are better able to make informed, responsible choices, and have the inner strength and conviction to act firmly and decisively on them. These individuals tend to become highly politicized, especially through participation in the formulation and implementation of policies affecting their well-being at the local level. As a result, a transformation is taking place in the form and content of politics in many countries, prompted by the heightened awareness and increased mobilization of these new historical subjects.

New Politics and the State

If social stability, democratization, and sustainability are to be serious development goals, it is clear that alternative approaches must be found to neoliberal SAPs and other top-down strategies. Conditions must be created in which strong social partners can participate in decision-making at the local, regional, and national levels, enabling a consensus or 'social contract' to be constructed over how development should proceed. This means strengthening popular organizations and other associational groups so that they can play an active and responsible part in the decision-making process. A widely acknowledged and respected social contract cannot be achieved, especially in highly politicized Third World societies, if important social groups are unable to exercise a decisive influence on governments to ensure their concerns are taken into account by the political system. Since SAPs, or any other development program, necessarily involve difficult choices over how the costs and benefits of development are to be distributed, any meaningful development strategy must obviously be based on a fair degree of social consensus if it is to be successfully sustained without resort to authoritarianism.

Especially under SAPs, the development efforts of most popular organizations have turned toward 'filling the gaps' in larger development programs created by macroeconomic policy reforms. However, as poverty and other social problems mount, there is a growing perception that these bottom-up initiatives are fighting a losing battle – for every 'gap' filled, many more appear.

Through their neglect of larger concerns, local development projects have become focused on isolated problems of survival, ignoring more fundamental issues concerning the systems that generate poverty and inequalities in the first place. Because they remain isolated and uncoordinated, grassroots projects and movements are unable to step beyond the micro-level, marked by individual local responses to specific problems, into the macro-realm of creating effective national popular organizations capable of mounting cohesive responses to pressing development issues. As a result, they run the risk of creating societies composed of disconnected apolitical individuals.

To avoid such problems, local popular movements need to 'scale up' their operations by building micro–macro links that can affect the overall framework of development policies. Genuine empowerment and participation in development decision-making cannot remain dependent solely on local pressure groups or the whims of NGO project funding. Instead, popular democratic structures need to be created that can translate decisions from below into needed changes in macro-policy. Designing development policies to serve majority needs and interests necessitates removing barriers to popular participation inherent in existing top-down methods of policy formulation. But simply improving communications and information flows through administrative decentralization is not sufficient; there also needs to be a political commitment to ensure that this process takes place, and this can only be ensured through the constant vigilance of a politically organized grassroots.

In most countries, relations between grassroots organizations and the state are marked by feelings of mutual suspicion and mistrust. Popular movements often view the state as an adversary, dominated by elite groups aligned against the interests of the majority. For its part, the state commonly ignores bottom-up development efforts, unless they threaten to provoke popular unrest. The state's attitude toward alternative development initiatives might typically be described as one of benign indifference. Local development projects are regarded as a cheap and easy way of maintaining social order. But if conscientization and empowerment become major components of such initiatives, the state usually intervenes to exert control.

In recent years, however, both popular organizations and the state have begun to modify their stances toward one another. Many grassroots groups have realized that progressive development is unlikely without state support. Individual development efforts that are poorly coordinated and not integrated with national plans have often produced duplication, confusion and disarray, with projects operating at cross-purposes. In some instances, the state may also offer useful support for popular organizations against reactionary local elites. Moreover, more equitable development generally requires a fair measure of resource distribution, which can most effectively be accomplished with state assistance. For all of these reasons, an increasing number of popular movements feel the need to influence the state more effectively – through a combination of grassroots political action, the electoral process, and radical political reforms that will render state structures more responsive to majority needs and interests.

At the same time, a growing number of Third World governments are realizing the value of bottom-up development strategies, especially because they normally offer private responses to what have traditionally been considered public responsibilities. If managed properly, increased cooperation with NGOs and popular organizations could make social services more efficient and less costly, without the state simply having to push responsibility for the well-being of the poor onto the informal economy. Local groups are often able to respond better to changing conditions than can state institutions. In some cases, state agencies can use knowledge, technologies, and innovations developed at the local level by NGOs and other groups. Given all of these advantages, governments might want to treat bottom-up development projects as a type of 'infant industry' which, if nurtured by state support, could help to generate more equitable growth by relatively inexpensive means.

Nevertheless, cooperation between the state and popular organizations frequently remains a difficult process. The establishment of closer links between state agencies and popular organizations normally demands a reconciliation of radically different organizational styles and approaches to development, as well as the negotiation of clear areas of jurisdiction. Whether relations between the state and popular organizations can be institutionalized in an effective and mutually agreeable manner remains an open question in most countries. On the one hand, popular organizations must increase their visibility and establish their own niche within national development agendas by scaling up their operations and becoming more politically involved, but without losing their autonomy and representativeness. On the other hand, the state must provide the resources and environment for this process, but without becoming overbearing.

In the end, processes of popular development, if they are to occur in more than a handful of isolated communities, require the support of a strong state. However, a strong state is not top heavy with bureaucracy dominated by elitist interests; instead, it is agile, responsive, and accountable to the majority. It is a state over which social control is exercised both indirectly via representative democratic institutions and directly through the involvement of pluralistic popular organizations. It is a state, therefore, that receives its vision and strength from the principles of a popular or inclusive democracy, in which a division of power permits development problems to be primarily managed by decentralized units of governance in cooperation with the people themselves, organized in their own communities and popular organizations. This new way of 'doing politics' may take on specific forms according to each country's individual circumstances, but it should also contain the following broad elements: the creation of more direct and varied styles of popular participation, the expansion and deepening of the realm of the 'political' beyond the state and old-style party politics, the recovery of the 'social' from control by top-down institutions, and the exercise of popular social control via representative and direct participatory democracy.

Global Institutions and International Cooperation

The need to scale-up popular development initiatives implies turning attention not only to national political structures, but also to international institutions and forums. This involves creating new methods for popular organizations and related groups to exchange information and provide mutual support. It also entails building novel, fresh forms of international cooperation via the reform/restructuring of existing global institutions and the creation of new, popularly based counterparts. Although forms of collaboration will necessarily vary according to the actors involved and the specific aspects of development being addressed, there is much potential to increase international cooperation, including North–South, South–South, and intra-regional relationships.

Studies of international relations within postwar development frameworks have typically been marked by biases of either endogenism or exogenism. Both stances, however, if carried to their logical extremes, are equally contradictory and impractical. The obvious remedy is to find a realistic synthesis which will transcend this dichotomy. Especially in today's interdependent world, there are no countries that can be completely autonomous and self-reliant. At the same time, countries do not develop (or underdevelop) merely as a reflection of events beyond their national borders. Moreover, the effects of interactions between more developed and less developed countries may be either positive or negative, and are normally socially differentiated by class, gender, and other factors. The most important question concerning international relations is, therefore, not whether Third World countries benefit or lose from global interrelationships, but how the popular sectors can pursue selective policies allowing them to maximize the benefits of the positive forces of such relations, while minimizing the harm caused by negative forces.

Given the different historical and geographical conditions faced by Third World countries, strategies of selective international participation should stress flexibility. Selective forms of global participation and interdependence should be shaped by the varying potentials of the countries involved, according to the needs and interests of their populations. This requires knowledge of the effects of development processes at a variety of levels. It is particularly important to understand the conditions under which global linkages (e.g., trade, capital flows, migration) are productive or counterproductive to particular social groups. Neither the North nor the South are monolithic blocks. Accordingly, development strategies need to eschew universalism in favor of flexible ideas and methods that can address particularities in social structures and relations which influence the impact of trade and other global linkages on individual groups. Such diversity cannot be accommodated by any single reconstruction of the development agenda; instead, 'polycentric' reconstructions are required, based on the varying objective conditions and subjective concerns of different peoples.

There is nothing inherently progressive about strategies that either expand or restrict global linkages such as trade. As we saw in chapter 2, even traditionally exploitative models based on primary exports can be modified to provide for more broadly based, sustainable development. Research into the functioning of international markets should form a key part of any new trade strategy – to find methods for countries to exploit new trade opportunities and to overcome constraints and limitations in potentially promising sectors. Third World producers may often improve their market position via collective negotiation along international marketing chains – perhaps with the support of producer associations, state agencies, and/or regional trading bodies. In concert with producer associations, countries may also increase their shares of selected global markets by cooperating in various aspects of commercialization (e.g., advertising, improving export services, certifying export quality).

In addition to exploring possibilities for expanding and transforming trade with the North, economic diversification should also be encouraged by seeking new ways to stimulate South–South trading links. In many cases this goal might be furthered by strengthening regional trading blocs and common markets which have frequently existed only 'on paper' or in quite limited form. Finding methods to facilitate freer trade within Southern regions opens up possibilities of rapid market expansion for firms previously restricted to relatively small domestic markets. Moreover, enhanced regional economic cooperation should allow for mutually beneficial exchange of products. Trade ties with local NICs, for example, might prove especially important for opening up new growth opportunities in more underdeveloped, rural countries. A growing body of evidence shows that enhanced cooperation among developing countries has an enormous and, as yet, largely unexplored potential for overcoming common problems in a variety of areas from research and development to production, and to commercialization.

To explore these opportunities, Third World countries need to develop better means for pooling their competence and resources to find answers to their own problems. Cooperation and interdependence can be envisioned at several levels, encompassing sub-regional and regional groupings as well as the entire South. At the same time, new mechanisms for global cooperation and interdependence must be created. These should include international networks of information, aimed at connecting various popular organizations and related groups, to assist in spreading knowledge, technical advice, and resources. Such networks might be accompanied by independent development institutes or learning centers, perhaps under the auspices of the United Nations or some other equivalent organization, to promote more meaningful North–South and South–South dialogue. In contrast to the one-way, top-down flow of information that presently characterizes neoliberal development planning, these centers might create a more cooperative atmosphere and level 'playing field' for the exchange of development ideas and methods. Toward this end, stress should be placed on enhancing the indigenous capabilities of popular development institutes and learning centers in the South itself. This would go a long way toward overcoming the South's debilitating dependence on Northern-based development frameworks and would help to instill a sense of self-confidence among Third World peoples to use their indigenous knowledge to define distinct paths of development appropriate to their own needs and interests.

Finally, it should be emphasized that no one should expect popular development to suddenly appear everywhere in some sort of finished form. Given the difficult circumstances faced by the popular sectors throughout the South, such expectations would be idealistic. Moreover, there is no single model of popular development which might be applied to all countries and regions. It is only possible to formulate some general elements or principles, which should be put into practice according to the needs and interests of specific groups of people. Elitist attempts to impose particular forms of development on local people must be rejected. Instead, development should be redefined from the bottom-up to meet local conditions and subjective concerns. Popular development should be seen as a process which empowers people to take control of their destinies. It is a process which can only take place through prolonged political struggle at various levels. Although some progress has recently been made in this area, especially through the efforts of a broad range of popular movements, in most countries these struggles have barely begun.

Further Reading

Comparative development theories

Apter, D. (1987) *Rethinking Development: Modernization, Dependency and Postmodern Politics*. Newbury Park, CA: Sage.

Harrison, D. (1988) *The Sociology of Modernization and Development*. London: Unwin Hyman.

Hettne, B. (1990) *Development Theory and the Three Worlds*. Essex: Longman.

Hulme, D. and Turner, M. (1990) *Sociology and Development: Theories, Policies and Practices*. New York: Harvester Wheatsheaf.

Preston, P. (1986) *Making Sense of Development: An Introduction to Classical and Contemporary Theories of Development and their Application to Southeast Asia*. London: Routledge.

Somjee, A. (1991) *Development Theory: Critiques and Explorations*. Basingstoke: Macmillan.

Comparative development strategies and practices

Black, J. (1991) *Development Theory and Practice: Bridging the Gap*. Boulder: Westview.

Dietz, J. and James, D. (eds) (1990), *Progress Toward Development in Latin America: From Prebisch to Technological Autonomy*. Boulder: Lynne Reinner.

Gereffi, G. and Wyman, D. (eds) (1990) *Manufacturing Miracles: Paths of Industrialization in Latin America and East Asia*. Princeton: Princeton University Press.

Thorp, R. (1992) A reappraisal of the origins of import-substituting industrialization, 1930–50, *Journal of Latin American Studies*, 24, 181–95.

Torres-Rivas, E. (1980) The Central American model of growth: crisis for whom? *Latin American Perspectives*, 7 (2/3), 24–44.

Williams, R. (1986) *Export Agriculture and the Crisis in Central America*. Chapel Hill: University of North Carolina Press.

The Asian NICs

Bello, W. and Rosenfeld, S. (1990) *Dragons in Distress: Asia's Miracle Economies in Crisis*. San Francisco: Institute for Food and Development Policy.

Burmeister, L. (1990) State industrialization and agricultural policy in Korea, *Development and Change*, 21, 197–223.

Jenkins, R. (1991) The political economy of industrialization: a comparison of Latin American and East Asian newly industrializing countries, *Development and Change*, 22, 197–231.

Onis, Z. (1991) The logic of the developmental state, *Comparative Politics*, 24, 109–26.

Vogel, E. (1991) *The Four Little Dragons: The Spread of Industrialization in East Asia*. Cambridge, MA: Harvard University Press.

Wade, R. (1992) East Asia's economic success: conflicting perspectives, partial insights, shaky evidence, *World Politics*, 44, 270–320.

World Bank (1993) *The East Asian Miracle: Economic Growth and Public Policy*. New York: Oxford University Press/World Bank.

Neoliberalism and structural adjustment programs

Cheru, F. (1992) Structural adjustment, primary resource trade and sustainable development in sub-Saharan Africa, *World Development*, 20, 497–512.

Helleiner, G. (1992) The IMF, the World Bank and Africa's adjustment and external debt problems: an unofficial view, *World Development*, 20, 779–92.

Rausser, G. and Thomas, S. (1990) Market politics and foreign assistance, *Development Policy Review*, 8, 365–81.

Ruccio, P. (1991) When failure becomes success: class and the debate over stabilization and adjustment, *World Development*, 19, 1315–34.

Schoenholtz, A. (1987) The IMF in Africa: unnecessary and undesirable Western restraints on development, *Journal of Modern African Studies*, 25, 403–33.

Streeten, P. (1993) Markets and states: against minimalism, *World Development*, 21, 1281–98.

Toye, J. (1993) *Dilemmas of Development: Reflections on the Counter-Revolution in Development Theory and Policy*. Oxford: Blackwell.

Alternative development approaches

de Janvry, A. (1981) *The Agrarian Question and Reformism in Latin America*. Baltimore: Johns Hopkins University Press.

Friedmann, J. (1992) *Empowerment: The Politics of an Alternative Development*. Oxford: Blackwell.

Griffin, K. (1989) *Alternative Strategies for Economic Development*. New York: St. Martin's.

Samoff, J. (1990) Decentralization: the politics of interventionism, *Development and Change*, 21, 513–30.

Simon, D. (ed.) (1990) *Third World Regional Development: A Reappraisal*. London: Chapman.

Slater, D. (1989) Territorial power and the peripheral state: the issue of decentralization, *Development and Change*, 20, 501–31.

Wisner, B. (1988) *Power and Need in Africa: Basic Human Needs and Development Policies*. London: Earthscan.

Popular participation and empowerment

Chambers, R. (1991) The search of professionalism, bureaucracy and sustainable livelihoods for the 21st century, *IDS Bulletin*, 22, 5–11.

Escobar, A. (1995) *Encountering Development: The Making and Unmaking of the Third World*. Princeton: Princeton University Press.

—— and Alvarez, S. (eds) (1992) *The Making of Social Movements in Latin America: Identity, Strategy and Democracy*. Boulder: Westview.

Fenster, T. (1993) Settlement planning and participation under principles of pluralism, *Progress in Planning*, 39, 171–242.

Moser, C. (1989) Gender planning in the Third World: meeting practical and strategic gender needs, *World Development*, 17, 1799–825.

Vivian, J. (1994) NGOs and sustainable development in Zimbabwe: no magic bullets, *Development and Change*, 25, 167–93.

Women, gender relations, and development

Chinchilla, N. (1992) Marxism, feminism, and the struggle for democracy. In Escobar, A. and Alvarez, S. (eds), *The Making of Social Movements in Latin America: Identity, Strategy, and Democracy*. Boulder: Westview, 37–51.

Geisler, G. (1993) Silences speak louder than claims: gender, household, and agricultural development in Southern Africa, *World Development*, 21, 1965–80.

Greenhalgh, S. (1985) Sexual stratification: the other side of 'growth with equity' in East Asia, *Population and Development Review*, 11, 265–314.

Joekes, S. (1994) Gender, environment and population, *Development and Change*, 25, 137–65.

Koopman, J. (1993) Neoclassical household models and modes of household production: problems in the analysis of African agricultural households, *Review of Radical Political Economics*, 23 (3/4), 148–73.

Parpart, J. (1993) Who is the 'other'? A postmodern feminist critique of women and development theory and practice, *Development and Change*, 24, 439–64.

Radcliffe, S. and Westwood, S. (eds) (1993) *Viva: Women and Popular Protest in Latin America*. New York: Routledge.

Environmental relations and sustainable development

Adams, P. (1992) The World Bank and the IMF in sub-Saharan Africa: undermining development and environmental sustainability, *Journal of International Affairs*, 46, 125–44.

Colchester, M. (1994) Sustaining the forests: the community based approach in South and South-East Asia, *Development and Change*, 25, 69–100.

Davies, S. and Leach, M. (1991) Globalism versus villagism: national and international issues in food security and the environment, *IDS Bulletin*, 22, 43–50.

Gadgil, M. and Guha, R. (1994) Ecological conflicts and the environmental movement in India, *Development and Change*, 25, 101–36.

Peet, R. and Watts, M. (1993) Introduction: development theory and environment in an age of market triumphalism, *Economic Geography*, 69, 227–53.

Redclift, M. (1987) *Sustainable Development: Exploring the Contradictions*. London: Methuen.

—— (1991) The multiple dimensions of sustainable development, *Geography*, 76, 36–42.

References

Abegglen, J. (1980) *Japan, the United States and Asia's Newly Industrializing Countries.* New York: Columbia University East Asian Institute.

Adams, P. (1992) The World Bank and the IMF in sub-Saharan Africa: undermining development and environmental sustainability, *Journal of International Affairs,* 46, 125–44.

Addison, T. and Demery, L. (1988) Wages and labour conditions in East Asia: a review of case-study evidence, *Development Policy Review,* 6, 371–93.

Adelman, I. (1975) Growth and income distribution and equity oriented development studies, *World Development,* 3 (2–3), 67–76.

—— and Morris, C. (1973) *Economic Growth and Social Equity in Developing Countries.* Stanford: Stanford University Press.

Afshar, H. and Dennis, C. (eds) (1992) *Women and Adjustment Policies in the Third World.* Basingstoke: Macmillan.

Agarwal, B. (1989) Rural women, poverty, and natural resources: sustenance, sustainability, and struggle for change, *Economic and Political Weekly,* 24 (43), WS46–65.

Agnew, J. (1987) *Place and Politics: The Geographical Mediation of State and Society.* Winchester, MA: Allen and Unwin.

Agnew, J. and Duncan, J. (eds) (1989) *The Power of Place: Bringing Together the Geographical and Sociological Imaginations.* Boston: Unwin Hyman.

Ahmad, Z. (1984) Rural women and their work: dependence and alternatives for change, *International Labour Review,* 123, 71–86.

Ahmed, I. (1987) Technology, production linkages and women's employment in South Asia, *International Labour Review,* 126, 21–40.

—— (1988) The bio-revolution in agriculture: key to poverty alleviation in the Third World? *International Labour Review,* 127, 53–72.

Alger, K. (1991) Newly and lately industrializing exporters: LDC manufactures exports to the United States, 1977–84, *World Development,* 19, 885–901.

Allison, C. (1985) Women, land, labour and survival: getting some basic facts straight, *IDS Bulletin,* 16 (3), 24–31.

—— and Green, R. (1985) Editorial: Toward getting some facts snarled? *IDS Bulletin* , 16 (3), 1–8.

Allor, D. (1984) Venezuela: from doctrine to dialogue to participation in the process of regional development, *Studies in Comparative International Development*, 19, 86–97.

Amadeo, E. and Banuri, T. (1991) Policy, governance, and the management of conflict. In Banuri, T. (ed.), *Economic Liberalization: No Panacea. The Experiences of Latin America and Asia*. New York: Oxford University Press, 29–55.

Amanor, K. (1994) Ecological knowledge and the regional economy: environmental management in the Aseewa District of Ghana, *Development and Change*, 25, 41–67.

Amin, S. (1990) Colonialism and the rise of capitalism: a comment, *Science and Society*, 54, 67–72.

—— (1993) Can environmental problems be subject to economic calculations? *Monthly Review*, 45, 16–32.

Amirahmadi, H. (1989) Development paradigms at a crossroad and the South Korean experience, *Journal of Contemporary Asia*, 19, 167–85.

Amsden, A. (1989) *Asia's Next Giant*. New York: Oxford University Press.

Appelbaum, R. and Henderson, J. (1992) *States and Development in the Asian Pacific Rim*. Newbury Park, CA: Sage.

Apter, D. (1987) *Rethinking Development: Modernization, Dependency and Postmodern Politics*. Newbury Park, CA: Sage.

Apthorpe, J. (1972) *Rural Cooperatives and Planned Change in Africa*, Vol. 5. Geneva: UNRISD.

Ariff, M. and Hill, H. (1986) *Export Oriented Industrialization: the ASEAN Experience*. London: Allen and Unwin.

Athukorala, P. (1989) Export performance of 'new exporting countries': how valid is the optimism? *Development and Change*, 20, 89–120.

Ayres, R. (1983) *Banking on the Poor: The World Bank and World Poverty*. Cambridge, MA: MIT Press.

Babb, F. (1990) Women and work in Latin America, *Latin American Research Review*, 25, 236–47.

Bagchi, A. (1990) Some fundamental issues in the analysis of technical change in developing countries, *Social Science Information*, 29, 399–415.

—— (1993) Rent seeking, new political economy and negation of politics, *Economic and Political Weekly*, 28, 1729–36.

Bajpai, N. (1993) Stabilization with structural reforms: can the two be pursued simultaneously? *Economic and Political Weekly*, 28, 990–4.

Balassa, B. (1981) *Structural Adjustment Policies in Developing Economies*. Washington: World Bank Working Paper 464.

—— (1988) The lessons of East Asian development: an overview, *Economic Development and Cultural Change*, 36, S273–90.

—— (1991) *Economic Policies in the Pacific Area Developing Countries*. London: Macmillan.

Bandarage, A. (1984) Women in development: Liberalism, Marxism, and Marxist-feminism, *Development and Change*, 15, 495–515.

Banuri, T. (ed.) (1991) *Economic Liberalization: No Panacea. The Experiences of*

Latin America and Asia. New York: Oxford University Press.

Baquedano, M. (1989) Socially appropriate technologies and their contribution to the design and implementation of social policies in Chile. In Downs, C., Solimano, G., Vergara, C. and Zuniga, L. (eds), *Social Policy from the Grassroots: Nongovernmental Organizations in Chile*. Boulder: Westview, 135–48.

Barham, B., Clark, M. and Katz, E. (1992) Nontraditional agricultural exports in Latin America, *Latin American Research Review*, 27, 43–82.

Barkin, D. (1990) *Distorted Development: Mexico in the World Economy*. Boulder: Westview.

Barnes, T. (1988) Rationality and relativism in economic geography: an interpretative review of the homo economicus assumption, *Progress in Human Geography*, 12, 473–96.

Barraclough, S. (1982) *A Preliminary Analysis of the Nicaraguan Food System*. Geneva: UNRISD.

Barry, T. (1987) *Roots of Rebellion: Land and Hunger in Central America*. Boston: South End.

—— (1992) *Mexico: A Country Guide*. Albuquerque: Inter-Hemispheric Education Resource Center.

Bartone, C. (1991) Environmental challenge in Third World cities, *Journal of the American Planning Association*, 57, 411–15.

Bartra, R. and Otero, G. (1987) Agrarian crisis and social differentiation in Mexico, *Journal of Peasant Studies*, 14, 334–62.

Basu, A. (1987) Grassroots movements and the state: reflections on radical change in India, *Theory and Society*, 16, 647–74.

Bates, R. (1981) *Markets and States in Tropical Africa: The Political Basis of Agricultural Policies*. Berkeley: University of California Press.

—— (1993) Urban bias: a fresh look, *Journal of Development Studies*, 29, 219–28.

Batie, S. (1989) Sustainable development: challenges to the agricultural economics profession, *American Journal of Agricultural Economics*, 71, 1083–101.

Bauer, P. (1972) *Dissent on Development*. London: Weidenfeld and Nicolson.

Baumeister, E. (1984) Estructura y reforma agraria en el proceso sandinista, *Desarrollo Económico*, 24, 187–202.

Baylies, C. and Wright, C. (1993) Female labour in the textile and clothing industry of Lesotho, *African Affairs*, 92, 577–91.

Bebbington, A. and Farrington, J. (1993) Governments, NGOs and agricultural development: perspectives on changing inter-organisational relationships, *Journal of Developmental Studies*, 29, 199–219.

Becker, G. (1983) A theory of competition among pressure groups for political influence, *Quarterly Journal of Economics*, 98, 371–400.

Behrman, J. (1990) *Human Resource Led Development? Review of Issues and Evidence*. New Delhi: ILO-ARTEP.

Bello, W. and Rosenfeld, S. (1990) *Dragons In Distress: Asia's Miracle Economies in Crisis*. San Francisco: Institute for Food and Development Policy.

Bentley, J. and Andrews, K. (1991) Pests, peasants, and publications: anthropological and entomological views of an integrated pest management program for small-scale Honduran farmers, *Human Organization*, 50 (2), 113–24.

Bernal, J. (1965) *Science of History*. London: Watts.

Bernstein, H. (1979) African peasantries: a theoretical framework, *Journal of Peasant Studies*, 6, 421–43.

—— (1990) Agricultural 'modernization' and the era of structural adjustment: observations on sub-Saharan Africa, *Journal of Peasant Studies*, 18, 3–35.

—— and Campbell, B. (eds) (1985) *Contradictions of Accumulation in Africa*. Beverly Hills, CA: Sage.

Bhagwati, J. (1986) Rethinking trade strategy. In Lewis, J. and Kallab, V. (eds), *Development Strategies Reconsidered*. New Brunswick, NJ: Transaction Books, 91–104.

—— (1988) *Protectionism*. Cambridge, MA: MIT Press.

—— (1991) *The World Trading System at Risk*. New York: Harvester Wheatsheaf.

Bhaskar, V. (1991) Export promotion, exchange rates and commodity prices, *Economic and Political Weekly*, 26, 1277–88.

Bhatt, E. (1989) Toward empowerment, *World Development*, 17, 1059–65.

Bienefeld, M. (1985) The lessons of Africa's industrial failure, *IDS Bulletin*, 16, 69–77.

—— (1988) The significance of the newly industrializing countries for the development debate, *Studies in Political Economy*, 25, 7–40.

—— (1989) The lessons of history and the developing world, *Monthly Review*, 41, 9–41.

Bienen, H. and Gersovitz, M. (1985) Economic stabilization, conditionality, and political stability, *International Organization*, 39, 729–54.

—— and Waterbury, J. (1989) The political economy of privatization in developing countries, *World Development*, 17, 617–32.

——, Kapur, D., Parks, J. and Riedinger, J. (1990) Decentralization in Nepal, *World Development*, 18, 61–75.

Biersteker, T. (1990) Reducing the role of the state in the economy: a conceptual exploration of the IMF and World Bank prescriptions, *International Studies Quarterly*, 34, 477–92.

Biggs, G. (1987) *La Crisis de la Deuda Latinoamericana Frente a los Precedentes Históricos*. Buenos Aires: Grupo Editor Latinoamericano.

Bina, C. and Yaghmaian, B. (1988) Import substitution and export promotion within the context of the internationalization of capital, *Review of Radical Political Economics*, 20, 234–40.

Bitar, S. (1988) Neo-conservatism versus neo-structuralism in Latin America, *CEPAL Review*, 34, 45–62.

Black, C. (1966) *The Dynamics of Modernization: A Study in Comparative History*. New York: Harper and Row.

Black, J. (1985) Ten paradoxes of rural development: an Ecuadorian case study, *Journal of Developing Areas*, 19, 527–56.

—— (1991) *Development in Theory and Practice: Bridging the Gap*. Boulder: Westview.

Blaikie, P. and Brookfield, H. (1987) *Land Degradation and Society*. New York: Methuen.

Bondi, L. (1990) Progress in geography and gender: feminism and difference, *Progress in Geography*, 14, 438–45.

Boserup, E. (1970) *Women's Role in Economic Development*. New York: St. Martin's.

Bossert, T. (1987) The promise of theory. In Klarén, P. and Bossert, T. (eds) *Promise of Development: Theories of Change in Latin America*. Boulder: Westview, 303–34.

Bourguignon, F., De Melo, J. and Morrisson, C. (1991) Poverty and income distribution during adjustment: issues and evidence from the OECD project, *World Development*, 19, 1485–508.

Boyer, W. and Ahn, B. (1991) *Rural Development in South Korea: A Sociopolitical Analysis*. Newark: University of Delaware Press.

Bradford, C. (1986) East Asian 'models': myths and lessons. In Lewis, R. and Kallab, V. (eds), *Development Strategies Reconsidered*. New Brunswick, NJ: Transaction Books, 115–28.

—— (1987) Trade and structural change: NICs and next tier NICs as transitional economies, *World Development*, 15, 299–316.

Bradshaw, V. and Wahl, A. (1991) Foreign debt, the International Monetary Fund, and regional variation in Third World poverty, *International Studies Quarterly*, 35, 251–72.

Brett, E (1987) States, markets and private power in the developing world: problems and possibilities, *IDS Bulletin*, 18, 31–8.

Brinkerhoff, D. (1988) Implementing integrated rural development in Haiti: the World Bank's experience in the northern region, *Canadian Journal of Development Studies*, 9, 63–79.

Broad, R. (1993) *Plundering Paradise*. Berkeley: University of California Press.

—— and Cavanaugh, J. (1989) Marcos's ghost, *The Amicus Journal*, 11, 18–29.

Brohman, J. (1989) Development theory and prerevolutionary Nicaragua. University of California at Los Angeles, unpublished dissertation.

—— (1995a) Universalism, Eurocentrism, and ideological bias in development studies: from modernisation to neoliberalism, *Third World Quarterly*, 16, 121–40.

—— (1995b) Economism and critical silences in development studies: a theoretical critique of neoliberalism, *Third World Quarterly*, 16, 297–318.

—— (1996) New directions in tourism for Third World development, *Annals of Tourism Research*, 23, 48–70.

Brookfield, H. (1975) *Interdependent Development*. London: Methuen.

Browett, J. (1985) The newly industrializing countries and radical theories of development, *World Development*, 13, 789–803.

Brown, L. (1981) *Building a Sustainable Society*. New York: Norton.

—— (1988) Reflections on Third World development: ground level reality, exogenous forces and conventional paradigms, *Economic Geography*, 64 (3), 155–278.

Bryant, C. (ed.) (1988) *Poverty, Policy, and Food Security in Southern Africa*.

Boulder: Lynne Rienner.

Bryant, R. (1992) Political ecology: an emerging research agenda in Third World studies, *Political Geography*, 11, 12–36.

Bulmer-Thomas, V. (1988) The economics of Central America, *Latin American Research Review*, 23, 154–69.

Burback, R. and Flynn, P. (1980) *Agribusiness in the Americas*. New York: New Left.

Burger, J. (1987) *Report from the Frontier: The State of the World's Indigenous People*. London: Zed.

Burgess, R. (1987) Lot of noise and no nuts: a reply to Alan Gilbert and Jan van der Linden, *Development and Change*, 18, 137–46.

Burkey, S. (1993) *People First: A Guide to Self-Reliant Participatory Rural Development*. London: Zed.

Burmeister, L. (1990) State industrialization and agricultural policy in Korea, *Development and Change*, 21, 197–223.

Cameron, J. (1994) Redemocratization and popular development alternatives in Chile. Unpublished manuscript.

Cardoso, F. and Faletto, E. (1969) *Dependencia y Desarrollo en América Latina: Ensayo de Interpretación Sociológica*. Mexico City: Siglo XXI.

Carter, M. and Mesbah, D. (1993) Can land market reform mitigate the exclusionary aspects of rapid agro-export growth? *World Development*, 21, 1085–100.

Cebotarev, E. (1988) Women, human rights and the family in development theory and practice (with reference to Latin America and the Caribbean), *Canadian Journal of Development Studies*, 9, 187–200.

Cebotarev, E. (1991) Women's actions in Third World communities: why popular education is not enough, *Canadian Journal of Development Studies*, 12, 421–30.

Cecelski, E. (1987) Energy and rural women's work: crisis, response and policy alterations, *International Labour Review*, 126, 41–64.

CEPAL (1983) *The Crisis in Central America: Its Origins, Scope, and Consequences*. Doc. E/CEPAL/G.1261 September.

Chakravarty, S. (1990) Development strategies for growth with equity: the South Asian Experience, *Asian Development Review*, 8, 133–59.

—— (1991) Development planning: a reappraisal, *Cambridge Journal of Economics*, 15, 5–20.

Chambers, R. (1991) The search of professionalism, bureaucracy and sustainable livelihoods for the 21st century, *IDS Bulletin*, 22, 5–11.

—— and Conway, G. (1992) Sustainable rural livelihoods: practical concepts for the 21st century. Institute for Development Studies Discussion Paper no. 296.

Chandra, R. (1992) *Industrialization and Development in the Third World*. London: Routledge.

Chang, H. (1993) The political economy of industrial policy in Korea, *Cambridge Journal of Economics*, 17, 131–57.

Cheema, G. (1985) *Rural Development in Asia: Case Studies on Programme Implementation*. New Delhi: Sterling.

Chenery, H., Ahluwalia, M., Bell, C., Duloy, J. and Jolly, R. (1974) *Redistribution with Growth*. New York: Oxford University Press.

Cheru, F. (1992) Structural adjustment, primary resource trade and sustainable development in sub-Saharan Africa, *World Development*, 20, 497–512.

Chidzero, B. (1987) A comment on professor Ghai's paper. In Toye, J. (ed.), *Development Policies and the Crisis of the 1980s*. Paris: Development Center for the OECD, 130–9.

Chinchilla, N. (1992) Marxism, feminism, and the struggle for democracy. In Escobar, A. and Alvarez, S. (eds), *The Making of Social Movements in Latin America: Identity, Strategy, and Democracy*. Boulder: Westview, 37–51.

Chisholm, A. and Tyers, R. (1982) *Food Security: Theory, Policy, and Perspectives from Asia and the Pacific Rim*. Lexington: Lexington Books.

Cho, Y. and Khatkhate, D. (1989) Financial liberalization: issues and evidence, *Economic and Political Weekly*, 24, 1105–14.

Choi, C. (1989) *Labor and the Authoritarian State*. Seoul: Korea University Press.

Chow, P. (1987) Causality between export growth and industrial development: empirical evidence from the NICs, *Journal of Development Economics*, 26, 55–63.

Clark, G. and Kim, W. (1993) Commentary: Industrial restructuring and regional adjustment in Asian NIE's, *Environment and Planning A*, 25, 1–4.

Cleveland, H. (1990) *The Global Commons: Policy for the Planet*. Aspen: Aspen Institute.

Coats, W. and Khatkhate, D. (1991) Money and monetary policy in LDCs in the 1990s, *Economic and Political Weekly*, 26, 1538–44.

Cobbe, J. (1990) Africa's economic crisis: the roles of debt, the state, and international economic institutions, *Journal of Modern African Studies*, 28, 351–8.

Cocoyoc Declaration (1974) *Development Dialogue*, 2, 88–96 (Uppsala: Dag Hammarskjöld Foundation).

Cohen, J. (1980) Integrated rural development: clearing out the underbrush, *Sociological Review*, 20, 195–212.

Cohen, S. (1978) *Agrarian Structures and Agrarian Reform*. Boston: Martinus Nijhoff.

Colchester, M. (1994) Sustaining the forests: the community based approach in South and South-East Asia, *Development and Change*, 25, 69–100.

Colclough, C. and Green, R. (1988) Editorial: Do stabilization policies stabilize? *IDS Bulletin*, 19, 1–6.

Collier, P. (1991) From critic to secular god: the World Bank and Africa, *African Affairs*, 90, 111–17.

Collins, J. (1985) *Nicaragua: What Difference Could a Revolution Make?* San Francisco: Institute for Food and Development Policy.

Commonwealth Secretariat (1990) *International Economic Issues*. London: Commonwealth Secretariat.

Concoran-Nantes, Y. (1993) Female consciousness or feminist consciousness? Women's consciousness raising in community-based struggles in Brazil. In

Radcliffe, S. and Westwood, S. (eds), *Viva: Women and Popular Protest in Latin America.* New York: Routledge, 136–55.

Conger Lind, A. (1992) Power, gender, and development: popular women's organizations and the politics of needs in Ecuador. In Escobar, A. and Alvarez, S. (eds), *The Making of Social Movements in Latin America: Identity, Strategy, and Democracy.* Boulder: Westview, 134–49.

Conroy, M. (1973) Rejection of growth center strategy in Latin American regional development planning, *Land Economics,* 49, 371–80.

Conway, G. and Barbier, E. (1988) After the Green Revolution: sustainable and equitable agricultural development, *Futures,* 20, 651–70.

Conyers, D. (1985) Rural regional planning: towards an operational theory, *Progress in Planning,* 23, 3–66.

Conyers, D. (1986) Future directions in development studies: the case of decentralization, *World Development,* 14, 593–603.

Corbridge, S. (1988) The Third World in global context. In Pacione, M. (ed.), *The Geography of the Third World: Progress and Prospect.* London: Routledge, 29–76.

—— (1989) Urban rural relations and the counter-revolution in development theory and policy. In Potter, B. and Unwin, T. (eds), *The Geography of Urban–Rural Interaction in Developing Countries.* London: Routledge, 233–56.

Cornia, G. (1984) A summary and interpretation of the evidence, *World Development,* 12 (special issue on the impact of world recession on children), 381–91.

——, Jolly, R. and Stewart, F. (1987) *Adjustment with a Human Face: Protecting the Vulnerable and Promoting Growth.* Oxford: Oxford University Press.

Cox, M., Niño de Zepeda, A. and Rojas, A. (1990) *Política Agraria en Chile: Del Crecimiento Excluyente al Desarrollo Equitativo.* Santiago, Chile: Centro de Estudios para América Latina sobre Desarrollo Rural, Pobreza y Alimentación.

Cox Edwards, A. and Edwards, S. (1992) Markets and democracy: lessons from Chile, *World Economy,* 15, 203–19.

Craske, N. (1993) Women's political participation in colonias populares in Guadalajara, Mexico. In Radcliffe, S. and Westwood, S. (eds), *Viva: Women and Popular Protest in Latin America.* New York: Routledge, 112–35.

Culpeper, R. (1988) The debt crisis and the World Bank: adjustment workout and growth, *Canadian Journal of Development Studies,* 9, 131–6.

Cypher, J. (1990) Latin American structuralist economics: an evaluation, critique, and reformulation. In Dietz, J. and James, D. (eds), *Progress toward Development in Latin America: From Prebisch to Technological Autonomy.* Boulder: Lynne Rienner, 41–65.

Dag Hammarskjöld Foundation (1975) What now? Another development. Report prepared for the seventh special session of the United Nations General Assembly, special issue of *Development Dialogue,* 1/2.

Daniel, P., Green, R. and Lipton, M. (1985) A strategy for the rural poor, *Journal of Development Planning,* 15, 113–36.

Dattoo, B. and Gray, A. (1979) Underdevelopment and regional planning in the

Third World: a critical overview, *Canadian Journal of African Studies,* 13, 245–64.

Davies, S. and Leach, M. (1991) Globalism versus villagism: national and international issues in food security and the environment, *IDS Bulletin,* 22, 43–50.

Dearlove, J. (1987) Economists on the state, *IDS Bulletin,* 18, 5–11.

Deere, C. (1985) Rural women and state policy: the Latin American agrarian reform experience, *World Development,* 13, 1037–53.

de Gregorio, J. (1992) Economic growth in Latin America, *Journal of Development Economics,* 39, 59–84.

de Janvry, A. (1981) *The Agrarian Question and Reformism in Latin America.* Baltimore: Johns Hopkins University Press.

——, Sadoulet, E. and Wilcox, L. (1989) Rural labour in Latin America, *International Labour Review,* 128, 701–30.

—— and Sadoulet, E. (1993) Market, state, and civil organizations in Latin America beyond the debt crisis: the context for rural development, *World Development,* 21, 659–74.

Dell, S. (1982) Stabilization: the political economy of overkill, *World Development,* 10, 597–612.

Derosa, D. (1992) Increasing export diversification in commodity exporting countries, *International Monetary Fund Staff Papers,* 39, 572–95.

Desanti, A. (1988) *Agricultura de Cambio.* San José, Costa Rica: Imprenta Nacional.

de Valk, P. and Wekwete, K. (eds) (1990) *Decentralizing for Participatory Planning?* Aldershot: Avebury.

Dewitt, R. (1987) Policy directions in international lending, 1961–1984: the case of the Inter-American Development Bank, *Journal of Developing Areas,* 21, 277–84.

Dholakia, R., Dholakia, B., and Kumar, G. (1992) Issues in strategy for export promotion: an inter-industry analysis, *Economic and Political Weekly,* 27 (48), M149–55.

Diaz-Alejandro, C. (1985) Goodbye financial repression, hello financial crash, *Journal of Development Economics,* 19, 1–24.

Dietz, F. and van der Straaten, J. (1992) Rethinking environmental economics: missing links between economic theory and environmental policy, *Journal of Economic Issues,* 26, 27–51.

Dietz, J. (1992) Overcoming underdevelopment: what has been learned from the East Asian and Latin American experiences? *Journal of Economic Issues,* 26, 373–83.

—— and James, D. (1990) Trends in development theory in Latin America: from Prebisch to the present. In Dietz, J. and James, D (eds), *Progress toward Development in Latin America: From Prebisch to Technological Autonomy.* Boulder: Lynne Rienner, 1–11.

Dinh, D. (1993) Complementarity – a new trend in the international division of labour, *International Social Science Journal,* 45, 91–6.

Dodaro, S. (1991) Comparative advantage, trade and growth: export-led growth revisited, *World Development,* 19, 1153–65.

Dollar, D. (1992) Outward-oriented developing economies really do grow more rapidly: evidence from 95 LDCs, 1976–85, *Economic Development and Cultural Change*, 40, 523–44.

Domar, E. (1957) *Essays in the Theory of Economic Growth*. London: Oxford University Press.

Dore, E. (1992) Debt and ecological disaster in Latin America, *Race and Class*, 34, 73–87.

Douglass, M. (1993) Social, political and spatial dimensions of Korean industrial transformation, *Journal of Contemporary Asia*, 23, 149–72.

Downs, C. and Solimano, G. (1989) Toward an evaluation of the NGO experience in Chile: implications for social policy and future investigation. In Downs, C., Solimano, G., Vergara, G., and Zuniga, L. (eds), *Social Policy from the Grassroots: Nongovernmental Organizations in Chile*. Boulder: Westview, 199–212.

Drabek, A. (1987) Development alternatives: the challenge for NGOs, *World Development*, 15 (supplement), ix–xv.

Drakakis-Smith, D. (ed.) (1990) *Economic Growth and Urbanization in Developing Areas*. London: Routledge.

Dube, S. (1988) *Modernization and Development: The Search for Alternative Paradigms*. London: Zed.

Duncan, S. (1989) Uneven development and the difference that space makes, *Geoforum*, 20, 131–9.

—— and Savage, M. (1991) New perspectives on the locality debate, *Environment and Planning A*, 23, 155–63.

Durkheim, E. (1984) *The Division of Labour in Society*. London: Macmillan.

Durning, A. (1989) Action at the grassroots: fighting poverty and environmental decay, *Worldwatch Paper*, 88, Washington: Worldwatch Institute.

Dutt, A. (1990) Sectoral balance in development: a survey, *World Development*, 18, 915–30.

Duvvury, N. (1989) Women in agriculture: a review of the Indian Literature, *Economic and Political Weekly*, 24 (43), WS 96–112.

Eckstein, S. (1988) *The Poverty of Revolution: The State and the Urban Poor in Mexico*. Princeton: Princeton University Press.

—— (ed.) (1989) *Power and Popular Protest: Latin American Social Movements*. Berkeley: University of California Press.

Edgington, D. (1993) The globalization of Japanese manufacturing corporations, *Growth and Change*, 24, 87–106.

Edwards, M. (1989) The irrelevance of development studies, *Third World Quarterly*, 11, 116–35.

Egger, P. (1986) Banking for the rural poor: lessons from some innovative saving and credit schemes, *International Labour Review*, 125, 447–62.

—— (1992) Rural organizations and infrastructure projects: social investment comes before material investment, *International Labour Review*, 131, 45–62.

Eisenstadt, S. (1970) Breakdowns of modernization. In Eisenstadt, S. (ed.), *Readings in Social Evolution and Development*. Oxford: Pergamon, 421–52.

El-Ghonemy, M. (1990) *The Political Economy of Rural Poverty: The Case for*

Land Reforms. London: Routledge.

Ellison, C. and Gereffi, G. (1990) Explaining strategies and patterns of industrial development. In Gereffi, G. and Wyman, D. (eds), *Manufacturing Miracles: Paths of Industrialization in Latin America and East Asia*. Princeton: Princeton University Press, 368–403.

El-Naggar, S. (1987) *Adjustment Policies and Development Strategies in the Arab World*. Washington: International Monetary Fund.

Elson, D. (1989) How is structural adjustment affecting women? *Development*, 1, 67–74.

—— (ed.) (1991) *Male Bias in the Development Process*. New York: Manchester University Press.

Emmerij, L. (ed.) (1987) *Development Policies and the Crisis of the 1980s*. Paris: Development Centre for the OECD.

Enge, K. and Martinez-Enge, P. (1991) Land, malnutrition, and health: the dilemmas of development in Guatemala. In Whiteford, S. and Ferguson A. (eds), *Harvest of Want: Hunger and Food Security in Central America and Mexico*. Boulder: Westview, 75–101.

Escobar, A. (1989) Social science discourse and new social movements research in Latin America: trends and debates. Paper presented at the Annual Meetings of the Latin American Studies Association, Miami, December.

—— (1992) Reflections on development: grassroots approaches and alternative politics in the Third World, *Futures*, 24, 411–34.

—— and Alvarez, S. (eds) (1992) *The Making of Social Movements in Latin America: Identity, Strategy and Democracy*. Boulder: Westview.

Esfahani, S. (1991) Exports, imports and economic growth in semi-industrialized countries, *Journal of Development Economics*, 35, 93–116.

Eshag, E. (1989) Some suggestions for improving the operation of IMF stabilization programmes, *International Labour Review*, 128, 297–320.

—— (1991) Successful manipulation of market forces: case of South Korea, 1961–78, *Economic and Political Weekly*, 26, 629–44.

Esman, M. and Uphoff, N. (1984) *Local Organizations: Intermediaries in Rural Development*. Ithaca, NY: Cornell University Press.

Esteva, G. (1983) *The Struggle for Rural Mexico*. South Handley, MA: Bergin and Garvey.

Eswaran, M. and Kotwal, A. (1993) Export led development: primary vs. industrial exports, *Journal of Development Economics*, 41, 163–72.

Evans, A. (1991) Gender issues in rural household economics, *IDS Bulletin*, 22, 51–9.

Evans, P. and Gereffi, G. (1982) Foreign investment and dependent development: comparing Brazil and Mexico. In Hewlett, S. and Weinert, R. (eds), *Brazil and Mexico: Patterns in Late Development*. Philadelphia: Institute for the Study of Human Issues, 111–68.

Faini, R., de Melo, J., Senhadji, A. and Stanton, J. (1991) Growth-oriented adjustment programs: a statistical analysis, *World Development*, 19, 957–67.

Fajnzylber, F. (1990) The United States and Japan as models of industrialization. In Gereffi, G. and Wyman, D. (eds), *Manufacturing Miracles: Paths of*

Industrialization in Latin America and East Asia. Princeton: Princeton University Press, 323–52.

Fals Borda, O. (1971) *Cooperatives and Rural Development in Latin America: An Analytic Report.* Geneva: UNRISD.

Fass, S. (1985) Book review of Rondinelli, D. (1985), *Applied Methods of Regional Analysis: The Spatial Dimensions of Development Policy, Economic Geography,* 61, 376–9.

Faulkner, A. and Lawson, V. (1991) Employment versus empowerment: a case study of the nature of women's work in Ecuador, *Journal of Development Studies,* 27 (4), 16–47.

Fenster, T. (1993) Settlement planning and participation under principles of pluralism, *Progress in Planning,* 39, 171–242.

Firebaugh, G. and Bullock, B. (1987) Export upgrading, export concentration and economic growth in less developed countries: a cross-national study, *Studies in Comparative International Development,* 22, 87–109.

Fishlow, A. (1972) Brazilian size distribution of income, *American Economic Review,* 60, 391–402.

—— (1984) Summary comment on Adelman, Balassa and Streeten, *World Development,* 12, 979–82.

—— (1985) External debt: crisis and adjustment, in Inter-American Development Bank, *Economic and Social Progress in Latin America,* Washington: IDB, 123–48.

—— (1991) Some reflections on comparative Latin American economic performance and policy. In Banuri, T. (ed.), *Economic Liberalization: No Panacea. The Experiences of Latin America and Asia.* New York: Oxford University Press, 149–70.

FitzGerald, E. (1985) Planned accumulation and income distribution in the small peripheral economy. In Irvin, G. and Gorostiaga, X. (eds), *Towards an Alternative for Central America and the Caribbean.* London: Allen and Unwin, 95–110.

—— (1991) Introduction: the Central American agro-export economy – issues and debates. In Pelupessy, W. (ed.), *Perspectives on the Agro-Export Economy in Central America.* London: Macmillan, 1–10.

Fleming, S. (1991) Between the household: researching community organization and networks, *IDS Bulletin,* 22, 37–43.

Forbes, D. (1984) *The Geography of Underdevelopment: A Critical Survey.* London: Croom Helm.

Fowler, A. (1991) The role of NGOs in changing state–society relations: perspectives from Eastern and Southern Africa, *Development Policy Review,* 9, 53–84.

Fox, J. (1990) Editor's introduction, *Journal of Development Studies,* 26 (4), 1–18.

Foxley, A. (1982) *Experimentos Neoliberales en América Latina.* Santiago: Colección Estudios Cieplan.

Fraser, N. and Nicholson, L. (1990) Social criticism without philosophy: an encounter between feminism and postmodernism. In Nicholson, L. (ed.), *Feminism/Postmodernism.* London: Routledge, 19–38.

Friedmann, J. (1981) The active community: towards a political-territorial framework for rural development in Asia, *Economic Development and Cultural Change*, 29, 226–61.

—— (1988) *Life Space and Economic Space: Essays in Third World Planning*. New Brunswick, NJ: Transaction.

—— (1992) *Empowerment: The Politics of an Alternative Development*. Oxford: Blackwell.

—— and Douglass, M. (1978) Agropolitan development: towards a new strategy for regional planning in Asia. In Lo, F. and Salih, K. (eds), *Growth Pole Strategy and Regional Development Policy*. Oxford: Pergamon, 163–92.

—— and Weaver, C. (1979) *Territory and Function: The Evolution of Regional Planning*. London: Edward Arnold.

Fröbel, F., Heinrichs, J. and Kreye, O. (1980) *The New International Division of Labour*. Cambridge: Cambridge University Press.

Fuentes, M. and Frank, A. (1989) Ten theses on social movements, *World Development*, 17, 179–81.

Furtado, C. (1987) Underdevelopment: to conform or reform. In Meier, G. (ed.), *Pioneers in Development*. Oxford: Oxford University Press, 205–27.

Gadgil, M. and Guha, R. (1994) Ecological conflicts and the environmental movement in India, *Development and Change*, 25, 101–36.

García, A., Infante, R. and Tokman, V. (1989) Paying off the social debt in Latin America, *International Labour Review*, 128, 467–84.

GATT (1985–6, 1990–1) *International Trade* (annual). Geneva: General Agreement on Tariffs and Trade.

Geisler, G. (1992) Who is losing out? Structural adjustment, gender and the agricultural sector in Zambia, *Journal of Modern African Studies*, 30, 113–39.

—— (1993) Silences speak louder than claims: gender, household, and agricultural development in Southern Africa, *World Development*, 21, 1965–80.

George, S. (1988) *A Fate Worse than Debt: The Financial Crisis and the Poor*. New York: Grove.

Gereffi, G. and Wyman, D. (eds) (1990) *Manufacturing Miracles: Paths of Industrialization in Latin America and East Asia*. Princeton: Princeton University Press.

Ghai, D. (1977) What is the basic-needs approach to development all about? In International Labour Organisation, *The Basic-needs Approach to Development: Some Issues Regarding Concepts and Methodology*. Geneva: ILO.

—— (1987) Successes and failures in growth in sub-Saharan Africa. In Emmerij, L. (ed.), *Development Policies and the Crisis of the 1980s*. Paris: Development Center for the OECD, 106–29.

—— (1989) Participatory development: some perspectives from grass-roots experiences, *Journal of Development Planning*, 19, 215–46.

—— and Hewitt de Alcántara, C. (1990) The crisis of the 1980s in sub-Saharan Africa, Latin America and the Caribbean: economic impact, social change and political implications, *Development and Change*, 21, 389–426.

Ghose, A. (ed.) (1983) *Agrarian Reform in Contemporary Developing Countries*. London: Croom Helm.

—— (1991) Political economy of structural readjustment, *Economic and Political Weekly*, 26, 2727–9.

Ghosh, S. (1990) Redefining concepts of science, technology, and development, *Economic and Political Weekly*, 25, 1343–8.

Gibson, M. and Ward, M. (1992) Export orientation: pathway or artifact? *International Studies Quarterly*, 36, 331–44.

Giddens, A. (1984) *The Constitution of Society: Outline of the Theory of Structuration*. Berkeley: University of California Press.

Gilbert, A. (1982) Urban and regional systems: a suitable case for treatment? In Gilbert, A. and Gugler, J. (eds), *Cities, Poverty and Development: Urbanization in the Third World*. Oxford: Oxford University Press, 162–97.

Gladwin, C. and McMillan, D. (1989) Is a turnaround in Africa possible without helping African women to farm? *Economic Development and Cultural Change*, 37, 345–69.

Glaeser, B. and Vyasulu, V. (1984) The obsolescence of ecodevelopment? In Glaeser, B. (ed.), *Ecodevelopment: Concepts, Projects, Strategies*. Oxford: Pergamon.

Glover, D. (1991) A layman's guide to structural adjustment, *Canadian Journal of Development Studies*, 12, 173–86.

Gómez, S. and Echenique, J. (1988) *La Agricultura Chilena: Las dos Caras de la Modernización*. Santiago, Chile: Facultad Latinamericana de Ciencias Sociales y Agraria.

Gore, C. (1984) *Regions in Question: Space, Development and Regional Policy*. London: Methuen.

—— (1991) The spatial separatist theme and the problem of representation in location-allocation models, *Environment and Planning A*, 23, 939–53.

Grabowski, R. (1988) Taiwanese economic development: an alternative interpretation, *Development and Change*, 19, 53–67.

—— (1989) Development as displacement: a critique and alternative, *Journal of Developing Areas*, 23, 505–18.

Grais, W., de Melo, J. and Urata, S. (1986) A general equilibrium estimation of the effects of reductions in tariffs and quantitative restrictions in Turkey in 1978. In Srivivasan, T. and Whalley, J. (eds), *General Equilibrium Trade Policy Modelling*. Cambridge, MA: MIT Press, 61–88.

Gray, J. (1988) Contemporary issues in East Asia. In Pacione, M. (ed.) *Geography of the Third World*. London: Routledge, 408–26.

Green, R. (1985) IMF stabilization and structural adjustment in sub-Saharan Africa: are they technically compatible? *IDS Bulletin*, 16, 61–8.

—— (1991) Structural adjustment and national environmental strategies: what interactions? Notes from Namibia, *IDS Bulletin*, 22, 38–44.

Greenaway, D. and Morrissey, O. (1993) Structural adjustment and liberalisation in developing countries: what lessons have we learned? *Kyklos*, 46, 241–61.

Greenhalgh, S. (1985) Sexual stratification: the other side of 'growth with equity' in East Asia, *Population and Development Review*, 11, 265–314.

Grice, K. and Drakakis-Smith, D. (1985) The role of the state in shaping development: two decades of growth in Singapore, *Transactions, Institute of British Geographers*, 10, 347–59.

Griffin, K. (1969) *Underdevelopment in Spanish America*. London: Allen and Unwin.

—— (1976) *Land Concentration and Rural Poverty*. New York: Holmes and Meier.

—— (1981) Economic development in a changing world, *World Development*, 9, 221–6.

—— (1989) *Alternative Strategies for Economic Development*. New York: St. Martin's.

—— and Khan, A. (eds) (1972) *Growth and Inequality in Pakistan*. London: Macmillan.

Griffith, W. (1987) Can CARICOM countries replicate the Singapore experience? *Journal of Development Studies*, 24, 60–81.

Grindle, M. (1986) *State and Countryside: Development Policy and Agrarian Politics in Latin America*. Baltimore: Johns Hopkins University Press.

—— (1988) *Searching for Rural Development: Labor Migration and Employment in Mexico*. Ithaca: Cornell University Press.

Gulati, U. (1992) The foundations of rapid economic growth: the case of the Four Tigers, *American Journal of Economics and Sociology*, 51, 161–72.

Haberler, G. (1987) Liberal and illiberal development policy. In Meier, G. (ed.), *Pioneers in Development*. Oxford: Oxford University Press, 49–83.

Haggard, S. (1986) The newly industrializing countries in the international system, *World Politics*, 38, 343–70.

Hahner, J. (1985) Recent research on women in Brazil, *Latin American Research Review*, 20, 163–79.

Hamilton, C. (1983) Capitalist industrialisation in East Asia's Four Little Tigers, *Journal of Contemporary Asia*, 13, 35–73.

Handelman, H. and Baer, W. (1989) *Paying the Costs of Austerity in Latin America*. Boulder: Westview.

Harris, L. (1989) The Bretton Woods system and Africa. In Onimode, B. (ed.) *The IMF, the World Bank and the African Debt, vol. 1*. London: Zed, 19–24.

Harris, N. (1986) *The End of the Third World: Newly Industrializing Countries and the Decline of an Ideology*. London: Tauris.

—— (1991) Export processing in Mexico, *Journal of Development Studies*, 27, 117–25.

Harrison, D. (1988) *The Sociology of Modernization and Development*. London: Unwin Hyman.

Harriss, J. (ed.) (1982) *Rural Development: Theories of Peasant Economy and Agrarian Change*. London: Hutchinson.

Harrod, R. (1948) *Towards a Dynamic Economics*. London: Macmillan.

Havnevik, K. (ed.) (1987) *The IMF and the World Bank in Africa*. Uppsala: Scandinavian Institute of African Studies.

Hawkins, J. (1991) Understanding the failure of IMF reform: the Zambian case, *World Development*, 19, 839–49.

Haynes, J., Parfill, T. and Riley, S. (1987) Debt in sub-Saharan Africa: the local politics of stabilization, *African Affairs*, 86, 343–66.

Hazell, P. and Ramasamy, C. (eds) (1991) *Importance of a Macroeconomic*

Perspective: The Green Revolution Reconsidered. Baltimore: Johns Hopkins University Press.

Heilbroner, R. (1990) After Communism, *The New Yorker*, 10 (September), 91–100.

Helleiner, G. (1979) 'Intrafirm trade and the developing countries: an assessment of the data, *Journal of Development Economics*, 6, 391–406.

—— (1986) Balance of payments experience and growth prospects of developing countries: a synthesis, *World Development*, 14, 877–908.

—— (1987) Stabilization, adjustment, and the poor? *World Development*, 15, 1499–513.

—— (1989) Conventional foolishness and overall ignorance: Current approaches to global transformation and development, *Canadian Journal of Development Studies*, 10, 107–20.

—— (1990) *The New Global Economy and the Developing Countries: Essays in International Economics and Development.* Brookfield, VT: Gower.

—— (1992) The IMF, the World Bank and Africa's adjustment and external debt problems: an unofficial view, *World Development*, 20, 779–92.

——, Cornia, G. and Jolly, R. (1991) IMF adjustment policies and approaches and the needs of children, *World Development*, 19, 1823–34.

Hellman, J. (1992) Making women visible: new works on Latin American and Caribbean women, *Latin American Research Review*, 27, 182–91.

Herbst, J. (1990) The structural adjustment of politics in Africa, *World Development*, 18, 949–58.

Herrera, A., Scolnik, H., Chichilinisky, G., Gallopin, G., Hardoy, J., Mosovich, D., Oteiza, E., Best, G., Suarez, C and Talavera. L. (1976) *Catastrophe or New Society? A Latin American World Model.* Ottawa: International Development Research Center.

Hettne, B. (1990) *Development Theory and the Three Worlds.* Essex: Longman.

Hewlett, S. and Weinert, R. (eds) (1982) *Brazil and Mexico: Patterns in Late Development.* Philadelphia: Institute for the Study of Human Issues.

Hiemenz, U. (1989) Development strategies and foreign aid policies for low-income countries in the 1990s. Kiel Discussion Papers, 152, Institut für Weltwirtschaft, Kiel.

Higgot, R. (1983) *Political Development Theory.* London: Croom Helm.

Hindess, B. (1991) Imaginary presuppositions of democracy, *Economy and Society*, 20, 173–95.

Hirschman, A. (1982) The rise and decline of development economics. In Gersovitz, M., Diaz-Alejandro, C., Ranis, G. and Rosengweig, M. (eds), *The Theory and Experience of Economic Development: Essays in Honour of Sir W. Arthur Lewis.* London: Allen and Unwin, 372–90.

—— (1985) Against parsimony: three easy ways of complicating some categories of economic discourse, *Economics and Discourse*, 1, 7–21.

—— (1987) The political economy of Latin American development: seven exercises in retrospection, *Latin American Research Review*, 22, 7–36.

Hirschmann, D. (1991) Women and political participation in Africa: broadening the scope of research, *World Development*, 19, 1679–94.

Hirway, I. (1988) Reshaping IRDP: some issues, *Economic and Political Weekly*, 23 (26), A89–96.

Hon, W. (1992) Exploiting information technology: a case study of Singapore, *World Development*, 20, 1817–28.

Honadle, G., Morss, E., Van Sant, J. and Gow, D. (1980) *Integrated Rural Development: Making it Work*. Washington: Development Alternatives Inc.

Hoogvelt, A. (1990) Extended review: Rethinking development theory, *Sociological Review*, 38, 352–61.

Horowitz, I. (1985) The 'Rashomon' effect: ideological proclivities and political dilemmas of the IMF, *Journal of Interamerican Studies and World Affairs*, 27 (4), 37–55.

Hoselitz, B. (1957) Generative and parasitic cities, *Economic Development and Cultural Change*, 3, 278–94.

—— (1960) *Sociological Aspects of Economic Growth*. Glencoe, IL: Free Press.

House-Midamba, B. (1990) The United Nations decade: political empowerment or increased marginalization for Kenyan Women? *Africa Today*, 37, 37–48.

Hughes, A. and Singh, A. (1991) The world economic slowdown and the Asian and Latin American economies: a comparative analysis of economic structure, policy, and performance. In Banuri, T. (ed.), *Economic Liberalization: No Panacea. The Experiences of Latin America and Asia*. New York: Oxford University Press, 57–98.

Hughes, H. (ed.) (1988) *Achieving Industrialization in East Asia*. Cambridge: Cambridge University Press.

Hugon, P. (1991) Politiques d'ajustement et repartition des effets, *Journal of African Studies*, 25, 12–33.

Hulme, D. and Turner, M. (1990) *Sociology and Development: Theories, Policies and Practices*. New York: Harvester Wheatsheaf.

Huntington, S. (1968) *Political Order in Changing Societies*. New Haven: Yale University Press.

Hyden, G. (1980) *Beyond Ujamaa in Tanzania: Underdevelopment and an Uncaptured Peasantry*. Nairobi: Heinemann.

Ibrahim, B. (1989) Policies affecting women's employment in the formal sector: strategies for change, *World Development*, 17, 1097–107.

Iglesias, E. (1985) Address to the meeting of experts on crisis and development in Latin America and the Caribbean. Santiago, April 29–May 3.

International Foundation for Development Alternatives (1980) Dossier no. 17.

International Labor Organization (1976) *Employment, Growth, and Basic Needs: A One-World Problem*. Geneva: ILO.

IMF (1993) *International Financial Statistics Yearbook* (Annual). Washington: International Monetary Fund.

Irwan, A. (1987) Real wages and class struggle in South Korea, *Journal of Contemporary Asia*, 17, 385–408.

Islam, N. (1992) Poverty in South Asia: approaches to its alleviation, *Food Policy*, 17, 108–28.

Islam, R. (1990) Rural poverty, growth and macroeconomic policies: the Asian experience, *International Labour Review*, 129, 693–714.

Jackson, C. (1993) Doing what comes naturally? Women and environment in development, *World Development*, 21, 1947–63.

Jackson, P. (1989) *Maps of Meaning: An Introduction to Cultural Geography*. London: Unwin Hyman.

—— (1991) Mapping meanings: a cultural critique of locality studies, *Environment and Planning A*, 23, 215–28.

Jaeger, W. and Humphreys, C. (1988) The effect of policy reforms on agricultural incentives in sub-Saharan Africa, *American Journal of Agricultural Economics*, 70, 1036–43.

Jahan, R. (1987) Women in South Asian politics, *Third World Quarterly*, 9, 848–70.

Jain, L., Krishnamurthy, B. and Tripathi, P. (1986) *Grass without Roots: Rural Development under Government Auspices*. New Delhi: Sage.

James, S. (1992) Transgressing fundamental boundaries: the struggle for women's human rights, *Africa Today*, 39 (4), 35–46.

Jansen, R. and van Hoof, P. (1990) Regional development planning for rural development in Botswana. In D. Simon (ed.), *Third World Regional Development: A Reappraisal*. London: Chapman, 190–209.

Jaquette, J. (1987) United Nations decade for women: its impact and legacy, *World Development*, 15, 419–27.

Jelin, E. (1986) Otros silencios, otras voces; el tiempo de la democratización en Argentina. In Calderón, F. (ed.), *Los Movimientos Sociales ante la Crisis*. Buenos Aries: CLACSO/UNO, 17–44.

—— (ed.) (1990) *Women and Social Change in Latin America*. London: Zed.

Jenkins, R. (1985) Internationalization of capital and the semi-industrialized countries: the case of the motor industry, *Review of Radical Political Economics*, 17, 59–81.

—— (1991) The political economy of industrialization: a comparison of Latin American and East Asian newly industrializing countries, *Development and Change*, 22, 197–231.

Jiggins, J. (1989) How poor women earn income in sub-Saharan Africa and what works against them, *World Development*, 17, 953–63.

Joekes, S. (1994) Gender, environment and population, *Development and Change*, 25, 137–65.

Johnson, E. (1970) *The Organization of Space in Developing Countries*. Cambridge, MA: Harvard University Press.

Johnson, J. (1958) Political change in Latin America: the emergence of the middle sectors. In Klarén, P. and Bossert, T. (eds), *Promise of Development: Theories of Change in Latin America*. Boulder: Westview, 88–99.

Johnston, B. and Clark, W. (1982) *Redesigning Rural Development: A Strategic Perspective*. Baltimore: Johns Hopkins University Press.

—— and Kilby, P. (1975) *Agriculture and Structural Transformation: Economic Strategies in Late-Developing Countries*. New York: Oxford University Press.

Jolly, R. (1988) From speeches to action: implementing what is agreed, *IDS Bulletin*, 19, 75–80.

—— and van der Hoeven, R. (1991) Editors' introduction: Adjustment with a human face – record and relevance, *World Development,* 19, 1801–5.

Jones, S., Joshi, P. and Murmis, M. (eds) (1982) *Rural Poverty and Agrarian Reform.* Norfolk: Durham Books.

Jorgenson, D. (1961) The development of a dual economy, *Economic Journal,* 71, 309–34.

Kabeer, N. and Joekes, S. (1991) Editorial, *IDS Bulletin,* 22, 1–4.

Kandiyoti, D. (1990) Women and rural development policies: the changing agenda, *Development and Change,* 21, 5–22.

Kaplan, T. (1982) Female consciousness and collective action: the case of Barcelona, 1910–1918, *Signs,* 7, 545–66.

Kaufman, R. (1989) The politics of economic adjustment policy in Argentina, Brazil and Mexico: experiences in the 1980s and challenges for the future, *Policy Sciences,* 22, 395–413.

—— (1990) How societies change development models or keep them: reflections on the Latin American experience in the 1930s and the postwar world. In Gereffi, G. and Wyman, D. (eds), *Manufacturing Miracles: Paths of Industrialization in Latin America and East Asia.* Princeton: Princeton University Press, 110–38.

Kay, C. (1985) The monetarist experiment in the Chilean countryside, *Third World Quarterly,* 7, 301–22.

—— (1993) For a renewal of development studies: Latin American theories and neoliberalism in the era of structural adjustment, *Third World Quarterly,* 14, 691–702.

Kearney, R. (1990) Mauritius and the NIC model redux, or how many cases make a model? *Journal of Developing Areas,* 24, 195–216.

Keller, B. and Mbewe, D. (1991) Policy and planning for the empowerment of Zambia's women farmers, *Canadian Journal of Development Studies,* 12, 75–88.

Kellman, M. and Chow, P. (1989) The comparative homogeneity of the East Asian NIC exports of similar manufacturers, *World Development,* 17, 267–73.

Khan, A. (1987) A comment on Professor Chakravarty's paper. In Toye, J. (ed.), *Development Policies and the Crisis of the 1980s.* Paris: Development Centre for the OECD, 96–100.

Khan, M. (1990) The macroeconomic effects of Fund-supported adjustment programs, *International Monetary Fund Staff Papers,* 37, 195–231.

Killick, T. (1986) Twenty-five years in development: the rise and impending decline of market solutions, *Development Policy Review,* 4, 99–116.

—— (1989) *A Reaction Too Far: Economic Theory and the Role of the State in Developing Countries.* London: Overseas Development Institute.

—— and Stevens, C. (1991) Eastern Europe: lessons on economic adjustment from the Third World, *International Affairs,* 67, 679–96.

Kim, C., Kim, Y. and Yoon, C. (1992) Korean telecommunications development: achievements and cautionary lessons, *World Development,* 20, 1829–41.

Kim, W. (1993) Industrial restructuring and the dynamics of city–state adjustments, *Environment and Planning A,* 25, 27–46.

Kishwar, M. (1988) Nature of women's mobilization in rural India: an exploratory essay, *Economic and Political Weekly,* 23, 2754–63.

Kitching, G. (1982) *Development and Underdevelopment in Historical Perspective: Populism, Nationalism, and Industrialism.* New York: Methuen.

Klarén, P. (1986) Lost promise: explaining Latin American underdevelopment. In Klarén, P. and Bossert, T. (eds), *Promise of Development: Theories of Change in Latin America.* Boulder: Westview, 1–35.

Kohli, A. (1989) Politics of economic liberalization in India, *World Development,* 17, 305–28.

Kolko, J. (1988) *Restructuring the World Economy.* New York: Pantheon.

Koo, H. (1991) Middle classes, democratization, and class formation: the case of South Korea, *Theory and Society,* 20, 485–509.

Koopman, J. (1993) Neoclassical household models and modes of household production: problems in the analysis of African agricultural households, *Review of Radical Political Economics,* 23 (3/4), 148–73.

Kraus, J. (1991) The struggle over structural adjustment in Ghana, *Africa Today,* 38, 19–37.

Kreye, O. and Schubert, A. (1988) Social implications of Third World debt, *Dialectical Anthropology,* 12, 261–70.

Krishnaswamy, K. (1991) On liberalisation and some related matters, *Economic and Political Weekly,* 26, 2415–22.

Krueger, A. (1986) Changing perspectives on development economics and World Bank research, *Development Policy Review,* 4, 195–210.

—— (1990) Economists' changing perceptions of government, *Weltwirtschaftliches Archiv,* 126, 417–31.

Krugman, P. (1986) Introduction: New thinking about trade policy. In Krugman, P. (ed.), *Strategic Trade Policy and the New International Economics.* Cambridge, MA: MIT Press, 1–22.

—— (1987) The narrow moving band, the Dutch disease, and the competitive consequences of Mrs Thatcher: notes on trade in the presence of dynamic scale economies, *Journal of Development Economics,* 27, 40–55.

Kuah, K. (1990) Confucian ideology and social engineering in Singapore, *Journal of Contemporary Asia,* 20, 371–83.

Kuczynski, P. (1988) *Latin American Debt.* Baltimore: Johns Hopkins University Press.

Kung, L. (1984) Taiwan garment workers. In Sheridan, M. and Salaff, J. (eds), *Lives: Chinese Working Women.* Bloomington: Indiana University Press, 109–22.

Kuznets, S. (1955) Economic growth and income inequality, *American Economic Review,* 45, 1–28.

—— (1965) *Economic Growth and Structure: Selected Essays.* New York: Norton.

Kyle, S. and Cunha, A. (1992) National factor markets and the macroeconomic context for environmental destruction in the Brazilian Amazon, *Development and Change,* 23, 7–33.

Laclau, E. and Mouffe, C. (1985) *Hegemony and Socialist Strategy: Towards a Radical Democratic Politics.* London: Verso.

LaFeber, W. (1983) *Inevitable Revolutions: The United States in Central America.* New York: Norton.

Lal, D. (1983) *The Poverty of Development Economics.* Cambridge, MA: Harvard

University Press.

—— (1984) The political economy of the predatory state. Discussion Paper DRD 105, Washington: World Bank Development Research Department.

Landell-Mills, P., Agarwala, R. and Please, S. (1989) *Sub-Saharan Africa: From Crisis to Sustainable Growth: A Long-term Perspective Study*. Washington: World Bank.

Lawless, R. (1988) Contemporary issues in the Middle East. In Pacione, M. (ed.) *The Geography of the Third World: Progress and Prospect*. London: Routledge, 365–90.

Lawrence, R. (1993) Japan's different trade regime: an analysis with particular reference to Keiretsu, *Journal of Economic Perspectives*, 7, 3–19.

Lazreg, M. (1988) Feminism and difference: the perils of writing as a woman on women in Algeria, *Feminist Studies*, 14, 81–107.

Lea, D. and Chaudhri, D. (eds) (1983) *Rural Development and the State*. London: Methuen.

Leach, M. and Mearns, R. (1991) Editorial, *IDS Bulletin*, 22, 1–4.

Leamer, E. (1990) Latin America as a target of trade barriers erected by the major developed countries in 1983, *Journal of Development Economics*, 32, 337–68.

Lebeau, A. and Salomon, J. (1990) Science, technology and development, *Social Science Information*, 29, 841–58.

LeBeuf, A. (1991) The role of women in the political organization of African societies. In Paulme, D. (ed.), *Women of Tropical Africa*. Berkeley: University of California Press, 93–119.

Lederer, K. (ed.) (1980) *Human Needs: A Contribution to the Current Debate*. Oelgeschlager: Gunn & Hain/Verlag Anton.

Lee, A. (1993) Culture shift and popular protest in South Korea, *Comparative Political Studies*, 26, 63–80.

Lee, S. (1993) Transitional politics of Korea, 1987–1992: activation of civil society, *Pacific Affairs*, 66, 351–67.

Lele, U. (1981) Cooperatives and the poor: a comparative perspective, *World Development*, 9, 55–72.

—— (1984) The role of risk in an agriculturally-led strategy in sub-Saharan Africa, *American Journal of Agricultural Economics*, 66, 677–83.

—— (1990) Structural adjustment, agricultural development and the poor: some lessons from the Malawian experience, *World Development*, 18, 1207–19.

—— and Adu-Nyako, K. (1992) Approaches to uprooting poverty in Africa, *Food Policy*, 17, 95–108.

Leonard, A. (ed.) (1989) *Seeds: Supporting Women's Work in the Third World*. New York: City University of New York Press.

Lerner, D. (1958) *The Passing of Traditional Society: Modernizing the Middle East*. New York: Free Press.

Leung, Y. (1987) The uncertain phoenix: Confucianism and its modern fate, *Asian Culture*, 10, 85–94.

Levitt, K. (1990) Debt, adjustment and development: looking to the 1990s, *Economic and Political Weekly*, 25, 1585–94.

Lewis, J. (1989) Government and national economic development, *Daedalus*, 118, 69–88.

—— and Kallab, V. (eds) (1986) *Development Strategies Reconsidered*. New Brunswick, NJ: Transaction.

Lewis, W. (1950) Industrialization of the British West Indies, *Caribbean Economic Review*, 2, 1–61.

—— (1954) Economic development with unlimited supplies of labour, *Manchester School of Economics and Social Studies*, 22, 139–91.

Liang, N. (1992) Beyond import substitution and export promotion: a new typology of trade strategies, *Journal of Development Studies*, 28, 447–72.

Lie, J. (1991) The state, industrialization and agricultural sufficiency: the case of South Korea, *Development Policy Review*, 9, 37–51.

Lim, L. (1983) Capitalism, imperialism and patriarchy: the dilemma of Third World women workers in multinational factories. In Nash, J. and Fernández-Kelly, M. (eds), *Women, Men and the International Division of Labor*. Albany: SUNY, 70–91.

Lin, C. (1988) East Asia and Latin America as contrasting models, *Economic Development and Cultural Change*, 36, S153–97.

—— (1989) *Latin America vs East Asia: A Comparative Development Perspective*. New York: Sharpe.

Lin, J. (1989) Beyond neoclassical shibboleths: a political-economic analysis of Taiwanese economic development, *Dialectical Anthropology*, 14, 283–300.

Lipset, S. (1967) Values, education, and entrepreneurship. In Lipset, S. and Solari, A. (eds), *Elites in Latin America*. New York: Oxford University Press, 3–60.

Lipton, M. (1977) *Why Poor People Stay Poor: Urban Bias in World Development*. Cambridge, MA: Harvard University Press.

—— (1993) Land reform as commenced business: the evidence against stopping, *World Development*, 21, 641–57.

List, F. (1844) *The National System of Political Economy*. Translated by S. Lloyd, 1916; London: Longmans.

Logan, K. (1990) Women's participation in urban protest. In Foweraker, J. and Craig, A. (eds), *Popular Movements and Political Change in Mexico*. Boulder: Lynne Rienner, 150–9.

López, R. (1992) Environmental degradation and economic openness in LDCs: the poverty linkage, *American Journal of Agricultural Economics*, 74, 1138–43.

Loufti, M. (1987) Development with women: action not alibis, *International Labour Review*, 126, 111–20.

Lowder, S. (1990) Development policy and its effects on regional inequality: the case of Ecuador. In Simon, D. (ed.), *Third World Regional Development: A Reappraisal*. London: Chapman, 73–93.

Loxley, J. (1987) The IMF, the World Bank, and sub-Saharan Africa: politics and politics. In Havnevik, K. (ed.), *The IMF and the World Bank in Africa: Conditionality, Impact and Alternatives*. Uppsala: Scandinavian Institute of African Studies, 47–63.

Lücke, M. (1993) Developing countries' terms of trade in manufactures, 1967–87: a note, *Journal of Development Studies*, 29, 588–95.

Luedde-Neurath, R. (1986) *Import Controls and Export-Directed Development: A Reassessment of the South Korean Case.* Boulder: Westview.

Mackenzie, F. (1991) Political economy of the environment, gender and resistance under colonialism: Murang's District, Kenya, 1910–1950, *Canadian Journal of African Studies*, 25, 226–62.

—— (1993) Exploring the connections: structural adjustment, gender and the environment, *Geoforum*, 24, 71–87.

Mackintosh, M. (1981) Gender and economics: the sexual division of labour and the subordination of women. In Young, K., Wolkowitz, C. and McCullagh, R. (eds), *Of Marriage and the Market.* London: CSE Books, 1–15.

Maddock, N. (1987) Privatizing agriculture: policy options in developing countries, *Food Policy*, 12, 295–8.

Mahon, J. (1992) Was Latin America too rich to prosper? Structural and political obstacles to export led industrial growth, *Journal of Development Studies*, 28, 241–63.

Maizels, A. (1987) Commodities in crisis: an overview of the main issues, *World Development*, 15, 537–57.

Maralidharan, S. (1991) Food studies: IMF prescriptions could spell trouble, *Economic and Political Weekly*, 26, 1083–5.

Marien, M. (1992) Environmental problems and sustainable futures: major literature from WCED to UNCED, *Futures*, 24, 731–57.

Maro, P. (1990) The impact of decentralization on spatial equity and rural development in Tanzania, *World Development*, 18, 673–93.

Martin, J. (1989) Motherhood and power: the production of a women's culture of politics in a Mexican community, *American Ethnologist*, 17, 470–90.

Martinez-Alier, J. (1990) Poverty as a cause of environmental degradation. Manuscript prepared for the World Bank, Washington, DC.

Maxfield, S. and Nolt, J. (1990) Protectionism and the internationalization of capital: US sponsorship of import substitution industrialization in the Philippines, Turkey and Argentina, *International Studies Quarterly*, 34, 49–81.

Max-Neef, M. (1986) Human scale economics: the Challenges ahead. In Ekins, P. (ed.), *The Living Economy: A New Economics in the Making.* London: Routledge, 43–54.

Mayoux, L. (1992) From idealism to realism: women, feminism, and empowerment in a Nicaraguan tailoring cooperative, *Development and Change*, 23, 91–114.

McCall, M. and Skutsch, M. (1983) Strategies and contradictions in Tanzania's rural development: which path for peasants? In Lea, D. and Chaudhri, D. (eds), *Rural Development and the State.* London: Methuen, 241–72.

McClelland, D. (1961) *The Achieving Society.* Princeton: Van Nostrand.

Meadows, D. H. and Meadows, D. L. (1972) *The Limits to Growth.* New York: Universe.

Mehta, S. (1984) *Rural Development Policies and Programs: A Sociological Perspective.* Beverly Hills, CA: Sage.

Meier, G. and Seers, D. (eds) (1984) *Pioneers in Development.* New York: Oxford University Press.

Meister, E. (1991) Selected problems of the EEC market for Central American coffee. In Pelupessy, W. (ed.) *Perspectives on the Agro-Export Economy in Central America*. London: Macmillan, 54–70.

Mengisteab, K. and Logan, B. (1990) Implications of liberalization policies for agricultural development in sub-Saharan Africa, *Comparative Political Studies*, 22, 437–57.

Merchant, C. (1980) *The Death of Nature: Women, Ecology and the Scientific Revolution*. New York: Harper and Row.

—— (1992) *Radical Ecology: The Search for a Livable World*. New York: Routledge.

Michaely, M., Papageorgiou, D. and Choksi, A. (1991) *Liberalizing Foreign Trade, vol. 7: Lessons of Experience in the Developing World*. Oxford: Blackwell.

Midgley, J. (1986) *Community Participation, Social Development and the State*. London: Methuen.

Migdal, J. (1974) *Peasants, Politics and Revolution: Pressures toward Political and Social Change in the Third World*. Princeton: Princeton University Press.

Minocha, A. (1991) Indian development strategy, *Economic and Political Weekly*, 26, 2488.

Mohanty, C. (1988) Under Western eyes: feminist scholarship and colonial discourses, *Feminist Review*, 30, 61–88.

Molyneux, M. (1985) Mobilization without emancipation? Women's interests, the state, and revolution in Nicaragua. In Fagen, R., Deere, C. and Coraggio, J. (eds), *Transition and Development: Problems of Third World Socialism*. New York: Monthly Review, 280–302.

Moore, D. (1993) Contesting terrain in Zimbabwe's eastern highlands: political ecology, ethnography, and peasant resource struggles, *Economic Geography*, 69, 380–401.

Moran, C. (1989) Economic stabilization and structural transformation: lessons from the Chilean experience, 1973–87, *World Development*, 17, 491–502.

Morishima, M. (1982) *Why Has Japan 'Succeeded'?* Tokyo: Tokyo University Press.

Moser, C. (1987) Mobilization is women's work: struggles for infrastructure in Guayquil, Ecuador. In Moser, C. and Peake, L. (eds), *Women, Human Settlements and Housing*. London: Tavistock, 166–94.

—— (1989) Gender planning in the Third World: meeting practical and strategic gender needs, *World Development*, 17, 1799–1825.

—— and Peake, L. (1987) *Women, Human Settlements and Housing*. London: Tavistock.

Mosley, P., Harrigan, J. and Toye, J. (1991) *Aid and Power: The World Bank and Policy-Based Lending*. New York: Routledge.

Munasinghe, M. (1993) Environmental issues and economic decisions in developing countries, *World Development*, 21, 1729–48.

Murphy, A. (1991) Regions as social constructs: the gap between theory and practice, *Progress in Human Geography*, 15, 23–35.

Myint, J. (1987) The neoclassical resurgence in development economics: its strength and limitations. In Meier, G. (ed.), *Pioneers in Development (Second Series)*.

Oxford: Oxford University Press, 107–36.

Nash, M. (1984) *Unfinished Agenda: The Dynamics of Modernization in Developing Nations*. Boulder: Westview.

Nayar, B. (1972) *The Modernization Imperative of Indian Planning*. Delhi: Vikas.

Nelson, J. (1989) *Fragile Coalitions: The Politics of Economic Adjustment*. New Brunswick, NJ: Transaction.

Nerfin, M. (ed.) (1977) *Another Development: Approaches and Strategies*. Uppsala: Dag Hammarskjöld Foundation.

Nesmith, C. (1991) Gender, trees and fuel – social forestry in West Bengal, India, *Human Organization*, 50, 337–48.

Neumann, R. (1992) Political ecology of wildlife conservation in the Mt. Meru area of northeast Tanzania, *Land Degradation and Society*, 3, 85–98.

Ngau, P. (1987) Tensions in empowerment: the experience of the Harambee (self-help) movement in Kenya, *Economic Development and Cultural Change*, 35, 523–38.

Niskanen, W. (1971) *Bureaucracy and Representative Government*. Chicago: Aldine-Atherton.

Nolan, P. (1990) Assessing economic growth in the Asian NICs, *Journal of Contemporary Asia*, 20, 41–63.

Nuñez, O. (1980) *El Somocismo: Desarrollo y Contradicciones del Modelo Capitalista Agroexportador en Nicaragua, 1950–1975*. Havana: Centro de Estudios sobre América.

Nurkse, R. (1953) *Problems of Capital Formation in Underdeveloped Countries*. Oxford: Blackwell.

Nyaga, M. (1986) Against many odds: the dilemmas of women's self-help groups in Mbeere, Kenya, *Africa*, 56, 210–28.

Oakley, P. and Marsden, D. (1984) *Approaches to Participation in Rural Development*. Geneva: International Labour Organization.

Ocampo, J. (1990) New economic thinking in Latin America, *Journal of Latin American Studies*, 22, 169–82.

O'Donnell, G. (1975) *Modernization and Bureaucratic Authoritarianism: Studies in South American Politics*. Berkeley: University of California Press.

Ogle, G. (1990) *South Korea: Dissent within the Economic Miracle*. London: Zed.

Oliveira, J. (1986) Trade policy, market 'distortions', and agriculture in the process of economic development: Brazil, 1950–1974, *Journal of Development Economics*, 24, 91–109.

Olsen, M. (1965) *The Logic of Collective Action*. Cambridge MA: Harvard University Press.

—— (1982) *The Rise and Decline of Nations: Economic Growth, Stagflation, and Social Rigidities*. New Haven: Yale University Press.

Oman, C. and Wignaraja, G. (1991) *The Postwar Evolution of Development Thinking*. London: Macmillan.

Onis, Z. (1991) The logic of the developmental state, *Comparative Politics*, 24, 109–26.

Organski, A. (1965) *The Stages of Political Development*. New York: Knopf.

Ortega, E. (1988) *Transformaciones Agrarias y Campesinado: De la Participación a la Exclusión*. Santiago, Chile: Corporación de Investigación para Latinoamérica.

Ovitt, G. (1989) Appropriate technology: development and social change, *Monthly Review*, 40 (9), 22–34.

Paasi, A. (1991) Deconstructing regions: notes on the scales of social life, *Environment and Planning A*, 23, 239–56.

Pacione, M. (1988) Introduction. In Pacione, M. (ed.), *The Geography of the Third World: Progress and Prospect*. London: Routledge, 1–25.

Page, D. (1986) Growing hope in Santiago's urban organic gardens, *Grassroots Development*, 10 (2), 38–44.

Painter, M. (1987) Spatial analysis and regional inequality: some suggestions for development planning, *Human Organization*, 46, 318–29.

Palloix, C. (1977) The self-expansion of capital on the world scale, *Review of Radical Political Economics*, 9, 1–28.

Parayil, G. (1992) Social movements, technology and development – a query and an instructive case from the Third World, *Dialectical Anthropology*, 17, 339–52.

Park, K. (1993) Women and development: the case of South Korea, *Comparative Politics*, 25, 127–45.

Parpart, J. (1993) Who is the 'other'? A postmodern feminist critique of women and development theory and practice, *Development and Change*, 24, 439–64.

Parsonage, J. (1992) Southeast Asia's 'growth triangle': a subregional response to global transformation, *International Journal of Urban and Regional Research*, 16, 307–20.

Parsons, T. (1951) *The Social System*. Chicago: Free Press.

Pastor, M. (1987) The effects of IMF programs in the Third World: debate and evidence from Latin America, *World Development*, 15, 249–62.

—— (1989) Latin America, the debt crisis, and the International Monetary Fund, *Latin American Perspectives*, 16, 79–109.

Patel, S. (1992) Social technology: a new factor in development, *Economic and Political Weekly*, 27, 1871–5.

Paul, E. (1993) Prospects for liberalization in Singapore, *Journal of Contemporary Asia*, 23, 291–305.

Paul, S. (1986) Community participation in development projects: the World Bank experience. World Bank Discussion Papers, 6.

Paus, E. (1989) The political economy of manufactured export growth: Argentina and Brazil in the 1970s, *Journal of Developing Areas*, 23, 173–200.

Peek, P. (1988) How equitable are rural development projects? *International Labour Review*, 127, 73–90.

Peet, R. and Watts, M. (1993) Introduction: Development theory and environment in an age of market triumphalism, *Economic Geography*, 69, 227–53.

Pelupessy, W. (1991a) Developments in the coffee and cotton sectors of El Salvador and perspectives for agrarian policy in the 1980s. In Pelupessy, W. (ed.), *Perspectives on the Agro-Export Economy in Central America*. London: Macmillan, 136–63.

—— (ed.) (1991b) *Perspectives on the Agro-Export Economy in Central America*. London: Macmillan.

Penna, D., Mahoney-Norris, K., McCarthy-Arnolds, E., Sanders, T and Campbell, P. (1990) Africa rights monitor: a women's right to political participation in Africa, *Africa Today*, 37, 49–64.

Petras, J. and Hui, P. (1991) State and development in Korea and Taiwan, *Studies in Political Economy*, 34, 179–98.

—— and Morley, M. (1990) *US Hegemony Under Siege: Class, Politics, and Development in Latin America*. London: Verso.

—— and Morley, M. (1992) *Latin America in the Time of Cholera: Electoral Politics, Market Economics, and Permanent Crisis*. New York: Routledge.

Pettman, R. (1992) Labor, gender and the balance of productivity: South Korea and Singapore, *Journal of Contemporary Asia*, 22, 45–56.

Phillips, L. (1990) Rural women in Latin America: directions for future research, *Latin American Research Review*, 25 (3), 89–107.

Pinstrup-Andersen, P. (1988) Macroeconomic adjustment and human nutrition, *Food Policy*, 13, 37–46.

Pires de Rio Caldeira, T. (1990) Women, daily life and politics. In Jelin, E. (ed.), *Women and Social Change in Latin America*. London: Zed, 41–78.

Porpora, P., Lim, M. and Rommas, U. (1989) The role of women in the international division of labour: the case of Thailand, *Development and Change*, 20, 269–94.

Prager, J. (1992) Is privatization a panacea for LDCs? Market failure versus the public sector failure, *Journal of Developing Areas*, 26, 301–22.

Pratt, G. and Hanson, S. (1994) Geography and the construction of difference, *Gender, Place and Culture*, 1, 5–29.

PREALC (Regional Development Program for Latin America and the Caribbean) (1988) *Adjustment and Social Debt: A Structural Approach*. Geneva: International Labour Organization.

Prebisch, R. (1950) *The Economic Development of Latin America and its Principal Problems*. New York: United Nations Department of Economic Affairs.

Preibisch, K. (1994) The personal is political: structural adjustment, women and the politics of everyday life. Unpublished manuscript.

Preston, P. (1986) *Making Sense of Development: An Introduction to Classical and Contemporary Theories of Development and their Application to Southeast Asia*. London: Routledge.

Prosterman, R., Temple, M. and Hamstad, T. (eds) (1990) *Agrarian Reform and Grassroots Development*. Boulder: Lynne Rienner.

Radcliffe, S. and Westwood, S. (eds) (1993) *Viva: Women and Popular Protest in Latin America*. New York: Routledge.

Raghuram, P. and Momsen, J. (1993) Domestic service as a survival strategy in Delhi, India, *Geoforum*, 24, 55–62.

Rakodi, C. (1990) Policies and preoccupations in rural and regional development planning in Tanzania, Zambia and Zimbabwe. In Simon, D. (ed.), *Third World Regional Development: A Reappraisal*. London: Chapman, 129–53.

Randolph, S. and Sanders, R. (1988) Constraints to agricultural production in Africa: a survey of female farmers in the Ruhengeri Prefecture of Rwanda, *Studies in Comparative International Development*, 23, 79–98.

Ranis, G. (1989) The role of institutions in transition growth: the East Asian newly industrializing countries, *World Development*, 17, 1443–53.

—— (ed.) (1992) *Taiwan: From Developing to Mature Economy*. Boulder: Westview.

—— and Fei, J. (1961) A theory of economic development, *American Economic Review*, 51, 533–65.

—— and Stewart, F. (1993) Rural nonagricultural activities in development: theory and application, *Journal of Development Economics*, 40, 75–101.

Rao, C. (1988) Agricultural development and ecological degradation: an analytical framework, *Economic and Political Weekly*, 23 (52/53), A142–6.

Rausser, G. and Thomas, S. (1990) Market politics and foreign assistance, *Development Policy Review*, 8, 365–81.

Razeto, L. (1991) Popular organizations and the economy of solidarity. In Aman, K. and Parker, C. (eds), *Popular Culture in Chile: Resistance and Survival*. Boulder: Westview, 81–96.

Redclift, M. (1984) *Development and the Environmental Crisis: Red or Green Alternatives?* London: Methuen.

—— (1987) *Sustainable Development: Exploring the Contradictions*. London: Methuen.

—— (1988) Sustainable development and the market: a framework for analysis, *Futures*, 20, 635–50.

—— (1991) The multiple dimensions of sustainable development, *Geography*, 76, 36–42.

Reusse, E. (1987) Liberalization and agricultural marketing, *Food Policy*, 12, 299–317.

Richards, P. (1985) *Indigenous Agricultural Revolution: Ecology and Food Production in West Africa*. Boulder: Westview.

Richardson, H. (1977) Growth centers, rural development, and national urban policy: a defense, *International Journal of Regional Science*, 3 (2), 131–52.

Riddell, J. (1992) Things fall apart again: structural adjustment programmes in sub-Saharan Africa, *Journal of Modern African Studies*, 30, 53–68.

Riedel, J. (1988) Economic development in East Asia: doing what comes naturally? In Hughes, H. (ed.), *Achieving Industrialization in East Asia*. Cambridge: Cambridge University Press, 1–38.

Riveros, L. (1990) Recession, adjustment and the performance of urban labour markets in Latin America, *Canadian Journal of Development Studies*, 11, 33–60.

Roarty, M. (1993) Trade in manufactures and the developing countries: the impact of new protectionism, *Kyklos*, 46, 105–19.

Robertson, C. (1988) Invisible workers: African women and the problem of the self-employed in labour history, *Journal of Asian and African Studies*, 23, 180–98.

Robison, R. (1989) Structures of power and the industrialisation process in Southeast Asia, *Journal of Contemporary Asia*, 19, 371–97.

Rodan, G. (1989) *The Political Economy of Singapore's Industrialization: National State and International Capital*. London: Macmillan.

Rodan-Rodenstein, P. (1943) Problems of industrialization in Eastern and South-Eastern Europe, *Economic Journal*, 52, 202–11.

Rodda, A. (ed.) (1993) *Women and the Environment*. London: Zed.

Rodrik, D. (1990) How should structural adjustment programs be designed? *World Development*, 18, 933–47.

Romein, A. and Schuurman, J. (1990) Regional strategy in Costa Rica and its impact on the northern region. In Simon, D. (ed.), *Third World Development: A Reappraisal*. London: Chapman, 94–108.

Rondinelli, D. (1981) Government decentralization in comparative perspective: theory and practice in developing countries, *International Review of Administrative Sciences*, 2, 133–45.

—— (1983) *Secondary Cities in Developing Countries: Policies for Diffusing Urbanization*. Beverly Hills, CA: Sage.

—— (1985) *Applied Methods of Regional Analysis: The Spatial Dimensions of Development Policy*. Boulder: Westview.

—— (1990) Decentralization, territorial power and the state: a critical response, *Development and Change*, 21, 491–500.

—— and Evans, H. (1983) Integrated regional development planning: linking urban centres and rural areas in Bolivia, *World Development*, 11, 31–53.

——, McCullough, J. and Johnson, R. (1989) Analysing decentralization policies in developing countries: a political economy framework, *Development and Change*, 20, 57–87.

—— and Montgomery, J. (1990) Managing economic reform: an alternative perspective on structural adjustment policies, *Policy Sciences*, 23, 73–93.

—— and Nellis, J. (1986) Assessing decentralization policies in developing countries: the case for cautious optimism, *Development Policy Review*, 4, 3–23.

—— and Ruddle, K. (1978) *Urbanization and Rural Development: A Spatial Policy for Equitable Growth*. New York: Praeger.

Rosales, O. (1988) An assessment of the structuralist paradigm for Latin American development and the prospects for its renovation, *CEPAL Review*, 34, 19–39.

Rose, G. (1993) *Feminism and Geography: The Limits of Geographical Knowledge*. Minneapolis: University of Minnesota Press.

Rosene, C. (1990) Modernization and rural development in Costa Rica: a critical perspective, *Canadian Journal of Development Studies*, 11, 367–74.

Rosset, P. (1989) Interview by Martha Honey. Unpublished paper, San José, Costa Rica.

Rostow, W. (1956) The take-off into self-sustained growth, *Economic Journal*, 66, 25–48.

Routledge, P. (1992) Putting politics in its place: Baliapal, India as a terrain of resistance, *Political Geography*, 21, 588–611.

Roxborough, I. (1979) *Theories of Underdevelopment*. London: Macmillan.

—— (1988) Modernization theory revisited: a review article, *Comparative Studies in Society and History*, 30, 753–61.

Ruccio, P. (1991) When failure becomes success: class and the debate over stabilization and adjustment, *World Development*, 19, 1315–34.

Rutten, M. (1990) The district focus policy for rural development in Kenya:

the decentralization of planning and implementation, 1983–9. In Simon, D. (ed.), *Third World Regional Development: A Reappraisal.* London: Chapman.

Sachs, I. (1974) Ecodevelopment, *Ceres,* 17 (4), 17–21.

—— (1988) Work, food and energy in urban ecodevelopment, *Economic and Political Weekly,* 23, 425–34.

—— and Silk, D. (1990) *Food and Energy: Strategies for Sustainable Development.* Tokyo: United Nations University Press.

Sack, R. (1974) The spatial separatist theme in geography, *Economic Geography,* 50, 1–19.

Safa, H. and Butler Flora, C. (1992) Production, reproduction and the polity: women's strategic and practical gender issues. In Stepan, A. (ed.), *Americas: New Interpretive Essays.* Oxford: Oxford University Press, 109–36.

Saha, S. (1991) Role of industrialisation in development of sub-Saharan Africa: a critique of the World Bank's approach, *Economic and Political Weekly,* 26, 2753–62.

Samoff, J. (1990) Decentralization: the politics of interventionism, *Development and Change,* 21, 513–30.

Samuels, W. (1988) An essay on the nature and significance of the normative nature of economics, *Journal of Post-Keynesian Economics,* 10, 347–54.

Samuelson, P. (1948) International trade and the equalization of factor prices, *Economic Journal,* 48, 163–84.

Sanford, J. (1988) The World Bank and poverty: the plight of the world's impoverished is still a major concern of the international agency, *American Journal of Economic Sociology,* 47, 257–74.

Santos, M. (1975) Underdevelopment, growth poles and social justice, *Civilizations,* 25, 18–30.

Sarkar, P. (1991) IMF/World Bank stabilization programmes: a critical assessment, *Economic and Political Weekly,* 26, 2307–10.

—— and Singer, H. (1991) *Debt Crisis, Community Prices, Transfer Burden and Relief.* Institute of Development Studies, University of Sussex.

Sazanami, H. and Newels, R. (1990) Subnational development and planning in Pacific island countries. In Simon, D. (ed.), *Third World Regional Development: A Reappraisal.* London: Chapman, 56–70.

Scarpaci, J. and Frazier, L. (1993) State terror: ideology, protest and the gendering of landscapes, *Progress in Human Geography,* 17, 1–21.

Schirmer, J. (1993) The seeking of truth and the gendering of consciousness: the comrades of El Salvador and the *Conaviuga* widows of Guatemala. In Radcliffe, S. and Westwood, S. (eds), *Viva: Women and Popular Protest in Latin America.* New York: Routledge, 30–64.

Schive, C. and Majumdar, B. (1990) Direct foreign investment and linkage effects: the experience of Taiwan, *Canadian Journal of Development Studies,* 11, 325–42.

Schoenholtz, A. (1987) The IMF in Africa: unnecessary and undesirable Western restraints on development, *Journal of Modern African Studies,* 25, 403–33.

Schuh, G. and McCoy, J. (eds) (1986) *Food, Agriculture, and Development in the Pacific Basin.* Boulder: Westview.

Schultz, T. (1980) Nobel lecture: The economics of being poor, *Journal of Political Economy*, 88, 639–51.

Scitovsky, T. (1954) Two concepts of external economies, *Journal of Political Economy*, 62, 143–51.

—— (1984) Comment on Adelman, *World Development*, 12, 953–4.

Sebstad, J. (1989) Introduction: Toward a wider perspective on women's employment, *World Development*, 17, 937–52.

Seidman, A. (1989) Towards ending IMF-ism in Southern Africa: an alternative development strategy, *Journal of Modern African Studies*, 27, 1–22.

Sen, A. (1981) *Poverty and Famines: An Essay on Entitlement and Deprivation*. Oxford: Clarendon.

—— (1987) *Gender and Cooperative Conflicts*. Helsinki: World Institute of Development Economics Research.

—— (1989) Development as capability expansion, *Journal of Development Planning*, 19, 41–58.

Senghaas, D. (1984) *The European Experience: A Historical Critique of Development Theory*. Dover, NH: Berg.

Sengupta, J. (1991) Rapid growth in the NICs in Asia: tests of new growth theory for Korea, *Kyklos*, 44, 561–79.

—— (1993) Growth in the NICs in Asia: some tests of new growth theory, *Journal of Development Studies*, 29, 342–57.

Senses, F. (1991) Turkey's stabilization and structural adjustment program in retrospect and prospect, *Developing Economies*, 29, 210–34.

Shao, J. (1986) The villagization program and the disruption of the ecological balance in Tanzania, *Canadian Journal of African Studies*, 20, 219–42.

Shaw, J. and Singer, H. (1988) Introduction: Food policy, food aid and economic adjustment, *Food Policy*, 13, 2–9.

Shaw, T. (1991) Reformism, revisionism and radicalism in African political economy during the 1990s, *Journal of Modern African Studies*, 29, 191–212.

Shiva, V. (1988) *Staying Alive: Women, Ecology and Development*. London: Zed.

—— (1991) *The Violence of the Green Revolution: Third World Agriculture, Ecology and Politics*. London: Zed.

Simon, D. (1990) The question of regions. In Simon, D. (ed.), *Third World Regional Development: A Reappraisal*. London: Chapman, 1–23.

—— and Rakodi, C. (1990) Conclusions and prospects: what future for regional planning? In Simon, D. (ed.), *Third World Regional Development: A Reappraisal*. London: Chapman, 249–60.

Singer, H. (1989) The World Bank: human face or facelift? Some comments in the light of the World Bank's annual report, *World Development*, 17, 1313–16.

Singh, A. (1987) Socio-cultural relevance: the missing link in development planning, *Journal of Social, Political and Economic Studies*, 12, 227–36.

Sklair, L. (1990) Regional consequences of open-door development strategies: export zones in Mexico and China. In Simon, D. (ed.), *Third World Regional Development: A Reappraisal*. London: Chapman, 109–26.

Slater, D. (1975) Underdevelopment and spatial inequality, *Progress in Planning*, 4, 97–167.

—— (1989) Territorial power and the peripheral state: the issue of decentralization, *Development and Change*, 20, 501–31.

—— (1993) The geopolitical imagination and the enframing of development theory, *Transactions, Institute of British Geographers*, 18, 419–37.

Smelser, N. (1963) Mechanisms of change and adjustment to change. In Hoselitz, B. and Moore, W. (eds), *Industrialization and Society*. The Hague: Mouton, 32–54.

Smil, V. (1989) Our changing environment, *Current History*, 88 (534), 9–12 and 47–8.

Smith, A. (1973) *The Concept of Social Change. A Critique of the Functionalist Theory of Social Change*. London: Routledge and Kegan Paul.

Smith, Adam. (1880) *An Inquiry into the Nature and Causes of the Wealth of Nations*. Oxford: Clarendon (originally published 1776).

Smith, B. (1985) *Decentralization – The Territorial Dimension of the State*. London: Allen & Unwin.

Smith, J. (1990) Zimbabwe's horticultural export trade, *Geography*, 75, 160–2.

Smith, S. (1990) Social geography: patriarchy, racism, nationalism, *Progress in Geography*, 14, 261–71.

Smith, T. (1985) Requiem or new agenda for Third World studies? *World Politics*, 37, 532–61.

So, A. (1986) The economic success of Hong Kong: insights from a world-system perspective, *Sociological Perspectives*, 29, 241–58.

—— (1990) *Social Change and Development: Modernization, Dependency and World System Theories*. Newbury, CA: Sage.

Soja, E. (1989) *Postmodern Geographies: The Reassertion of Space in Critical Social Theory*. London: Verso.

Sollis, P. and Moser, C. (1991) A methodological framework for analysing the social costs of adjustment at the micro-level: the case of Guayaquil, Ecuador, *IDS Bulletin*, 22, 23–30.

Somjee, A. (1991) *Development Theory: Critiques and Explorations*. Basingstoke: Macmillan.

South Commission (1990) *The Challenge to the South: The Report of the South Commission*. Oxford: Oxford University Press.

Spivak, G. (1988) Can the subaltern speak? In Nelson, C. and Grossberg, L. (eds), *Marxism and the Interpretation of Culture*. Urbana: University of Illinois Press, 271–313.

Sridharan, E. (1993) Economic liberalisation and India's political economy: towards a paradigm synthesis, *Journal of Commonwealth and Comparative Politics*, 31, 1–31.

Stallings, B. (1990) The role of foreign capital in economic development. In Gereffi, G. and Wyman, D. (eds), *Manufacturing Miracles: Paths of Industrialization in Latin America and East Asia*. Princeton: Princeton University Press, 55–89.

Standing, G. (1989) Global feminization through flexible labor, *World Development*, 17, 1077–95.

Stein, H. (1992) Deindustrialization, adjustment, the World Bank and the IMF in Africa, *World Development*, 20, 83–95.

—— and Nafziger, E. (1991) Structural adjustment, human needs, and the World Bank agenda, *Journal of Modern African Studies*, 29, 173–89.

Stephen, L. (1989) Not just one of the boys: from female to feminist in popular rural movements in Mexico. Paper delivered at the annual meetings of LASA (Latin American Studies Association), Puerto Rico.

—— (1992) Women in Mexico's popular movements: survival strategies against ecological and economic impoverishment, *Latin American Perspectives*, 72, 73–96.

Stewart, F. (1987) Should conditionality change? In Havnevik, K. (ed.), *The IMF and the World Bank in Africa: Conditionality, Impact and Alternatives.* Uppsala: Scandinavian Institute of African Studies, 29–45.

—— (1991) Are adjustment policies in Africa consistent with long-run development needs? *Development Policy Review*, 9, 413–36.

Stöhr, W. (1981) Development from below: the bottom-up and periphery-inward development paradigm. In Stöhr, W. and Taylor, D. (eds), *Development from Above or Below? The Dialectics of Regional Planning in Developing Countries.* Chichester: Wiley, 39–72.

—— and Taylor, D. (eds) (1981) *Development from Above or Below? The Dialectics of Regional Planning in Developing Countries.* Chichester: Wiley.

—— and Tödtling, F. (1978) Spatial equity – some antitheses to current regional development doctrine, *Papers of the Regional Science Association*, 38, 33–53.

Stokke, K., Dias, H. and Yapa, L. (1991) Growth linkages, the non-farm sector and rural inequality: a case study in southern Sri Lanka, *Economic Geography*, 67, 223–39.

Stone, L. (1989) Cultural crossroads of community participation in development: a case from Nepal, *Human Organization*, 48, 206–13.

Streeten, P. (1972) *The Frontiers of Development Studies.* London: Macmillan.

—— (1981) *Development Perspectives.* London: Macmillan.

—— (1987) Structural adjustment: a survey of the issues and options, *World Development*, 15, 1469–82.

—— (1989) Interests, ideology and institutions: a review of Bhagwati on protectionism, *World Development*, 17, 293–8.

—— (1993) Markets and states: against minimalism, *World Development*, 21, 1281–98.

—— and Burki, S. (1978) Basic needs: some issues, *World Development*, 6, 411–21.

Stren, R., White, R. and Whitney, J. (eds) (1992) *Sustainable Cities – Urbanization and the Environment in International Perspective.* Boulder: Westview.

Stromquist, N. (1992) Women and literacy: promises and constraints, *Annals of the American Academy of Political and Social Science*, 520, 54–65.

Summers, R. and Heston, A. (1988) A new set of international comparisons of real products and prices: estimates for 130 countries, *Review of Income and Wealth*, 34, 1–25.

Sunkel, O. (1973) Transnational capitalism and national disintegration in Latin America, *Social and Economic Studies*, 22, 132–76.

Sutton. F. (1989) Development ideology: its emergence and decline, *Daedalus*, 118,

35–67.

—— (ed.) (1990) *A World to Make: Development in Perspective.* New Brunswick, NJ: Transaction.

Tak-wing, N. (1993) Civil society and political liberalization in Taiwan, *Bulletin of Concerned Asian Scholars*, 25, 3–16.

Tan, G. (1993) The next NICs of Asia, *Third World Quarterly*, 14, 57–73.

Tang, A. (1988) Introduction: Why does overcrowded, resource-poor Asia succeed – lessons for the LDCs? *Economic Development and Cultural Change*, 36, S5–10.

Taplin, R. (1989) *Economic Development and the Role of Women.* Aldershot: Gower.

Taylor, D. and Mackenzie, F. (eds) (1992) *Development from Within: Survival in Rural Africa*, New York: Routledge.

Taylor, L. (1988) *Varieties of Stabilization Experience: Towards Sensible Macro-economics in the Third World.* Oxford: Clarendon.

Texier, J. (1974) Promotion of cooperatives in traditional areas, *Cooperative Information*, 1, 1–8.

Thiesenhusen, W. (1987) Rural development questions in Latin America, *Latin American Research Review*, 22, 171–203.

—— (ed.) (1989) *Searching for Agrarian Reform in Latin America.* London: Unwin Hyman.

—— (1991) Implications of the rural land tenure system for the environmental debate: three scenarios, *Journal of Developing Areas*, 26, 1–24.

Thomas-Slayter, B. (1992) Implementing effective local management of natural resources: new roles for NGOs in Africa, *Human Organization*, 51, 136–43.

Thorbecke, E. (1979) Agricultural development. In Galenson, W. (ed.), *Economic Growth and Structural Change in Taiwan: The Postwar Experience of the Republic of China.* Ithaca: Cornell University Press, 132–205.

Thorp, R. (1992) A reappraisal of the origins of import-substituting industrialization, 1930–50, *Journal of Latin American Studies*, 24, 181–95.

Thrift, N. (1983) On the determination of social action in space and time, *Environment and Planning D*, 1, 23–57.

Thrupp, L. (1990) Environmental initiatives in Costa Rica: a political ecology perspective, *Society and Natural Rescources*, 3, 243–56.

Timmer, P. (1973) Choice of techniques in rice milling on Java, *Bulletin of Indonesian Economic Studies*, 9, 57–76.

Timossi Dolinsky, G. (1990) Debt and structural adjustment in Central America, *Latin American Perspectives*, 17 (4), 76–90.

Tokman, V. (1986) Adjustment and employment in Latin America: the current challenges, *International Labour Review*, 125, 533–43.

Torres-Rivas, E. (1980) The Central American model of growth: crisis for whom? *Latin American Perspectives*, 7 (2/3), 24–44.

—— (1981) *Crisis del Poder en Centroamérica.* San José, Costa Rica: Editorial Universitaria Centroamericana.

Touraine, A. (1988) Discussions in modernity and identity: a symposium on culture, economy and development, *International Social Science Journal*, 40, 533–85.

Townsend, J. and Bain de Corcuera, J. (1993) Feminists in the rainforest in Mexico, *Geoforum*, 24, 45–54.

Toye, J. (1987) *Dilemmas of Development: Reflections on the Counter-Revolution in Development Theory and Policy*. Oxford: Blackwell.

Turnham, D. (1971) *The Employment Problem in Less Developed Countries: A Review of Evidence*. Paris: OECD Development Center.

ul Haq, M. (1976) *The Poverty Curtain: Choices for the Third World*. New York: Columbia University Press.

UNCHS (1984) *Community Participation in the Execution of Low Income Housing Projects*. Nairobi: United Nations Center for Human Settlements – HABITAT.

UNCTAD (1985) *Trade and Development Report, 1985*. New York: United Nations Conference on Trade and Development.

UNCTC (1988) *Transnational Corporations in World Development: Trends and Prospects*. New York: United Nations Conference on Transnational Corporations.

United Nations (1981) *Popular Participation as a Stategy for Promoting Community Level Action and National Development*. New York: United Nations Press.

—— (1989) *Report on the World Social Situation*. New York: United Nations Department of International Economic and Social Affairs.

Unwin, T. (1989) Urban–rural interaction in developing countries: a theoretical perspective. In Potter, R. and Unwin, T. (eds), *The Geography of Urban–Rural Interaction in Developing Countries*. London: Routledge, 11–32.

Uphoff, N. (1993) Grassroots organizations and NGOs in rural development: opportunities with diminishing states and expanding markets, *World Development*, 21, 607–22.

Urrutia, M. (1987) Latin America and the crisis of the 1980s. In Toye, J. (ed.), *Development Policies and the Crisis of the 1980s*. Paris: Development Center for the OECD, 55–69.

Utting, P. (1994) Social and political dimensions of environmental protection in Central America, *Development and Change*, 25, 231–59.

van der Laan, L. (1993) Boosting agricultural exports? A 'marketing channel' perspective on an African dilemma, *African Affairs*, 92, 173–201.

van Ginneken, W. (1990) Labour adjustment in the public sector: policy issues for developing countries, *International Labour Review*, 129, 441–58.

Vartiainen, P. (1987) The strategy of territorial integration in regional development: defining territoriality, *Geoforum*, 18, 117–26.

Velasco, B. and Leppe, A. (1989) Appropriate technologies: a way of meeting human needs. In Downs, C., Solimano, G., Vergara, G. and Zuniga, L. (eds), *Social Policy from the Grassroots: Nongovernmental Organizations in Chile*. Boulder: Westview, 93–112.

Vengroff, R. and Johnston, A. (1989) *Decentralization and the Implementation of Rural Development in Senegal: The View from Below*. New York: Edwin Mellen.

Vilas, C. (1984) *Perfiles de la Revolución Sandinista*. Havana: Ediciones Casa de las Américas.

Viner, J. (1953) *International Trade and Economic Development*. Oxford: Clarendon.

Viola, E. (1988) The ecologist movement in Brazil (1974–1986): from environmentalism to ecopolitics, *International Journal of Urban and Regional Research*, 12, 21–8.

Vivian, J. (1994) NGOs and sustainable development in Zimbabwe: no magic bullets, *Development and Change*, 25, 167–93.

Vogel, E. (1991) *The Four Little Dragons: The Spread of Industrialization in East Asia*. Cambridge, MA: Harvard University Press.

von Braun, J. and Paulino, L. (1990) Food in sub-Saharan Africa: trends and policy challenges for the late 1990s, *Food Policy*, 15, 505–17.

Wade, R. (1990) *Governing the Market: Economic Theory and the Role of Government in East Asian Industrialization*. Princeton: Princeton University Press.

—— (1992) East Asia's economic success: conflicting perspectives, partial insights, shaky evidence, *World Politics*, 44, 270–320.

—— (1993) Managing trade: Taiwan and South Korea as challenges to economics and political science, *Comparative Politics*, 25, 147–67.

Watkins, K. (1992) GATT and the Third World: fixing the rules, *Race and Class*, 34, 23–40.

Watts, M. (1983) *Silent Violence: Food, Famine and Peasantry in Northern Nigeria*. Berkeley: University of California Press.

Weaver, C. (1981) Development theory and the regional question: a critique of spatial planning and its detractors. In Stöhr, W. and Taylor, D. (eds), *Development from Above or Below? The Dialectics of Regional Planning in Developing Countries*. Chichester: Wiley, 73–105.

Weber, M. (1951) *The Religion of China*. New York: Free Press.

—— (1958) *The Protestant Ethic and the Spirit of Capitalism*. New York: Scribner.

Webster, A. (1990) *Introduction to the Sociology of Development*. Basingstoke: Macmillan.

Weil, C. (ed.) (1988) *Lucha: The Struggle of Latin American Women*. Minneapolis: Prisma Institute.

Weisskoff, R. (1992) The Paraguayan agro-export model of development, *World Development*, 20, 1531–40.

Weissman, S. (1990) Structural adjustment in Africa: insights from the experiences of Ghana and Senegal, *World Development*, 18, 1621–34.

Westwood, S. and Radcliffe, S. (1993) Gender, racism and the politics of identities in Latin America. In Radcliffe, S. and Westwood, S. (eds), *Viva: Women and Popular Protest in Latin America*. New York: Routledge, 1–29.

White, G. (ed.) (1988) *Developmental States in East Asia*. New York: St. Martin's.

White, L. (1990) Policy reforms in sub-Saharan Africa: conditions for establishing a dialogue, *Studies in Comparative International Development*, 25, 24–42.

Whiteford, S. and Ferguson, A. (eds) (1991) *Harvest of Want: Hunger and Food Security in Central America and Mexico*. Boulder: Westview.

Wickramasinghe, A. (1993) Development intervention and the changing status of rural women in Sri Lanka, *Geoforum*, 24, 63–9.

Wilken, G. (1987) *Good Farmers: Traditional Agricultural Resource Management in Mexico and Central America*. Berkeley: University of California Press.

Williams, J. (1992) Capitalist development and human rights: Singapore under Lee Kuan Yew, *Journal of Contemporary Asia*, 22, 360–72.

Williams, R. (1986) *Export Agriculture and the Crisis in Central America*. Chapel Hill: University of North Carolina Press.

Wilson, F. (1985) Women and agricultural change in Latin America: some concepts guiding research, *World Development*, 13, 1017–35.

Wilson, G. (1991) Thoughts on the cooperative conflict model of the household in relation to economic method, *IDS Bulletin*, 22, 31–6.

Winckler, E. and Greenhalgh, S. (eds) (1988) *Contending Approaches to the Political Economy of Taiwan*. Armonk, NY: Sharpe.

Winpenny, J. (1991) Environmental values and their implications for development, *Development Policy Review*, 9, 381–90.

Wipper, A. (1988) Reflections on the past sixteen years, 1972–1988, and future challenges, *Canadian Journal of African Studies*, 22, 409–21.

Wisner, B. (1988) *Power and Need in Africa: Basic Human Needs and Development Policies*. London: Earthscan.

Wolch, J. and Dear, M. (1989) How territory shapes social life. In Wolch, J. and Dear, M. (eds), *The Power of Geography*. London: Unwin Hyman, 3–18.

Wolf, D. (1990) Daughters, decisions and domination: an empirical and conceptual critique of household strategies, *Development and Change*, 21, 43–74.

Wood, G. (1985) The politics of developmental policy labelling, *Development and Change*, 16, 347–73.

World Bank (1982, 1983, 1985, 1987, 1992, 1993a) *World Development Report* (Annual). New York: Oxford University Press/ World Bank.

—— (1988) *Rural Development: World Bank Experience, 1965–86*. Washington: World Bank Operations Evaluation Department.

—— (1993b) *The East Asian Miracle: Economic Growth and Public Policy*. New York: Oxford University Press/ World Bank.

—— (1993c) *World Tables* (Annual). New York: Oxford University Press.

World Commission on Environment and Development (1987) *Our Common Future*. New York: Oxford University Press.

Yoon, B. (1992) Reverse brain drain in South Korea: state-led model, *Studies in Comparative International Development*, 27, 4–26.

Zarkovic, M. (1988) The effects of economic growth and technological innovation on the agricultural labour force in India, *Studies in Comparative International Development*, 22, 103–20.

Index